CABLE NETWORKS, SERVICES, AND MANAGEMENT

CABLE NETWORKS, SERVICES, AND MANAGEMENT

Edited by

Mehmet Toy

**IEEE Press
Series on
Networks and
Services Management**

**Thomas Plevyak and
Veli Sahin,** *Series Editors*

IEEE Press

For general information on our other products and services or for technical support, please contact our
Customer Care Department within the United States at (800) 762-2974, outside the United States at (317)
572-3993 or fax (317) 572-4002.

Wiley also publishes its books in a variety of electronic formats. Some content that appears in print may
not be available in electronic formats. For more information about Wiley products, visit our web site at
www.wiley.com.

Library of Congress Cataloging-in-Publication Data:

Cable networks, services and management / editor, Mehmet Toy.
    pages   cm
  Includes bibliographical references.
  ISBN 978-1-118-83759-7 (hardback)
1. Integrated services digital networks.   2. Internet service providers–Management.
3. Home computer networks.   4. Cable television.   I. Toy, Mehmet.
  TK5103.75.C33 2015
  384.068–dc23

                                                    2014027711

Printed in the United States of America

10 9 8 7 6 5 4 3 2 1

CONTENTS

Preface   vii

Contributors   ix

1 INTRODUCTION AND OVERVIEW   1

1.1 Introduction   1

1.2 Residential Network Architectures and Services   2

1.3 OAMPT (Operations, Administration, Maintenance, Provisioning, Troubleshooting) for Residential Services   4

1.4 Business Network Architectures and Services   6

1.5 OAMPT for Business Services   7

1.6 Future Directions in Cable Networks, Services and Management   8

2 RESIDENTIAL NETWORK ARCHITECTURES AND SERVICES   11

2.1 Introduction   11

2.2 DOCSIS 3.0/3.1 Architecture and Services   12

2.3 PacketCable Architecture and Services   22

2.4 IMS Architecture and Services   26

2.5 IPTV Architecture and Services   34

2.6 CDN Architecture and Services   45

2.7 CCAP Architectures and Services   53

2.8 Wi-Fi Architecture and Services   69

2.9 Conclusion   78

References   80

3 OPERATIONS, ADMINISTRATION, MAINTENANCE, PROVISIONING, AND TROUBLESHOOTING FOR RESIDENTIAL SERVICES   84

3.1 Introduction   84

3.2 Operational Systems and Management Architectures   85

3.3 Service Orders   89

3.4 Provisioning   91

3.5 Fault Management   92
3.6 Performance Management   94
3.7 Billing Systems and Formats   97
3.8 Security   99
3.9 Conclusion   131
References   132

4 BUSINESS NETWORK ARCHITECTURES AND SERVICES   134
4.1 Introduction   134
4.2 Metro Ethernet Architecture and Services   135
4.3 DPoE Architecture and Services   161
4.4 EPoC Architecture and Services   168
4.5 Business Voice   171
4.6 Conclusion   180
References   181

**5 OPERATIONS, ADMINISTRATION, MAINTENANCE, PROVISIONING,
AND TROUBLESHOOTING FOR BUSINESS SERVICES   183**
5.1 Introduction   183
5.2 Operations Systems and Management Architectures   184
5.3 Service Orders   196
5.4 Provisioning   205
5.5 Fault Management   224
5.6 Performance Management   230
5.7 Billing Systems and Formats   235
5.8 Security   236
5.9 Conclusion   241
References   243

**6 FUTURE DIRECTIONS IN CABLE NETWORKS,
SERVICES AND MANAGEMENT   246**
6.1 Introduction   246
6.2 Cloud Services   247
6.3 Virtualization   274
6.4 Network Functions Virtualization   295
6.5 Software-Defined Networks   297
6.6 Self-Managed Networks   318
6.7 Conclusion   337
References   338

Index   341

PREFACE

Cable operators (MSOs) provide various types of voice, video, and data services to residential and commercial customers. The number of customers per operator is in the millions while the operator's network consists of equipment from various vendors. Compliance of equipment with international standards becomes a necessity for their interworking. Quality of service becomes a part of the contractual agreement with customers in commercial services.

Operations Support Systems, tools, and procedures for automation developed by the operators play an important role in managing these complex networks. Vendor-developed Network Management Systems (NMSs) and Element Management Systems (EMSs) are not widely used.

This is the first book describing cable networks, services, and their management in greater detail. The book is written by 13 experts in various fields covering network architectures and services, operations, administration, maintenance, provisioning, and troubleshooting (OAMPT) for residential services; network architectures, services, and OAMPT for business services; cloud, software-defined networks (SDN) and virtualization concepts and their applications as part of the future directions of cable networks. Finally, a proposed self-managed network concept is described.

The book begins by describing architecture and services for Data over Cable Service Interface Specification (DOCSIS) 3.0/3.1, Converged Cable Access Platform (CCAP), content distribution networks (CDNs), IP TV, and PacketCable and Wi-Fi for residential services. The book then follows with operational systems and management architectures, service orders, provisioning, fault management, performance management, billing systems and formats, and security for residential services.

Similar to residential services, the book describes architecture and services for Carrier Ethernet, DOCSIS provisioning over Ethernet Passive Optical Networks (EPON-DPoE), EPON Protocol over Coax (EPOC), CCAP, and IP Multimedia Subsystem (IMS) for business services. Following this, operational systems and management architectures, service orders, provisioning, fault management, performance management, billing systems and formats, and security for business services are explained.

As MSO networks and services change rapidly, future directions for cable networks are projected in Chapter 6 by describing cloud services, virtualization, SDN, and a self-managed network concept with their applications. I hope that this book provides guidance to all those involved in designing and operating MSO networks, researchers, and students.

I thank the co-authors who worked diligently to make this book a valuable reference; editors of IEEE Network Management series, Tom Plevyak and Dr. Veli Sahin, who asked me to write a book on cable networks and management, and provided valuable comments during the editing process; Mary Hatcher of John Wiley & Sons who helped greatly throughout the publication process; and Francois PascalRaj of SPi Global and Danielle LaCourciere of John Wiley & Sons who led the production of this book.

MEHMET TOY
Allendale, NJ

CONTRIBUTORS

Alireza Babaei, CableLabs, Denver, CO, USA

Victor Blake, Consultant, Leesburg, VA, USA

Niem Dang, Time Warner Cable, Herndon, VA, USA

Kirk Erichsen, Time Warner Cable, Colorado Springs, CO, USA

Sergio Gambro, Comcast Cable, Philadelphia, PA, USA

Kenneth Gould, Time Warner Cable, Herndon, VA, USA

Belal Hamzeh, CableLabs, Denver, CO, USA

David Hancock, CableLabs, Denver, CO, USA

Brian Hedstrom, OAM Technology Consulting, Denver, CO, USA

Stuart Hoggan, CableLabs, Denver, CO, USA

Curtis Knittle, CableLabs, Denver, CO, USA

Vikas Sarawat, CableLabs, Denver, CO, USA

Mehmet Toy, Comcast Cable, Mount Laurel, NJ, USA

1

INTRODUCTION AND OVERVIEW

Mehmet Toy

1.1 INTRODUCTION

Cable companies (multiple system operators or MSOs) have been offering phone and TV services over copper medium as the basic residential services for a long time. With the proliferation of broadband, fiber, and wireless technologies, fiber is deployed to provide TV, voice, and Internet services (i.e., triple play) while serving rates over cable are increased with new techniques in supporting triple-play services.

MSOs have been providing residential services over coaxial cable using the Data Over Cable Service Interface Specification (DOCSIS) protocol that permits the addition of high-speed data transfer to an existing cable TV (CATV) system in Mbps currently and aims at supporting capacities of at least 10 Gbps downstream and 1 Gbit/s upstream.

Toward the end of 1990s, MSOs initiated PacketCable project to deliver real-time communication services, namely Voice over Internet Protocol (VoIP). Later, content distribution network (CDN) is created to deliver content to deliver video distribution over IP. IP television (IPTV) system delivered television using IP over a packet-switched network such as a LAN or the Internet. For delivering IP multimedia services, IP Multimedia Subsystem (IMS) architectural framework was introduced by 3rd Generation Partnership Project (3GPP). MSOs provide both residential and commercial services over IMS platform today.

Cable Networks, Services, and Management, First Edition. Edited by Mehmet Toy.
© 2015 The Institute of Electrical and Electronics Engineers, Inc. Published 2015 by John Wiley & Sons, Inc.

In recent years, in order to centralize data and video applications on a single platform to reduce overall system cost, MSOs introduced Converged Cable Access Platform (CCAP).

In addition to residential services, MSOs have been offering services to small- and medium-sized enterprises (SMEs), and large businesses. Metro Ethernet services in the form of private line, virtual private line, and multipoint-to-multipoint services are among them. In order to optimize MSO networks and improve quality and rates of service offerings, DOCSIS 3.1, DOCSIS provisioning of EPON (DPoE), EPON protocol over coax (EPoC), and Wi-Fi systems have been introduced. CCAP will support business services as well.

Chapter 2 describes architecture and services for DOCSIS 3.0/3.1, CCAP, CDN, IP TV, and PacketCable and Wi-Fi for residential services.

Chapter 3 describes operational systems and management architectures, service orders, provisioning, fault management, performance management, billing systems and formats, and security for residential services.

Chapter 4 describes architecture and services for Carrier Ethernet, DPoE, EPoC, CCAP, IMS for business services.

Chapter 5 describes operational systems and management architectures, service orders, provisioning, fault management, performance management, billing systems and formats, and security for business services.

Finally, Chapter 6 explains the future directions for cable networks by describing cloud services, virtualization, SDN, and the author's proposed self-managed network concepts with their applications.

1.2 RESIDENTIAL NETWORK ARCHITECTURES AND SERVICES

MSOs provide residential services over coaxial cable using the Data Over Cable Service Interface Specification (DOCSIS) protocol. PacketCable project was initiated to deliver real-time communications services, namely VoIP. Later, Content Distribution Networks (CDN) is created to deliver content to deliver video distribution over IP. The goal is to serve content to end-users with high availability and high performance. In order to deliver television using IP over a packet-switched network such as a LAN or the Internet, Internet Protocol television (IPTV) system was introduced. Many advanced communications services to users are delivered by IP Multimedia Subsystem (IMS).

In order to reduce cost by centralizing data and video applications on a single platform, MSOs introduced Converged Cable Access Platform (CCAP). CCAP is expected to dramatically increase system capacity and density to enable the MSOs the ability to protect and maximize existing infrastructure investment, and provide orderly migration of existing video services to IP video.

The DOCSIS system allows transparent bi-directional transfer of Internet Protocol (IP) traffic, which is called High Speed Data (HSD) Internet or Broadband service, between the cable system head-end and customer location, over an all-coaxial or hybrid-fiber/coax (HFC) cable network. The service is supported by a Cable Modem Termination

System (CMTS) or a CCAP at the head-end, and a Cable Modem (CM) at customer location. It is also possible to support this service with a PON system at head-end and Optical Networking Unit (ONU) at customer location.

The CMTS and CCAP head-end devices reside within the MSO's core network and are generally considered secure. The CM provides the demarcation point between the subscriber and the MSO's network. It is considered unsecure and untrusted. Due to the different nature of each of these devices as well as differing business and technical requirements, they are managed differently from the MSO's back office.

DOCSIS 3.0 protocol is the currently deployed within MSO networks today. DOCSIS 3.0 and DOCSIS 3.1 systems were designed to support high capacities with predictable Quality of Service (QoS). The DOCSIS 3.0 protocol supports 240 Mbps in the upstream and 1.3 Gbps in the downstream, while the DOCSIS 3.1 protocol increases the upstream and downstream capacities significantly promising 2 Gbps and 10 Gbps on the upstream and downstream respectively.

In 1990s, MSOs began looking for value-added services such as VoIP that could ride on top of their newly deployed DOCSIS technology. The PacketCable project was initiated for the delivery of real-time communications services over the two-way cable plant. PacketCable 1.5 was designed to provide traditional voice telephone service, and subsequently PacketCable 2.0 was designed to provide advanced multi-media services beyond basic voice.

In order to reach a large population of leased or consumer-owned devices, enable linear and on-demand video to non-technical subscribers, and provide a viewing experience similar to traditional TV, MSOs developed IPTV service delivery architecture. The video services comprise local linear video, national linear video, video-on-demand (VOD), and pay-per-view (PPV).

National VOD programming is typically delivered to MSOs over satellite as well, where the video asset and its associated metadata are placed on a satellite by the content programmers, beamed by the satellite to all of the MSOs who are authorized to distribute that content, received by the MSOs, and placed onto the appropriate VOD distribution servers and represented in a navigational client so that customers can find the video content.

Content Distribution Networks (CDN) are created to deliver content, whether video or web pages as quickly as possible, with the least latency, at the lowest cost, to the widest quantity of consumers who are spread over a disparate geographic area. A CDN is an interconnected network of servers used to deliver web assets to consuming devices in an efficient manner, taking advantage of caching of content to reduce or eliminate the retransmission of a single asset to multiple consumers. This chapter describes deploying a CDN for video distribution over IP, the architecture of a CDN, services supplied by a CDN and areas for future CDN research.

CCAP will dramatically increase system capacity and density to enable MSOs the ability to protect and maximize existing critical infrastructure investment, by delivering various access technologies from the same chassis. It will play a vital role in completing a smooth transition to IP video transport and an "all IP" service offering more generally, in lock-step with the ongoing evolution of the HFC plant and every increasing capacity needs per user.

Wi-Fi is becoming a network of choice for both service providers (SPs) and consumers alike. This chapter reviews some of the recent technological advances in the Wi-Fi domain and describes residential network architecture and services over DOCSIS, CDN, Packet Cable, IMS and CCAP.

1.3 OAMPT (OPERATIONS, ADMINISTRATION, MAINTENANCE, PROVISIONING, TROUBLESHOOTING) FOR RESIDENTIAL SERVICES

OAMPT capabilities are necessary to order and maintain the residential services described in Chapter II. In addition to OAMPT, services need to be billed to subscribers.

Service order process begins with sales. Once the service order is in, the provisioning process installs equipment at customer premises, sets-up equipment at customer premises and central offices, sets-up user accounts, and creates new circuits. The provisioning processes will also include checklists that need to be strictly adhered to and signed off, and integration and commissioning processes which will involve sign-off to other parts of the business life cycle.

Operations encompass automatic monitoring of the environment, detecting and determining faults and alerting network administration. Network administration collects performance statistics, accounts data for the purpose of billing, plans capacity based on usage data, maintains system reliability, administers network security, and maintains the service database for periodic billing.

Maintenance involves routine equipment checks, upgrading software and hardware, fixes, new feature enablement, backup and restore, standard network equipment configuration changes as a result of policy or design, and monitoring facilities. When there is a failure, failed components are identified by diagnostic tools and troubleshooting.

OAMPT for residential services are challenging mainly due to the involvement of a very large number of Customer Premises Equipment (CPE), facilities and connections (i.e., in the order of millions). As a result, automation for each OAMPT function is crucial for MSOs. Autoprovisioning of CPE and connections, administration of network security via DOCSIS protocol and back office systems are developed to resolve these scalability issues.

The DOCSIS service and device management approach uses FCAPS (Fault, Configuration, Accounting, Performance, and Security) model, to organize the requirements for the configuration and management of the CMTS/CCAP and CM devices.

Fault management is a proactive and on-demand network management function that allows abnormal operation on the network to be detected, diagnosed, and corrected. When an abnormal condition is detected, an autonomous event (often referred to as an alarm notification) is sent to the network operations center (NOC) to alert the MSO of a possible fault condition in the network affecting a customer's service. Once the MSO receives the event notification, further troubleshooting and diagnostics can be performed by the MSO to correct the fault condition and restore the service to proper operation.

Configuration Management provides a set of network management functions that enables system configuration building and instantiating, installation and system turn up, network and device provisioning, auto-discovery, backup and restore, software download, status, and control (e.g., checking or changing the service state of an interface). DOCSIS Configuration Management is typically performed at the network layer (e.g., device provisioning at the CMTS/CCAP and CM).

Accounting Management is a network management function that allows MSOs to measure the use of network services by subscribers for the purposes of cost estimation and subscriber billing. Subscriber Accounting Management Interface Specification (SAMIS), as defined in DOCSIS, is an example of an implemented Accounting Management function.

Performance Management is a proactive and on-demand network management function which is gathering and analyzing "statistical data for the purpose of monitoring and correcting the behavior and effectiveness of the network, network elements (NEs), or other equipment and to aid in planning, provisioning, maintenance and the measurement of quality." A Performance Management network layer and service-level use case might include the NOC performing periodic (15 min, for example) collections of Quality of Service (QoS) measurements from network elements to perform monitoring and identification of any potential performance issues that may be occurring with the service being monitored.

Security Management provides for the management of network and operator security, as well as providing an umbrella of security for the telecommunications management network functions including authentication, access control, data confidentiality, data integrity, event detection, and reporting.

The CM and CMTS reside within the Network Layer where services are provided to end subscribers and various metrics are collected about network and service performance, among other things. Various management servers reside in the Network Management Layer within the MSO back office to provision, monitor and administer the CM within the Network Layer.

The major service and network management features introduced in the DOCSIS 3.1 protocol specification include configuration, monitoring and reporting on the feature set DOCSIS Light Sleep Mode (DLS), Backup Primary Channels, Active Queue Management (AQM) and Proactive Network Maintenance (PNM). The DOCSIS service and device management approach used for the DOCSIS 3.1 network remains the same as the DOCSIS 3.0 network for the CM management model. With respect to DOCSIS 3.1 management of the CMTS, the management model is aligned with the CCAP management approach, where configuration management using XML-based configuration files is used.

Provisioning of CM is primarily performed using the configuration file download process. Binary configuration files are constructed using the TLV definitions from the DOCSIS specifications.

Fault management is a proactive and on-demand network management function including Alarm Surveillance, Fault Localization, Fault Correction, and Testing. When service-impacting abnormalities of an NE is detected, an autonomous event (often referred to as an alarm notification) is sent to the network operations center (NOC) to alert the MSO of a possible fault condition in the network affecting a customer service. Once the NOC receives the event notification, further troubleshooting and diagnostic testing can be performed on-demand by the NOC to correct the fault condition and restore the service to proper operation.

Performance Management is a proactive and on-demand network management function which is gathering and analyzing "statistical data for the purpose of monitoring and correcting the behavior and effectiveness of the network, NEs, or other equipment and to aid in planning, provisioning, maintenance and the measurement of quality. NOC periodically (15 min, for example) collects Quality of Service (QoS) measurements from network elements to perform monitoring and identification of any potential performance issues that may be occurring with the service being monitored. With the historical data that has been collected, trending analysis can be performed to identify issues that may be related to certain times of day or other corollary events.

For the DOCSIS 3.0 and 3.1 networks, identified performance metric data sets implemented in the CMTS and CCAP, both Simple Network Management Protocol (SNMP) and IP Detail Record/Streaming Protocol (IPDR/SP) protocols are available for bulk data collection of the data sets. MSO business policies and back office application architectures dictate which protocol is utilized for collecting the various performance metrics.

The core function of the "Billing" system is to generate an accurate bill. The name "billing" is a misnomer because it indicates just one of the many functions have been built into these software applications. Some of the biller functions are Product Catalogue, the CPE Inventory, the Order Entry system, the Service Order Manager, the Taxation Engine, the Biller as well as the Workforce Manager (determining the number of availability of installation technicians), and the Serviceable Homes Database of the operator (which premises are serviceable for what services).

When providing content and services over a network to subscribers it is important to apply proper security. There are many threats that exist that can negatively impact a subscriber's experience and service provider's business. These threats include spoofing, tampering and information disclosure.

DOCSIS 3.0/3.1 security features focus on preventing theft of service and loss of privacy. The main features are device authentication and traffic encryption, secure software download, and secure provisioning. A DOCSIS CM is provisioned by sending its configuration file to the CM which it then forwards to the CMTS. Since the CM is untrusted, it is important that security controls are in place to verify service settings.

This chapter further describes OAMPT including security and billing for residential services.

1.4 BUSINESS NETWORK ARCHITECTURES AND SERVICES

MSOs offer services to small and medium size enterprises (SMEs) and large businesses. Carrier Ethernet services in the form of private line, virtual private line and multipoint-to-multipoint services are among them.

As in residential services, business services can be voice, data and video services. The nature of services can differ depending on the size of the business. For example voice services for small and medium size enterprises (SMEs) might involve interfacing Private Branch Exchanges (PBXs) at customer premises and managing voice mailboxes while voice services for a large corporation might involve providing a private network serving many locations. Furthermore, services for large corporations require better availability, better performance and multiple classes of services.

MSOs offer various Carrier Ethernet commercial services. They use service OAM capabilities of Metro Ethernet extensively. Automation of Metro Ethernet equipment and service provisioning is underway.

Carrier Ethernet is connection-oriented by disabling unpredictable functions such as MAC learning, Spanning Tree Protocol and "broadcast of unknown." It is scalable from 10 Mbps to 100 Gbps with finer granularity as low as 56 Kbps and 1 Mbps. With QoS and synchronization capabilities, applications with strict performance requirements such as Mobile Backhaul are being supported. Port and connection level fault monitoring, detection and isolation, performance monitoring and statistics and protection failover are available.

Some service providers are concerned of the scalability of Carrier Ethernet. There is an effort by Cloud Ethernet Forum (CEF) to increase scalability of Carrier Ethernet by using Provider Backbone Bridge (PBB) and Provide Backbone Transport (PBT).

Carrier Ethernet services are delivered to users at fiber, copper and coax user interfaces. The services over fiber interfaces can employ point-to-point and point-to-multipoint architectures such as Passive Optical Network (PON) which is characterized by having a shared optical fiber network which supports a single downstream transmitter and multiple simultaneous upstream transmitters. The downstream transmitter, an Optical Line Terminal (OLT), is typically located in the head end or hub of a service provider's network. Optical Network Units (ONUs) located on the customer premise terminate the PON. A passive optical splitter/combiner enables the fiber to be divided to reach multiple customer locations. The OLT schedules each ONU for upstream transmission and guarantees the frames do not overlap.

Business Real-Time Communications Services include both legacy business voice-calling services such as hunt groups and paging, and advanced communications services such as video conferencing and desktop-sharing. In Hosted IP-Centrex service, the service provider owns and manages the service and the service delivery equipment. The business customer has no equipment to buy, and simply pays a monthly subscription fee for the service. With SIP Trunking service, the business customer owns and operates the service delivery equipment, while the service provider simply provides the SIP Trunk that connects the SIP-PBX to the global telecom network.

This chapter covers architecture and services for business customers over DOCSIS, IMS, Metro Ethernet, DPoE, and EPoC technologies.

1.5 OAMPT FOR BUSINESS SERVICES

In Chapter 3, we described OAMPT for residential services. OAMPT is also crucial for business services. In fact, the OAMPT requirements for business services are tighter than those for residential services. This is due the fact that degraded or lost services can have larger financial impact.

In order to process service orders, keep inventory, configure equipment, provision services, collect measurement and accounting data, secure network access, report service degradation and failures, troubleshoot the failures and fix them, and bill customers, MSOs have operations systems dedicated for business services.

Equipment manufacturers usually have their own Element Management System (EMS) to manage a Network Element (NE) such as a router and their own Network Management System (NMS) to manage a subnetwork of NEs. It is possible to combine these management systems with operators' operation systems to perform the OAMPT functions. In some cases, operators may choose to use only their operations systems to manage their network consisting of multiple vendor equipment.

Service providers have a need to tie together ordering, payment, and operation of systems. For example, when a typical enterprise service is provisioned, it would typically operate it at its maximum capacity. In contrast, a service provider selling a similar or even the same service to an enterprise may sell a fixed bandwidth service, a burstable service, or a billable service that must measure and count traffic.

The scale of service provider networks also dictates more automation. Service providers need to build processes to accept changes from business customers such as a move in location, additions of new locations or services at an existing location, or other changes such as capacity upgrades or downgrades.

Unlike most enterprise networks, service providers need to have means to control the traffic in their networks so as to avoid conflicts between the performance, quality, and reliability for both different services running on the same network and for different customers running on the same network.

In this chapter, operations systems, service order provisioning, fault management, performance management and security for business services over DOCSIS, CCAP, IMS, Metro Ethernet, DPoE, and EPoC have been described.

1.6 FUTURE DIRECTIONS IN CABLE NETWORKS, SERVICES AND MANAGEMENT

The previous chapters describe highly complex MSO residential and commercial networks and services, and their management. These networks and management systems have proven to be scalable and allow MSOs to create and offer new competitive services to their customers.

MSOs can reduce CAPEX and OPEX, and amount of time to create new services. The concepts of Cloud Services, Virtualization, Software Defined Networks (SDN) and Self-Managed will help MSOs to achieve these goals. In fact, there are substantial efforts in MSOs to take advantage of these new technologies. Some MSOs do provide Cloud based services today.

Self-Managed networks will be an ultimate goal for the industry and requires substantial changes in how equipment and management systems are built and operated.

Substantial growth in high speed personal devices such as phones, laptops and IPAD, and IP video and IPTV applications are driving huge bandwidth demand in networks. Applications such as storage networking, video streaming, collaborative computing, and online gaming and video sharing are driving not only bandwidth demand in networks, but also resources of various data centers connected with these networks.

The concepts of cloud computing and cloud-based services are expected to help service providers to deal with these challenges.

Cloud Computing technologies are emerging as infrastructure services for provisioning computing and storage resources on-demand. Multi-provider and multi-domain resources, and integration with the legacy services and infrastructures are involved.

Cloud based virtualization allows for easy upgrade and/or migration of enterprise application, including also the whole IT infrastructure segments. This brings significant cost saving comparing to traditional infrastructure development and management that requires lot of manual work.

Cloud based applications operate as regular applications in particular using web services platforms for services and applications integration, however their composition and integration into distributed cloud based infrastructure will require a number of functionalities and services.

Virtual infrastructure provides a layer of abstraction between computing, storage and networking hardware, and the applications running on it. Virtual infrastructure gives administrators the advantage of managing pooled resources across the enterprise, allowing IT managers to be more responsive to dynamic organizational needs and to better leverage infrastructure investments.

Virtualization separates a resource or request for a service from the underlying physical delivery of that service. With virtual memory, for example, computer software gains access to more memory than is physically installed, via the background swapping of data to disk storage. Similarly, virtualization techniques can be applied to networks, storage, server hardware, operating systems, applications, and so on.

Virtualization can be Management Virtualization, Network Virtualization, Hardware Virtualization, Storage Virtualization, Operating System Virtualization, Application Server Virtualization, Application Virtualization, Service Virtualization that are described in details.

Software-Defined Networking (SDN) is defined by Open Networking Foundation (ONF) as an emerging architecture that decouples the network control and forwarding functions enabling the network control to become directly programmable and the underlying infrastructure to be abstracted for applications and network services.

The switching plane can be heterogeneous, composed of network elements from multiple vendors, and it can provide distinct services with different characteristics, configurations, and control at the packet and/or optical layers. Abstracting the control plane from the network elements allows network-platform-specific characteristics and differences that do not affect services to be hidden. In addition, SDN is based on the principle that applications can request needed resources from the network via interfaces to the control plane. Through these interfaces, applications can dynamically request network resources or network information that may span disparate technologies.

As communication networks have grown in size and complexity, streamlining the architecture and implementation to reduce costs, simplify management, improve service provisioning time, and improve resource utilization has become increasingly important. Ideally, an application or service could be completely decoupled from the underlying network infrastructure, but this is not always realistic. Parameters that affect application performance, such as bandwidth, packet loss, and latency, are closely tied to the underlying network. To meet application performance objectives, it becomes necessary for the underlying network to be aware of the application requirements and provides the necessary services.

There is a great deal of optimism that SDN will make networks more flexible, dynamic, and cost-efficient, while greatly simplifying operational complexity. Vendors has begun unveiling its open network environment, extending network capabilities and extracting greater intelligence from network traffic through programmatic interfaces.

Rise of cloud services and virtualization, and large-scale computing and storage in huge data-centers require on-demand availability of additional network capacity (i.e. scalability), rapid service creation and delivery. SDN is expected to automate service provisioning at least and help service providers to deliver services much quicker.

The infrastructure consists of both physical and virtual network devices such as switches and routers. These devices implement the OpenFlow protocol as a standards-based method of implementing traffic forwarding rules. The control layer consists of a centralized control plane for the entire network to provide a single centralized view of the entire network. The control layer utilizes OpenFlow to communicate with the infrastructure layer.

The application layer consists of network services, orchestration tools, and business applications that interact with the control layer. These applications leverage open interfaces to communicate with the control layer and the network state.

Virtualization, Cloud, and SDN concepts are mainly focused on operational efficiency and maximum utilization of network resources. Further operational efficiencies can be achieved with self-managed networks. Network resources and services should be automatically provisioned, and faulty components should be automatically identified and fixed by the network itself. The future networks are very likely to be self-managed.

This chapter describes Cloud, Virtualization, SDN, and Self-Managed concepts, and provides examples.

2

RESIDENTIAL NETWORK ARCHITECTURES AND SERVICES

Belal Hamzeh, Brian Hedstrom, David Hancock,
Kenneth Gould, Kirk Erichsen, Niem Dang,
Vikas Sarawat, Alireza Babaei and Mehmet Toy

2.1 INTRODUCTION

Phone and TV services have been delivered over copper medium as the basic residential services for a long time. With the proliferation of broadband, fiber and wireless technologies, not only types of residential services but also rates of services and medium that these services delivered have been drastically changed. Rates have been increased from Kbps to Mbps. Wireless became a de facto medium for communications within homes and from home to outside world. Fiber is deployed to provide TV, voice, and Internet services (i.e., triple play), while serving rates over cable have been increased with new techniques in supporting triple-play services competitively. Home networking for appliances and communication devices is on the rise.

Cable companies (Multiple system operators or MSOs) worldwide have been providing residential services over coaxial cable using the Data over Cable Service Interface Specification (DOCSIS) protocol that is a set of telecommunications standards that permit the addition of high-speed data transfer to an existing cable TV(CATV) system.

Cable Networks, Services, and Management, First Edition. Edited by Mehmet Toy.
© 2015 The Institute of Electrical and Electronics Engineers, Inc. Published 2015 by John Wiley & Sons, Inc.

Toward the end of 1990s, MSOs initiated the PacketCable project to deliver real-time communications services, namely VoIP, over the two-way cable plant as an alternative to the existing circuit-switched telephony delivery platforms in terms of providing a richer feature set at lower cost Later, content distribution networks (CDNs) are created to deliver content to deliver video distribution over IP. The goal is to serve content to end-users with high availability and high performance.

In order to deliver television using IP over a packet-switched network such as a LAN or the Internet, instead of being delivered through traditional terrestrial, satellite signal, and cable television formats, Internet Protocol television (IPTV) system has been introduced. With IPTV services, MSOs aim to reach a large population of leased or consumer-owned devices for linear and on-demand video, and provide a viewing experience similar to traditional TV.

For delivering IP multimedia services, IP Multimedia Subsystem (IMS) that is an architectural framework was originally designed by the wireless standards body 3rd Generation Partnership Project (3GPP). Later, it was updated by 3GPP, 3GPP2, and ETSI TISPAN. IMS employs Session Initiation Protocol (SIP) for the establishment of IP multimedia sessions. The SIP enables the delivery of many advanced communications services to the user. MSOs provide both residential and commercial services over IMS platform today.

In order to centralize data and video applications on a single platform to reduce overall system cost, MSOs introduced Converged Cable Access Platform (CCAP). CCAP is expected to dramatically increase system capacity and density to enable the MSOs the ability to protect and maximize existing infrastructure investment, and provide orderly migration of existing video services to IP video.

In the following sections, we will describe MSO residential network architectures and services for the technologies introduced earlier.

2.2 DOCSIS 3.0/3.1 ARCHITECTURE AND SERVICES

The DOCSIS[1] system allows transparent bidirectional transfer of IP traffic, between the cable system headend, referred to as a Cable Modem Termination System (CMTS) or CCAP, and customer locations via a CM, over an all-coaxial or hybrid-fiber/coax (HFC) cable network. This type of service is generally offered to subscribers as high-speed data (HSD) Internet or broadband service. This is shown in a simplified form in Figure 2.1.

The CMTS and CCAP head-end devices reside within the MSO's core network and are generally considered secure. The CM resides within the subscriber's location and provides the demarcation point between the subscriber and the MSO's network. It is considered insecure and untrusted. Due to the different nature of each of these devices as well as differing business and technical requirements, they are managed differently from the MSO's back-office.

---

[1]CableLabs, DOCSIS, and PacketCable are trademarks of Cable Television Laboratories, Inc. and are used with permission throughout this book.

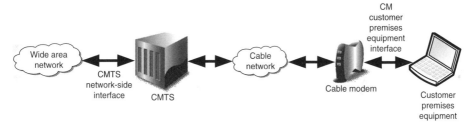

Figure 2.1. DOCSIS high-speed data or HSD services.

DOCSIS 3.0 protocol is the currently the latest deployed version of the DOCSIS protocol within MSOs' networks today. The major features introduced in the DOCSIS 3.0 protocol included the following:

- Upstream and downstream channel bonding.
- IPv6.
- Enhanced diagnostic tools.
- IP multicast.
- Enhanced security features.
- Enhanced statistics monitoring and reporting.

More recently, the DOCSIS 3.1 protocol has been introduced. The major features introduced in the DOCSIS 3.1 protocol include the following:

- OFDM downstream and OFDMA upstream channels.
- Scalable channel bandwidths.
- High-order modulations.
- Upstream and downstream profiles.
- Light sleep mode.
- Active queue management (AQM).
- Enhanced diagnostic tools for proactive network maintenance.
- Enhanced statistics monitoring and reporting.

The DOCSIS service and device management approach used for DOCSIS 3.1 system remains the same as DOCSIS 3.0 system for the CM management model. With respect to DOCSIS 3.1 management of the CMTS, the management model is aligned with the CCAP management approach, where configuration management using XML-based configuration files is used.

2.2.1 DOCSIS Evolution

DOCSIS 3.0 and DOCSIS 3.1 systems were designed with the long-term vision of an all IP service delivery in mind, thus achieving high capacities with predictable quality of service (QoS), concepts that are fundamental to the DOCSIS protocol design. For example,

DOCSIS 3.0 protocol supports 240 Mbps in the upstream and 1.3 Gbps in the downstream, while DOCSIS 3.1 protocol increases the upstream and downstream capacities to as much as 2 Gbps by 10 Gbps on the upstream and downstream, respectively. The DOCSIS 3.1 protocol allows expansion beyond 2 Gbps on the upstream and 10 Gbps on the down-stream by enabling support of multiple OFDMA and OFDM channels.

Additionally, the DOCSIS protocol provides QoS guarantees by using service flows which provide a mechanism for upstream and downstream QoS support. Service flows are characterized by a set of QoS Parameters such as latency, jitter, and throughput assurances. IP packets are routed to the appropriate service flow based on the required QoS; for example, a video stream would be routed to a service flow that has a guaranteed data rate with jitter limits, while an HTTP session would be routed to a best-effort service flow, given that HTTP packets are less delay sensitive than VoIP or video packets.

2.2.1.1 DOCSIS 3.0
DOCSIS 3.0 network is the currently deployed version of the DOCSIS protocol specification distinguished primarily by the addition of channel bonding, improved multicast capabilities (source-specific multicast or SSM) and incorporated IP Version 6 (IPv6), a more advanced bandwidth allocation method and other performance enhancements. The DOCSIS 3.0 protocol is now very mature technology. Its realistic limits will be reached in the coming years as upstream and downstream tier rates are pushed beyond what the protocol can realistically achieve over existing HFC plant (Figure 2.2). Data-over-Cable Reference Architecture from PHY v3.0 [1].

For all its virtues, DOCSIS 3.0 equipment becomes less efficient as you introduce more channels into the bonding groups of the upstream and to a lesser extent the downstream direction. Its forward error correction (FEC) and modulation techniques are limiting factors in achieving higher performance.

The DOCSIS 3.0 network's final generation of products will offer 8 channels upstream and 32 channels downstream for a raw capacity of roughly 240 Mbps in the upstream and 1.3 Gbps downstream with a MAC data rate of approximately 200 Mbit upstream by 1 Gbit downstream, assuming sufficient spectrum is provided. For example, eight 6.4 MHz channels will not fit within the cramped 5–42 MHz upstream band, requiring a change to the 85 MHz bandpass crossover point. Four channels at 6.4 MHz will consume the same spectrum and perform about the same as eight 3.2 MHz channels in a bonding group, all else being equal. Moving the passband to 85 MHz would allow for eight 6.4 MHz channels without interfering with any legacy upstream burst carriers, avoiding the dirtiest portion of the upstream's lower band edge and would be inherently "clean" spectrum. For more information on DOCSIS 3.0 systems, refer to MULPI v3.0 and PHY v3.0 [1, 2].

2.2.1.2 Next-Generation DOCSIS: DOCSIS 3.1
The next-generation DOCSIS system has improved on previous versions of DOCSIS in almost every way. Certain backward compatibility requirements have been relaxed, with full backward compatibility limited to the DOCSIS 1.1 specifications. The DOCSIS 3.1 protocol offers improved efficiency within the same spectrum, a more advanced and efficient bandwidth allocation scheme, an AQM algorithm with DOCSIS protocol

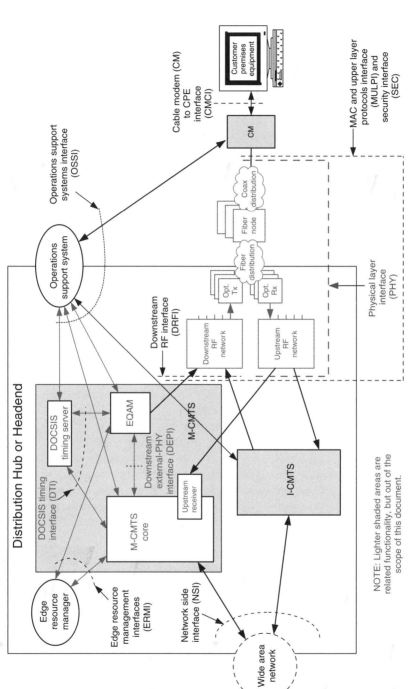

Figure 2.2. Data-over-cable reference architecture from PHY v3.0 [1].

NOTE: Lighter shaded areas are related functionality, but out of the scope of this document.

awareness that can improve the performance of latency sensitive applications and a new energy management capability to reduce power consumption.

The DOCSIS 3.1 protocol also established a set of technology targets, including 10 Gbps downstream and 2 Gbps upstream (raw PHY capacity) over the useful life of the protocol. First-generation DOCSIS 3.1 products will be required to support 5 Gbps raw downstream capacity per channel and as much as 21 Gbps raw upstream capacity depending on upstream diplex configuration and available spectrum.

The DOCSIS 3.1 protocol incorporates new modulation techniques such as orthogonal frequency division multiplexing (OFDM) to the physical layer (PHY) specifications using either a 25 kHz or 50 kHz (configurable) subcarrier spacing [3]. The OFDM modulation makes use of a new forward error correction (FEC) technique that greatly enhances the robustness of the DOCSIS system against interference on the plant.

The combination of new PHY capabilities provides a roughly 50% increase in spectral efficiency (expressed in terms of bits per second per Hertz (b/sec/Hz)) as compared to previous versions of DOCSIS systems within the same spectrum and operating conditions. The new scheduler's unique time and frequency division multiplexing (TaFDMA) design allows the narrow upstream spectrum band to be shared with legacy constant bit rate (CBR) voice services and set-top-box (STB) return, among others. Additionally, the upstream can rely on the OFDM protocol to selectively mute impaired or heavily utilized portions of the spectrum that are associated with legacy carriers or interference in order to avoid placing active subcarriers in those locations.

Additionally, the DOCSIS 3.1 specification specifically calls out mandatory support for radio frequency over glass (RFoG), which demands a scheduler configuration limiting simultaneous transmitters to one. This controls optical beat interference (OBI), which while statistically rare, is a unique aspect of the RFoG system.

2.2.1.3 Capacity Evolution in the DOCSIS Protocol

To meet the continuously increasing demand for capacity, the DOCSIS 3.1 system achieves higher capacities in comparison with the DOCSIS 3.0 system by enabling operation across larger bands of spectrum and improvement of spectral efficiency. In the following sections, we will address some of the key attributes that enable higher capacities in the DOCSIS 3.1 protocol.

2.2.1.3.1 Frequency Bands. Per Shannon's equation, the most effective way to increase capacity is to increase the utilized bandwidth, as the capacity is directly proportional to the signal bandwidth. A typical coaxial cable has a 3 dB bandwidth of more than 3 GHz, thus providing sufficient bandwidth to support higher capacities while keeping in mind that higher frequencies traversing through the coaxial cable suffer from higher attenuation in comparison to lower frequencies. The DOCSIS system utilizes frequency division duplexing (FDD), where the upstream operates in the lower bands, and the downstream operates in the higher bands, with a typical 26% transition region between both bands.

The DOCSIS 3.0 protocol mandates support of upstream operation bands from 5–42 MHz (5–65 MHz for Euro DOCSIS) and an optional upstream band of 5–85 MHz;

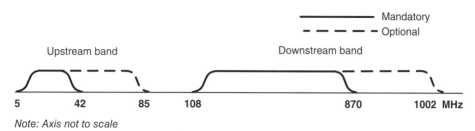

Note: Axis not to scale

Figure 2.3. DOCSIS 3.0 supported frequency bands.

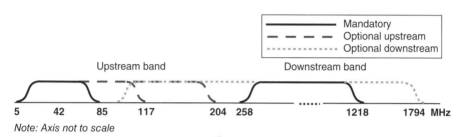

Note: Axis not to scale

Figure 2.4. DOCSIS 3.1 supported frequency bands.

while the downstream band operates in the 108–870 MHz with an optional downstream band of 108–1002 MHz (Figure 2.3).

The DOCSIS 3.1 protocol increases the supported bands for both the upstream and the downstream. On the upstream, DOCSIS 3.1 requires devices to support a minimum upper band edge of 85 MHz, and optional support for 117 MHz and 204 MHz. For backward compatibility, 42 and 65 MHz upper band edges are also supported. On the downstream, the DOCSIS 3.1 protocol supports 258–1218 MHz, and allows optional support for a downstream lower band edge of 108 MHz and an upper band edge of 1794 MHz, noting that using a downstream lower band edge of 108 MHz precludes using an upstream upper band edge greater than 85 MHz (Figure 2.4).

The transition to an 85 MHz or higher upstream upper band edge, although supported by the DOCSIS 3.1 system, is no simple task. MSOs will have to tackle some operational challenges starting with vacating the lower portion of the band of legacy analog channels, in addition to the out-of-band control channels used by some older set-top boxes (STBs).

2.2.1.3.2 Channel Bandwidths.
Physical layer communication in DOCSIS is performed in channels that transmit within the operational bands described in Section 2.2.1.1. In DOCSIS 3.0, upstream channels operate in 1.6, 3.2 or 6.4 MHz wide channels, and on the downstream operate in 6.4 MHz wide channels. DOCSIS 3.0 requires four upstream and four downstream channels with devices typically supporting a higher number of channels. Devices that support 8 upstream channels and 32 downstream channels are now commercially available.

In DOCSIS 3.1, the upstream channel bandwidth is configurable between 6.4 MHz (for 25 kHz subcarrier spacing or 10 MHz for 50 kHz subcarrier spacing) and 96 MHz. On the downstream, the channel bandwidth is configurable between 24 MHz and 192 MHz.

DOCSIS 3.1 channels are configurable at a subcarrier granularity, thus enabling a higher efficiency in utilizing the available spectrum and increasing spectral efficiency. In DOCSIS 3.1, a minimum of two upstream channels and two downstream channels are required, but these are significantly wider than what is found in previous versions of DOCSIS.

A DOCSIS 3.1 device that meets only the mandatory requirements will be able to support 192 MHz on the upstream and 384 MHz on the downstream; this is a huge increase in supported channel bandwidth in comparison with the latest commercially available DOCSIS 3.0 devices that support a modulated spectrum of 48 MHz on the upstream and 192 MHz on the downstream (8 upstream channels and 32 downstream channels). Also, DOCSIS 3.1 defines no limit on the maximum number of upstream and downstream channels, allowing devices to scale their capabilities to meet network capacity requirements.

To compensate for the reduction in power spectral density that results from using such wide upstream channels, the CM is required to provide a minimum of 65 dBmV of transmit output power, with additional power output permitted so long as the strict fidelity and spurious emissions requirements can be met. DOCSIS 3.1's PHY has much stricter fidelity and spurious noise emissions requirements than previous versions of DOCSIS, requiring a considerably cleaner amplifier implementation. Refer to PHY v3.1 (CM-SP-PHYv3.1-I01-131029) Sections 6.2.22, 7.4.12, and 7.5.9 for the updated fidelity requirements [3].

2.2.1.3.3 Modulation Techniques and FEC. DOCSIS 3.1 employs and extends low-density parity check (LDPC) as the forward error correction (FEC) code, a technique used in standards such as DVB-C2 (Digital Video Broadcasting—Cable 2), DVB-S2 (Digital Video Broadcasting—Satellite 2), long-term evolution (LTE), Multimedia over Coax Alliance (MoCA), and several others. However, DOCSIS 3.1 makes use of a more modern and efficient algorithm to simplify decoding operations. The LDPC provides roughly 3 dB higher SNR (signal-to-noise ratio) margin due to this improved coding, which translates into the ability to operate at higher modulation orders with greater robustness than previous versions of DOCSIS. Significantly, the upstream uses Orthogonal Frequency Division Multiple Access (OFDMA) to allow multiplexed access to the entire upstream band concurrent with legacy burst carriers. Additionally, DOCSIS 3.1 employs both time and frequency interleaving to increase robustness even further. Previous versions of DOCSIS employed only a downstream time interleaver that while configurable was very limited and was derived from the same ITU-T J.83 Annex B that governs generic QAM modulation requirements for the North American 6 MHz channel map. For more information on the time and frequency interleaver, refer to PHY v3.1 [3].

2.2.1.3.4 Subcarrier Bit Loading. DOCSIS 3.1 system supports higher and more granular bit loading (modulation orders) in comparison with previous versions of DOCSIS systems. DOCSIS 3.1 supports 4096 QAM (12 bits per symbol) on the downstream and 1024 QAM (10 bits per symbol) on the upstream with optional support of 8192 QAM and 16384 QAM on the downstream and 2048 QAM and 4096 QAM on the upstream. Additionally, the DOCSIS 3.1 supports the ability to reduce modulation orders down to 16 QAM in both upstream and downstream, as needed, with non-square modulation orders supported between the minimum and maximum modulations, on a per-subcarrier group basis. These modulation orders provide significantly higher spectral efficiency in comparison to DOCSIS 3.0, where up to 64 QAM was supported on the upstream and 256 QAM on the downstream.

Modulation orders (or bit loading) are assigned to CMs in terms of profiles, where a CM supports more than a single profile, and a profile may contain more than a single modulation order to enable even a finer granularity in modulation orders.

2.2.1.3.5 Upstream and Downstream Profiles.

Profiles are supported on a per Cable Modem (CM) basis so that CMs can benefit from the fastest modulation profile possible, which is particularly important on the upstream. On the upstream, the modulation order within a profile is set per mini-slot, and a profile can contain multiple mini-slots. The mini-slot is the building block representing the data grant supplied by the CMTS's scheduler to provide the CM with an upstream burst opportunity, as in previous versions of DOCSIS. While on the downstream, the modulation order within a profile can be set on a per subcarrier basis. In other words, the profile does not apply to the entire channel so much as it applies to the individual subcarrier groups within the channel, and all CMs are not forced to use the same modulation within the same service group unlike previous versions of DOCSIS.

Thus portions of the spectrum benefit from more robust modulation profiles as an alternative to not using that spectrum at all. A CM is required to support four concurrent profiles for live traffic on the downstream and two profiles on the upstream, with the CMTS always attempting to use the profile with the highest bit loading available to deliver the provisioned upstream and downstream data rate derived from the CM configuration file.

Profiles are channel independent which allows for great flexibility. For example, on the upstream, one channel could be allocated to the "dirty" portion of the spectrum between 6–25 MHz, while the second channel could be used between 26 MHz and 41 MHz (excluding the 1 MHz side lobes). The second channel could thus be configured with higher order profiles while the first channel could make use of more robust profiles appropriate to that portion of the spectrum. The new PHY will not only be equal to, but in fact exceed what was possible with Synchronous Code Division Multiple Access (S-CDMA) using Maximum Scheduled Codes (MSC) and Selectable Active Codes version 2 (SAC2) extensions which improved upon D3.0's S-CDMA's robustness when used in the lower portion of the return band.

DOCISS 3.1's profile assignment is flexible and is left up to the network operator on how to configure the profiles. For example, one option—using downstream profiles as an example—is for the profiles to be arranged between a base which is a minimum profile that is likely to work even in the most severely impaired plant conditions, and a top, high efficiency profile. Two additional profiles are situated between the lowest and highest profiles that the CM will use whenever the top profile cannot be used as a result of current operating conditions. Another option would be to have a base profile, a high-efficiency profile, and the remaining two profiles can be assigned for multicast and broadcast traffic. Additionally, profile definitions are not static, and a CMTS can choose to update the profile definitions as needed as plant conditions evolve. The CM will transparently switch between configured profiles to achieve the best balance of robustness and bit loading.

The DOCSIS 3.1 specification provides no specific requirements beyond the support for the profiles themselves for how vendors will implement switching. Vendors are free to implement intelligence in the switching of modulation profiles and the maintenance and automation of such profiles in both the upstream and downstream direction.

For example, a vendor could implement switching of profiles to occur after periodic probing and channel estimation determines a higher or lower profile should be used, or the testing of a proposed profile prior to the profile actually being assigned a portion of spectrum. Most vendors are expected to implement rapid switching of profiles to maintain the best possible bit loadings the HFC plant can support.

For more information on profiles, refer to PHY v3.1 (CM-SP-PHY v3.1-I01-131029) and MULPI v3.1 (CM-SP-MULPI-v3.1-I01-131029).

2.2.1.4 QoS in DOCSIS Systems

The varying QoS requirements of the applications transported over a data network require the network to be able to efficiently support various types of traffic. VoIP, online gaming, video streaming, and over-the-top services are just a few examples of such applications. Additionally, the migration to an all IP network, including the delivery of TV channels and video on demand over IP makes it critical the network to be able to support and guarantee QoS.

Applying QoS to traffic is a three step process as shown in Figure 2.5.

First the data packets go through a classifier, which is a set of packet matching criteria a classifier priority, and a reference to a service flow.

Various packet matching criteria can be used such as source or destination IP address, TCP or UDP port number, ToS/DSCP mask similar criterion.

If a packet matches the specified packet matching criteria of a QoS classifier, it is delivered to the referenced service flow. If the packet does not match any packet matching criterion, then it is delivered to the default (primary) service flow. Finally, the packet is scheduled for transmission using the scheduling service attached to the service flow.

2.2.1.5 DOCSIS Service Flows

A service flow (SF) is a MAC-layer transport service that provides unidirectional transport of packets on the upstream and the downstream. Service flows are characterized by a set of QoS parameters that define the traffic rates and latencies.

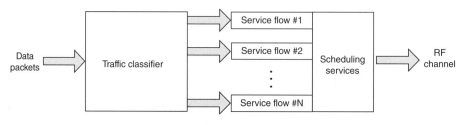

Figure 2.5. QoS process.

For each CM, a minimum of two SFs must be defined as one for the upstream and one for the downstream. Additional SFs can be defined to provide additional QoS services. The first SFs are the default SFs to which incoming packets are routed if the packet does not match any of the additional service flows. They are known as the primary service flows.

On the downstream, service flows can be either individual service flows or group service flows. Individual service flows apply to traffic forwarded by the CMTS to a single CM (unicast address), while group service flows apply to non-unicast destination traffic, such as Address Resolution Protocol (ARP).

Each service flow is assigned a scheduling service tailored to a specific type of data flow. A scheduling service is associated with a set of QoS parameters that define traffic rates and latency requirements such as maximum sustained traffic rate, minimum reserved traffic rate, maximum traffic burst, tolerated grant jitter, grants per interval, and so on. For more information on QoS traffic parameters, and their association with scheduling services, refer to MULPI v3.1 (CM-SP-MULPI-v3.1-I01-131029).

2.2.1.6 DOCSIS Scheduling Services

Data traffic patterns vary based on the type of traffic being carried, where the traffic pattern can vary in terms of periodicity and size of data packets. For example, VoIP without silence suppression has a periodic traffic pattern with constant size packets and low tolerance for latency and jitter, while streaming video traffic has a periodic traffic pattern with variable size packets and low tolerance for latency and jitter, while web browsing traffic has a random traffic pattern with varying packet sizes. The following scheduling services are supported in DOCSIS (CM-SP-MULPI-v3.1-I01-131029):

- **Unsolicited Grant Service (UGS):** Designed to support real-time service flows that generate fixed-size data packets on a periodic basis, such as Voice over IP. UGS offers fixed-size grants (in bytes) on a real-time periodic basis.
- **Real-Time Polling Service (rtPS):** Designed to support real-time service flows that generate variable size data packets on a periodic basis, such as MPEG video. rtPS offers real-time, periodic, unicast request opportunities, which meet the flow's real-time needs and allow the CM to specify the size of the desired grant.
- **Unsolicited Grant Service with Activity Detection (UGS-AD):** Designed to support UGS flows that may become inactive for substantial portions of time (i.e., tens of milliseconds or more), such as Voice over IP with silence suppression.
- **Non-Real-Time Polling Service (nrtPS):** Designed to support non real-time service flows that require variable-size data grants on a regular basis, such as high bandwidth FTP. The service offers unicast polls on a regular basis, which assures that the flow receives request opportunities even during network congestion.
- **Best-Effort (BE) Service:** Designed to provide efficient service to best-effort traffic. In order for this service to work correctly, the request/transmission policy setting should be such that the CM is allowed to use contention request opportunities.
- **Committed Information Rate (CIR):** Designed to guarantee a minimum data rate availability to the CM.

DOCSIS is the first access technology to directly incorporate an AQM into the standard. Previous versions of DOCSIS left the implementation of the CM and CMTS data forwarding queues entirely up to the vendor. Work within the IETF and at the university level had identified a phenomenon dubbed "buffer bloat," characterized by higher than optimal queue occupancy of data packets that in turn result in degraded performance of latency and jitter sensitive applications. In essence, too much buffering and an inability to dynamically alter such buffering can harm performance. This new field of study emerged in response to increasing demand for low latency flows to exist concurrently with high-capacity flows, where the high-capacity flows tend to bias the buffer queue depths up to a point that their negatively impacts the latency-sensitive application's responsiveness.

DOCSIS 3.1 incorporates an adaptation of the IETF's Proportional Integral Controller-Enhanced (PIE) algorithm that has been further extended to include unique aspects of the DOCSIS bandwidth allocation and token bucket scheme to further improve drop probability prediction accuracy. The DOCSIS PIE implementation can be applied on a per service flow basis in both the upstream and downstream direction and is required on both the CM and CMTS network elements. Additional AQM techniques are permitted if the implementer so chooses, allowing two or more AQMs in the CM or CMTS that are selectable for differing service flow requirements.

2.3 PacketCable ARCHITECTURE AND SERVICES

As the 1990s drew to a close, cable service providers (SP) began looking for value-added services that could ride on top of their newly deployed DOCSIS technology. An obvious choice was VoIP telephony service. Packet-based VoIP technology offered an intriguing alternative to the existing circuit-switched telephony delivery platforms in terms of providing a richer feature set at lower cost.

To kick this off, CableLabs and its member companies initiated the PacketCable project, whose task was to develop architectures and interface specifications for the delivery of real-time communications services over the two-way cable plant. The PacketCable project ultimately produced two service delivery architectures; first, PacketCable 1.5, which was designed to provide traditional voice telephone service, and subsequently PacketCable 2.0, a more complex and flexible architecture that was designed to provide advanced multimedia services beyond basic voice. PacketCable 1.5 and 2.0 both provided standards-based, scalable, and highly available end-to-end service delivery architectures that specified procedures for session establishment signaling, media transport with QoS, security, provisioning, billing, and other network management functions.

PacketCable 1.5 is primarily an IP-based replacement for the legacy circuit-switched telephone network deployed in the PSTN. It is designed to support a single service to a single type of device (i.e., residential telephone service to an analog telephone set). PacketCable 1.5 supports traditional telephony features, including local and long-distance calling with caller-ID, hold/conference/transfer, call-waiting, call-forwarding, and voice mail. Media support was originally limited to narrow-band voice

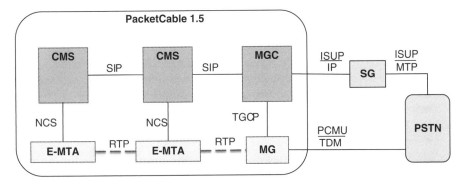

Figure 2.6. PacketCable 1.5 architecture.

and FAX (later, support for wide-band voice via DECT was added). PacketCable 1.5 also supports regulatory features such as emergency calling, electronic surveillance, and operator services.

The PacketCable 1.5 architecture shown in Figure 2.6 is commonly referred to as a "softswitch-based" architecture, where feature control is centralized in a single network-based entity called the Call Management Server (CMS). The CMS delivers phone service to a customer through a media terminal adapter (MTA) embedded in the CM at the customer's home (this endpoint device is called an E-MTA). The E-MTA supports a user facing RJ-11 interface to drive the metallic tip-ring pair of the analog phone, and a network-facing call-control signaling interface to the CMS.

A single CMS supports multiple E-MTAs, and so the PacketCable 1.5 network scales by adding more CMSs. A CMS can also play the role of a routing proxy to route calls between CMSs. The PacketCable CMSS protocol defined in Ref. [4] is used for call control signaling on the inter-CMS interface. CMSS is based on the Session Initiation Protocol (SIP) defined in by the IETF in RFC 3261 [5], and the Session Description Protocol defined in by the IETF in RFC 4566 [6].

The original framers of PacketCable 1.5 had to select a protocol for the call-control signaling interface between the CMS and E-MTA. One obvious choice was SIP, the same protocol used between CMSs. The downside of using SIP in this case is that it puts a lot of intelligence and hence complexity and cost in the endpoint. Therefore, the PacketCable 1.5 designers took a more lightweight approach by selecting an MGCP-based protocol called Network Call Signaling (NCS), defined in Ref. [7].

NCS is a master/slave protocol, where the CMS is the master and the E-MTA is the slave. The CMS tells the E-MTA to apply phone signals such as "apply dial tone," or "apply ringing." The CMS instructs the E-MTA to report phone events such as "the user went offhook," "the user dialed digit 2," or "the user flashed." Using a protocol that conveys these basic analog phone signals and events has two advantages. First, it minimizes the complexity and therefore the cost of the E-MTA. Second, it simplifies the specification work, since there is no need to mandate per-feature signaling flows; the CMS vendor is free to innovate how the CMS uses the NCS protocol to make each feature work.

The Media Gateway Controller (MGC) is a special purpose CMS that routes calls via the signaling gateway (SG) and the media gateway (MG) to SS7 trunks to the PSTN. The MGC controls the MG using a variation of the NCS protocol called TGCP [8], which is designed to control trunks. In addition to establishing the RTP media session to the MG, TGCP supports the special events and signals unique to legacy analog trunks used for operator services such as emergency call and Busy Line Verify.

Packetized media is carried between E-MTAs and between the MTS and the MG using the Real-Time Transport Protocol (RTP) profile defined in IETF RFC 3551 Ref. [9].

PacketCable 1.5 supports a dynamic QoS model where, at call establishment time, the E-MTA requests the DOCSIS bandwidth resources required to carry the media for the call. The CMS sends QoS policy information to the CMTS to ensure that the untrusted E-MTA can't ask for more bandwidth than the user is authorized to use.

PacketCable 1.5 specifies provisioning procedures and a data model for the E-MTA. Since it is embedded in the CM, MTA authentication (described in Chapter 3, Section 3.8.2.1) builds on the CM's authentication mechanisms. The PacketCable provisioning specifications describe the procedures for initial boot-up procedure, including identification and authentication, obtaining IP address, authentication, and downloading the configuration information to the E-MTA.

The PacketCable 2.0 project was initiated with the goal of developing a platform that would enable the delivery of new and advanced communications services to any type of user device, including mobile smart phones and tablets, laptops, STBs, and connected/smart TVs. Advanced services would include new types of media (e.g., hi-def voice, video, and text), new market segments (e.g., business services), and new features (unified communications, fixed/mobile convergence, and find-me/follow-me), plus new communication services yet to be invented.

The IP Multimedia System (IMS) architecture developed by the 3rd Generation Partnership Project (3GPP) was selected as the base architecture for PacketCable 2.0. 3GPP is a collaborative effort among multiple telecommunications standards development organizations (e.g., ARIB, ATIS, CCSA, ETSI, TTA, and TTC) that develop technologies and specifications for the telecommunications industry. IMS began as a mobile-centric service delivery platform over radio access, but has since evolved to become the dominant IP-based service delivery platform for real-time communications across all access technologies (see Section 2.4) for a more detailed description of IMS).

Some of the key design innovations of IMS over the centralized architecture of PC 1.5 are shown in Figure 2.7:

1. Decouples the application logic from the common application-agnostic core functions; that is, core functions that are common across all applications such as user authentication and authorization, message routing, and basic session establishment. The IMS core provides a well-defined SIP interface (the ISC interface) that the application logic can use to gain access to the common functions. This arrangement offers two advantages. First, it simplifies application development since the common core functions can be reused across multiple applications. Second, providing a standard application interface to the core enables SPs to select applications from different application vendors.

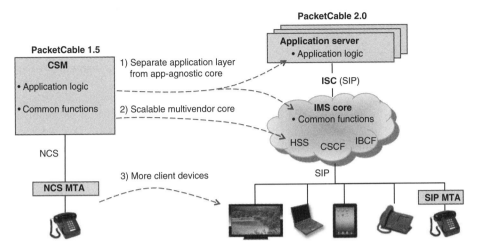

Figure 2.7. Comparing PacketCable 1.5 and 2.0.

2. Decomposes the IMS core into separate specialized functions with well-defined interfaces between each function. For example, IMS defines specific roles for functions such as the HSS (user database), X-CSCF (specialized types of SIP proxies), and IBCF (border element to peer networks). This more granular decomposition enables SPs to scale the IMS core in different dimensions (e.g., scale based on number of users vs. scale based on call capacity). It also enables SPs to deploy IMS with a mix of component solutions from different vendors.
3. Changes the client protocol from NCS to SIP. By moving away from the analog-phone-centric NCS protocol, SPs can offer more services to more types of client devices, including smart-TVs, mobile smart-phones and tablets, laptops, STBs, and SIP phones.

PacketCable 2.0 added some incremental extensions to IMS to support the unique technology, business, and operational requirements of the cable industry. Extensions include the use of additional or alternate functional components compared with the IMS architecture (e.g., the IMS MGCF is replaced with the PacketCable MGC), as well as enhancements to capabilities provided by the IMS functional components.

Some of the PacketCable 2.0 enhancements to the IMS include the following:

- Support for dynamic QoS for real-time multimedia on DOCSIS access networks, leveraging the PacketCable Multimedia architecture;
- Support for additional access signaling security and UE authentication mechanisms for endpoints with no SIM;
- Support for provisioning, activation, configuration, and management of UEs;
- Support for regulatory requirements such as number portability and PacketCable lawful interception;

- Support for signaling and media traversal of network address translation (NAT) and firewall (FW) devices, based on IETF mechanisms ICE, STUN, and TURN.
- Relaxation of certain mandatory procedures (e.g., signaling compression on the access network) that aren't required for cable.

The process for documenting the PacketCable 2.0 extensions to IMS was to first publish the extensions in PacketCable 2.0 "delta" specifications (e.g., Ref. [1]), and then contribute the changes to 3GPP so they could be incorporated into the actual IMS specifications. The extensions were added in IMS Release 7 and 8, so that all subsequent IMS releases now support the cable requirements.

Other PacketCable-only specifications were developed under the PacketCable 2.0 umbrella. One of the most important of these was the residential SIP telephony specification [2] that defines the SIP procedures to be supported by the UE and application server (AS) to provide residential phone service.

2.4 IMS ARCHITECTURE AND SERVICES

IMS defines an architecture and a set of open interfaces that provide a platform for the creation and delivery of IP-based real-time communication services.

Each user in an IMS network is identified by a globally unique identity, referred to as a public user identity. The public user identity is an SIP URI that identifies the domain of the SP and the user within that domain. IMS support both E.164-style URIs (e.g., sip:+13032241234@service-provider.com) and e-mail-style URIs (e.g., sip:john-doe@service-provider.com).

The public user identity is stored in the user's client device (i.e., the user equipment (UE) in IMS terminology). The public user identity is also stored in the IMS network, in the Home Subscriber Server (HSS) database function, along with the user's authentication credentials and other information required to provide services to the user's UE.

IMS leverages standard protocols defined in industry. One of the most important is the Session Initiation Protocol (SIP) defined by the IETF in RFC-3261 [3] for the establishment of IP multimedia sessions. This widely supported protocol enables the delivery of many advanced communications services to the user.

SIP supports many functions, including the following:

- **Registration:** The SIP REGISTER request enables the user equipment (UE) to attach to the SP network to obtain services. The REGISTER request contains the public user identity of the user, and the "location" of the UE on the network; that is, the contact IP address:port where the UE is listening for incoming SIP requests from the network. The network maintains the public user identity to location binding in a function called the SIP registrar.
- **Session establishment:** The SIP INVITE request enables the establishment of multimedia sessions between two UE devices. The INVITE request and its

response carry a message body encoded using the Session Description Protocol (SDP) defined in RFC 4566 [10] that enables the two endpoints to exchange media listening IP address:port, and negotiate a common set of media attributes, such as the media codecs (e.g., G.711, G.722, H.264, and VP8), using the offer/answer procedures defined in RFC 3264 [11].

- **Subscription:** A SIP entity issues a SIP SUBSCRIBE request to subscribe to events from another SIP entity using the procedures defined in IETF RFC 6665 Ref [12]. SIP defines a number of standard event packages, including "reg" to obtain the registration status of a UE (defined in by IETF in RFC 3680 [13]), "presence" to obtain the online/offline status of a user (defined in by IETF in RFC 3856 [14]), and "message-summary" to obtain the number of voice-mail messages stored for a user (defined by IETF in RFC 3842 [15]).

Another important protocol used by IMS is the Real-time Transport Protocol (RTP) documented in IETF RFC 3550 [16], which defines a standard format for carrying packetized media over IP networks.

IMS provides separate services to the originating and terminating users in a call. For example, traditional telephony includes originating services such as originating call blocking, speed dial, and autorecall/callback, and terminating services such as terminating call blocking, the various flavors of call-forwarding, and do-not-disturb. In addition, there are features such as call transfer that that can be invoked separately and independently by either the originating or terminating user in a call.

IMS supports separate originating and terminating services by adopting a half-call model, in which two instances of the IMS network are inserted in each call: one instance for the originating half of the call and a second instance for the terminating half of the call, as shown in Figure 2.8.

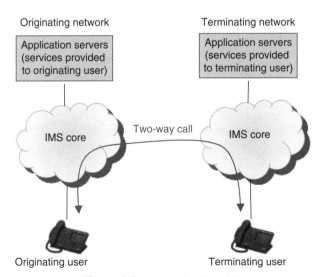

Figure 2.8. IMS half-call model.

2.4.1 IMS Architecture

The IMS architecture is defined in 23.228 "IP multimedia subsystem; Stage 2" [17]. A simplified version of the architecture is shown in Figure 2.9.

IMS is partitioned into the following four layers: local network, access network, IMS core, and application layer:

The local network contains the UE; the endpoint device that the customer uses to obtain services from the IMS network. The UE is basically a SIP user agent (UA) that is configured with the user's public user identity and authentication credentials, plus other data required to support services.

The access network connects the user's endpoint device (the UE) to the IMS Core. The access network could be any technology, for example, DOCSIS, 4G, DSL, Wi-Fi, and Fiber.

The IMS core contains common application-agnostic functions, such as user identification, authentication, and authorization, call routing, basic two-way session establishment, policy control, and many others. The IMS core provides a well-defined SIP interface, called the "ISC" interface, to the application layer.

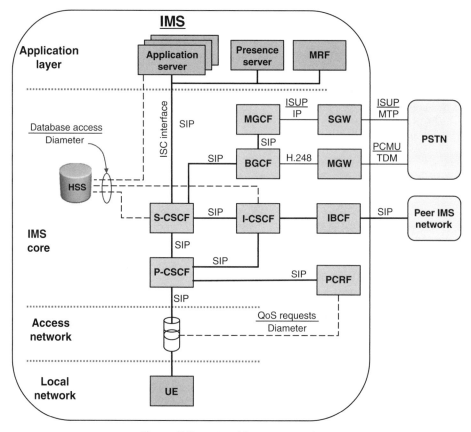

Figure 2.9. IMS architecture overview.

The Proxy Call Session Controller Function (P-CSCF), I-CSCF, and S-CSCF are all SIP proxies, each playing a different role in session control:

- **P-CSCF:** The P-CSCF is located at the SIP signaling trust boundary between the trusted core network and the untrusted access and local networks. The P-CSCF maintains a security association between itself and the UE of each authenticated user. When it receives an SIP request on an established security association, the P-CSCF asserts the public user identity of the sending user and routes the request to the correct S-CSCF. Likewise, when it receives a request from the network destined for an authenticated user, the P-CSCF routes the request to the target UE over the established security association.

- **Serving-CSCF (S-CSCF):** The S-CSCF serves as the "home proxy" for each user. The S-CSCF contains the SIP registrar function, and therefore knows the location of each registered user. All SIP requests received from or sent to the UE traverse the S-CSCF. Based on feature information obtained from the HSS, the S-CSCF invokes ASs via the ISC interface to provide originating and terminating services to its registered users.

- **Interrogating-CSCF (I-CSCF):** The I-CSCF provides two functions. First, it selects the S-CSCF/registrar for newly registering users; that is, when it receives the initial REGISTER request from a UE, the I-CSCF selects an available S-CSCF to process the REGISTER request. The selected S-CSCF becomes the new "home proxy" for the registering user. Second, when the I-CSCF receives an incoming SIP request from an originating network, it routes the request to the S-CSCF assigned to the target user.

- **Interconnection Border Control Function (IBCF):** The IBCF is located at the interconnect interface to a peer SP network. The IBCF provides multiple functions, including access control, topology hiding, signal interworking, and media transcoding.

- **Breakout Gateway Control Function (BGCF):** The BGCF selects the MGCF for calls destined to the PSTN.

- **Media GCF (MGCF):** The MGCF performs protocol conversion between the SIP and SS7 ISUP. It also provides connection control of the media channels in the MG.

- **Signaling Gateway (SGW):** The SGW performs signaling conversion at a transport layer between SS7-based transport and the IP-based transport used in the IMS network.

- **Media Gateway (MGW):** The MG provides bearer channel conversion between the circuit switch network and the IP RTP media streams in the IMS network.

2.4.2 Application Layer

The application layer contains one or more ASs. The AS contains the application logic that provides value-added communications services to the users. An AS may provide services to SIP sessions initiated by the UE, or it may initiate services or terminate services on behalf of a user.

2.4.3 Use Cases

The IMS core SIP procedures are specified in 3GPP TS 24.229 [18].The message flows for two mainline use cases—registration and session establishment—are described here.

The registration message sequence takes place when a UE first boots up and connects to the access network. Typically, two REGISTER requests are required to complete the registration process. The IMS network rejects the first REGISTER request with an authentication challenge response. The UE sends a second REGISTER request containing a challenge response based on the UE's stored credentials. The IMS network validates the challenge response to authenticate the user. The message flows for the first and second REGISTER requests are shown in Figures 2.10 and 2.11, respectively.

Initially, the UE contains Bob's public user identity and authentication credentials, plus the SIP signaling address of the P-CSCF (Figure 2.10). The HSS is configured with Bob's public user identity and credentials, the set of features and services that Bob is authorized to use, and the S-CSCF currently assigned to Bob (in this case no S-CSCF is assigned, since Bob hasn't registered yet).

1. The UE sends the initial REGISTER request (message [1] in Figure 2.10) to the P-CSCF, with a To header field containing Bob's public user identity, and a Contact header field containing the SIP signaling IP address: port of the UE (i.e., the contact address where the UE is listening for subsequent SIP requests from

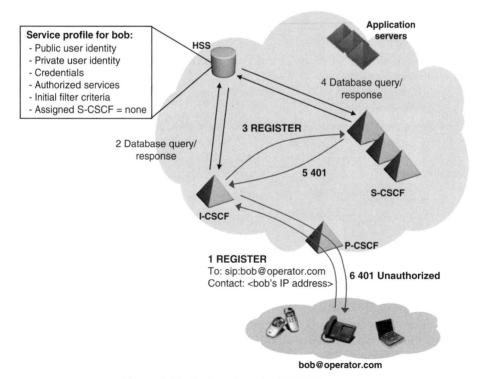

Figure 2.10. Registration—1st REGISTER request.

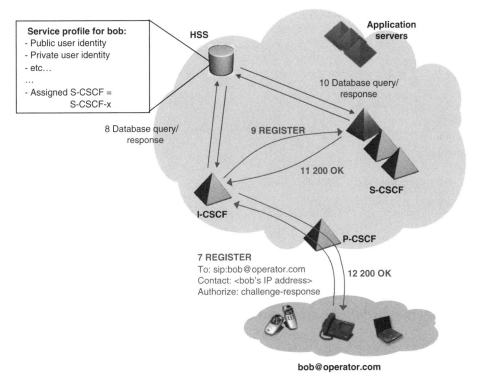

Figure 2.11. Registration—2nd REGISTER request.

the network). On receiving message [1] in Figure 2.10, the P-CSCF recognizes that this is an initial REGISTER since was not received over an established security association. The P-CSCF forwards the REGISTER request to the I-CSCF.

2. On receiving message [1], the I-CSCF queries the HSS via request/response [19] to determine whether the registering public user identity is currently assigned to an S-CSCF (i.e., whether the UE is currently registered or not). In this case, the HSS response indicates that there is no S-CSCF currently assigned (since this is an initial REGISTER request), and so the I-CSCF selects an S-CSCF. Before forwarding the REGISTER request to the selected S-CSCF, the I-CSCF records the public user identity to S-CSCF association in the HSS (i.e., the I-CSCF does not maintain state data across SIP messages).

3. On receiving message [3] in Figure 2.10, the S-CSCF recognizes that this is an initial REGISTER for an unregistered user (since there is no entry for this user in the S-CSCF's registrar database).

4. The S-CSCF queries the HSS via request/response [20] to verify that the registering user is authorized to receive services from this IMS network. In this case, the HSS response indicates that the registering public user identity identifies an authorized user.

The S-CSCF rejects the REGISTER request with a 401-Unauthorized response (message [5] in Figure 2.10). The S-CSCF maintains local state data that indicates it is waiting for subsequent REGISTER from the same user containing a challenge response.

5. The 401-unauthorized response (message [6]) is routed via the I-CSCF and P-CSCF to the UE.

6. On receiving message [6] as depicted in Figure 2.10, the UE generates a challenge response based on its stored credentials, and populates the challenge response in the Authorize header field of the second REGISTER request before sending the request to the P-CSCF, as shown in message [7] of Figure 2.11. The P-CSCF adds a Path header field defined in IETF RFC 3327 (Ref. [21]) to the REGISTER request identifying this specific P-CSCF, and forwards the request on to the I-CSCF as before.

7. On receiving message [7] in Figure 2.11, the I-CSCF queries the HSS via query/response [8] to determine whether the registering public user identity is assigned to an S-CSCF. The HSS responds with the S-CSCF that was assigned during initial REGISTER processing back in step [19]. The I-CSCF forwards the REGISTER request [9] to the assigned S-CSCF.

8. On receiving message [9], the S-CSCF recognizes that it is a challenge-response to a previous authentication challenge.

9. The S-CSCF obtains the authentication credentials for the registering public user identity from the HSS via query/response [1], and validates the challenge response contained in the Authorize header field of the received REGISTER request. In this case challenge-response validation check passes (meaning that the HSS and the UE contained the same shared-secret for this public user identity). The S-CSCF registrar function records a location binding entry mapping Bob's public user identity to his UE's contact address, and the P-CSCF path information contained in the received Path header field. The S-CSCF uses this saved path and contact information to route subsequent SIP requests to Bob's UE. At this time, the S-CSCF also fetches the initial filter criteria (IFC) for Bob from the HSS. This IFC identifies the ASs that must be invoked by the S-CSCF when it receives subsequent requests to or from Bob.

10. The S-CSCF accepts the REGISTER request by sending a 200 OK response [2]. The S-CSCF adds a Service-Route header field defined in IETF RFC 3608 (Ref. [22]) to the 200-OK response, containing a SIP URI that identifies this specific S-CSCF.

11. The 200 OK response is routed via the I-CSCF to the P-CSCF via message [12]. On receiving message [12], the P-CSCF creates a security association that contains Bob's public user identity, the contact address of Bob's UE, and other pertinent data. The P-CSCF uses this security association to recognize subsequent requests from Bob's public user identity. On receiving message [12], the UE saves the SIP-URI contained in the received Service-Route header field. The UE sends all subsequent SIP requests to this SIP URI (which is the SIP URI of its assigned S-CSCF), thus eliminating the need to route these requests via the I-CSCF.

The message sequence for establishing a basic two-way call is depicted in Figure 2.12.

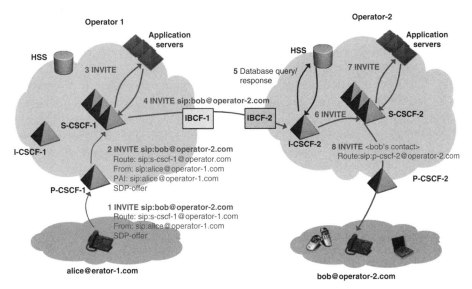

Figure 2.12. Session establishment.

Initially, Bob and Alice are both registered to their respective operator networks: Bob to Operator 1, and Alice to Operator 2.

1. As shown in Figure 2.12, Alice initiates a call to Bob from the contact list on her SIP phone (UE). Alice's UE creates an INVITE request with the following:
   - Request-URI containing Bob's public user identity;
   - A Route header field identifying the S-CSCF assigned to Alice, the S-CSCF SIP URI learned when Alice registered;
   - A From header field containing Alice's public user identity;
   - An SDP-offer describing the media capabilities of Alice's UE.

2. On receiving message [1], P-CSCF-1 first validates that the request is associated with the security association established by the P-CSCF when Alice registered. Once validated, P-CSCF-1 asserts that Alice is an authenticated user authorized to make this call, by adding a P-Asserted-Identity header containing Alice's public user identity. Downstream entities can trust that the P-Asserted-Identity contains the actual identity of the caller.

3. On receiving message [2], S-CSCF-1 uses the Initial Filter Criteria that was obtained when Alice previously registered to identify the set of ASs that must be invoked to provide services to Alice. In this case, the S-CSCF invokes the ASs identified by the IFC for originating services, since this is an "originating" request from Alice.

4. Once all originating ASs have been invoked via message exchange [3], S-CSCF-1 forwards the INVITE request [20] to the terminating network: Operator 2. The request is routed from IBCF-1 in the originating network to IBCF-2 in the terminating network.

5. On receiving message [4], I-CSCF-2 in Operator 2 network queries the HSS via query/response [23] to identify the S-CSCF assigned to Bob, and forwards the request to that S-CSCF (S-CSCF-2 in this example) via message [6].

6. On receiving message [6], the S-CSCF-2 uses the Initial Filter Criteria that was obtained when Bob previously registered to identify the set of ASs that must be invoked to provide services to Bob.

7. In this case the S-CSCF invokes the ASs identified by the IFC for terminating services, since this is a "terminating" request to Bob. Once all terminating ASs have been invoked via message transaction [7], S-CSCF-2 forwards the INVITE request [8] to all UEs currently registered for Bob; in this case, one UE. The S-CSCF retargets the INVITE Request-URI with the registered contact address of Bob, adds a Route header containing the path information that was saved by the S-CSCF when Bob registered, and sends the request.

8. On receiving message [8], Bob's UE alerts Bob of the newly arriving call, and displays the calling identity contained in the received P-Asserted-Identity header. When Bob answers the call, Bob's UE responds to the INVITE with a 200 OK response (not shown), containing an SDP answer describing the media capabilities of Bob's UE. Once this 200 OK response arrives at Alice's UE, both UEs have all the information they need to establish a bidirectional media stream between the two UEs.

2.5 IPTV ARCHITECTURE AND SERVICES

In this section, the term "IPTV" refers to the replication and enhancement of the traditional pay-TV viewing experience delivered to end users over an IP-based networks. While many people will refer to the same service as "streaming video," "video over IP," and "adaptive bit-rate video," many implementations of services that go by those terms do deliver video over the Internet. However, those services do not necessarily deliver the features one would associate with a pay-TV service such as closed captioning, parental controls, emergency alert messaging, and accessibility of services for the disabled; all of which are features that are expected of a cable TV company's video service whether delivered traditionally over QAM technologies or over IP.

MSO's goals for delivering IPTV services are to reach a large population of leased or consumer-owned devices, enable linear and on-demand video to non-technical subscribers, provide a viewing experience similar to traditional TV, generate continued revenue via video subscriptions, video rental and advertising, and meet regulatory requirements for video-related features.

The main components of the IPTV service delivery architecture are the following:

• Video acquisition;
• Video encoding;
• Video encryption and digital rights management;
• Video distribution;
• Video navigation and presentation.

2.5.1 Architecture

An MSO provides video services that comprise local linear video, national linear video, video-on-demand (VOD), and pay-per-view (PPV). The local linear video consists of channels only available for viewing in a limited geographic area such as your local ABC, NBC, and PBS affiliates along with local public, educational, and governmental channels (referred to as "PEG" channels). The national channels consist of channels that are typically made available to viewers from coast-to-coast such as HBO, TBS, TNT, and ESPN. Traditionally, the MSO would have off-air receivers or direct video connections to the local providers of linear content and would depend on satellite receivers for the reception of nationally available linear channels.

National VOD programming is typically delivered to MSOs over satellite as well, where the video asset and its associated metadata are placed on a satellite by the content programmers, beamed by the satellite to all of the MSOs who are authorized to distribute that content, received by the MSOs, and placed onto the appropriate VOD distribution servers and represented in a navigational client so that customers can find the video content. VOD programming distributors are being to transition to the delivery of lightly compressed, "Mezzanine-quality" VOD content over wired network connections [24]. Local VOD content, such as local sporting events are delivered by a variety of means, whether over an IP delivery or via videotape. PPV is typically of a live event, so is delivered over satellite just as a linear program is (Figure 2.13).

An MSO that served 20 different geographic areas might have 20 different video reception sites that received video destined for customers on the 20 different subregions. As the cost of fiber-based IP transport has fallen, it has become increasingly economical for large MSOs to centralize the video acquisition processes into one or two sites and to deliver the acquired nationally available video to the 20 subregions from that national site. The use of a second site is usually intended to provide a continuity of video services in a case of a disaster befalling the other video acquisition site (Figure 2.14).

IPTV services are destined for video-displaying devices that are either leased by the MSO to the subscriber or devices that are consumer-owned and managed (COAM). Typically, the goal of the IPTV service is to deliver similar linear, on-demand, and PPV services that a consumer enjoys on a traditional STB, but with the added convenience of device portability, varying screen form-factor, and perhaps, video-storage capabilities. The IPTV devices consist of desktop PCs, laptops, tablets, mobile phones, gaming consoles, IP STBs, and Internet-connected TVs (referred to here as "SmartTVs").

Given the variety of these devices, the varying in-home network connections that those devices use, and the fact that the throughput of those networks can change by the second, it is necessary to provide a means of delivering video over IP that can react to the varying network throughput. This is especially true for mobile devices utilizing a wireless network connection. Encoding the video as "adaptive bit-rate (ABR)" video is the solution to the varying-bandwidth problem [25]. ABR involves encoding video at multiple bit-rates and resolutions simultaneously, breaking the resulting video stream into small segments that each contain a few seconds-worth of video content, and allowing a video player on the video-rendering device to request the version of the video file

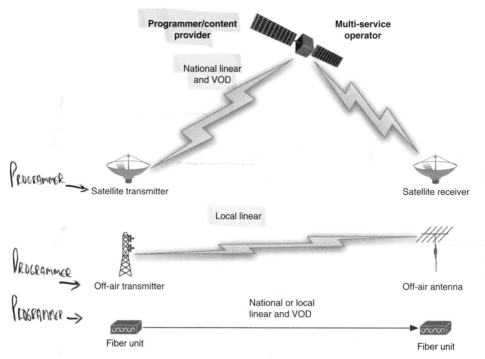

Programmer/content provider   **Multi-service operator**

National linear and VOD

PROGRAMMER → Satellite transmitter   Satellite receiver

Local linear

PROGRAMMER → Off-air transmitter   Off-air antenna

National or local linear and VOD

PROGRAMMER → Fiber unit   Fiber unit

Figure 2.13. Video delivery from programmers to MSOs.

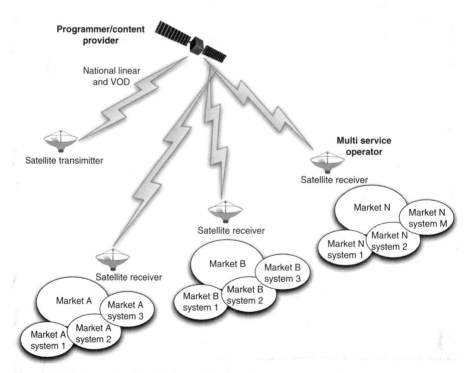

Figure 2.14. Example of multiple satellite-based reception sites for a single MSO (Adobe, 2012).

that can be delivered over a bandwidth-constrained network. The video would typically be encoded into H.264 with audio encoded with AAC, wrapped into MPEG-2 transport streams.

There are several proprietary adaptive bit-rate solutions: Apple has Apple HTTP Live Stream (HLS) [26] Adobe has Adobe HTTP Dynamic Streaming (HDS) [25], and Microsoft has Microsoft Smooth Streaming [27], as examples. Having a variety of video available on the Internet encoded in different video formats, delivered by different protocols and requiring different video player plugins leads to a inconsistent user experiences across devices for the same video services, requires MSOs to prepare and deliver video services in multiple incompatible formats, and increases the chances of service outages due to the need for multiple proprietary skill-sets by MSO personnel. As a result, international standards were created to provide a means to move away from proprietary IPTV delivery solutions to solutions that can interoperate. The 3GPP's Release 9 [28] and HTTP Adaptive Streaming Release 2 [29] from the Open IPTV Forum were created to specify the profiles for video data being delivered over HTTP to mobile/wireless devices. These specifications were followed by 3GPP Release 10 DASH and the MPEG-DASH specifications [30]. While MPEG-DASH is an international standard for adaptive bit-rate HTTP-based streaming, MPEG-DASH and the proprietary solutions listed earlier work in similar manners and the following descriptions are not specific to any one implementation.

For example, an MSO could elect to encode a linear video stream simultaneously in three different bit-rates: 5 Mbps, 1.5 Mbps, and 700 kbps. The video encoder converts the video stream into the appropriate format (i.e., H.264) and a segmenter divides the resulting video into a series of small files each representing a few seconds-worth of video. The segmenter also creates a manifest file or Media Presentation Description (MPD) file, in the vernacular of the Open IPTV Forum, that is delivered over HTTP to the video player, where the manifest file informs the player of the availability of the video stream at the different bit rates [29]. The player will request the lowest bit-rate version of the video stream, and then request the next higher bit-rate video if the network bandwidth is sufficient to allow its delivery. During times of network congestion, the video player will likewise request the lower bit-rate version until the player detects that the network has become uncongested.

In this ABR-based world, the client/player has responsibility for determining the best bit-rate/resolution of video to request. The client does so by monitoring the playout buffer, local resources (CPU and memory utilization), and the number of dropped frames.

The video player requests the URL of the manifest file. The video player continuously reads from the manifest file (which is a text file) where that manifest file contains a list of the files that represent the chunk of video in order. The manifest file is continuously updated as new video files are created by the segmenter. Note that the format of the manifest file varies based on the vendor or standard being used: Apple HLS uses a m3u8-based text file [26], Microsoft Smooth use a XML file [27], as does an MPEG DASH-based client [30] (Figure 2.15).

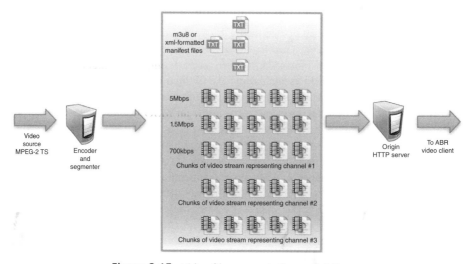

Figure 2.15. Video file segmentation and delivery.

There are benefits of delivering the video adaptively over HTTP. Generic HTTP web servers can be used instead of expensive and proprietary purpose-built video servers and the adaptive streaming over HTTP video content can traverse any router or firewall that regular HTTP-based web traffic traverses.

A drawback of using adaptive streaming is that, for now, all delivery is unicast. MSOs cannot take advantage of the network efficiencies as in the case where the video content is being multicast and delivered to multiple consumers of the video who were reachable by the same "last-mile" network segment.

2.5.2 Video Encryption and Digital Rights Management

Traditional cable TV services use "scrambling and conditional access (CA)" to ensure that only authorized subscribers and authenticated STBs can receive most video services. In North America, the most prevalent CA systems are provided by Cisco (for PowerKEY CA), and Arris (for DigiCipher CA). For both CA systems, video is encrypted using a strong encryption mechanism, and the video-rendering devices located in the consumers' homes has a hardware component that is capable of decrypting the video if the device is entitled to that video service. The business rules that need to be supported by these CA systems are relatively simple: either a device is entitled to a video service like HBO or it is not; either a device is entitled to consume a basic tier of video service or it is not.

For IPTV, there is a desire to not only replicate the simple traditional CA use cases but also expand upon them to allow a video service or individual piece of content to be viewed on a certain device (whether leased or consumer owned) at a certain location (whether from the customer's home network, or out of home), and at a certain time (usually until the customer stops paying their cable bill). There is also a desire to offer video service such as the ability to download content to a device for playback when the device is not on a data network, for example on a plane. This is referred to this as a "download-to-go" service.

All of these video services require some form of content protection to allow the MSO to implement content protection, rights management, and content consumption rules imposed upon the MSOs by their content-providing partners. The implementation of these content protection rules is referred to as "digital rights management (DRM)."

As examples, some content owners will allow an MSO to deliver content to any device in the world, to any device in the United States, to any device in the subscribers' home, or to any device more than 50 miles from the physical location of a sports arena. Likewise, MSOs have the responsibility to deny access to video from rendering devices that the content owners believe do not sufficient protect against un-authorized redistribution of that video.

Many of the major players in providing Internet-based video players have created DRM solutions: Apple has FairPlay DRM, Microsoft has PlayReady DRM [31], Google has Widevine [32], and Adobe has Adobe Access [33]. While the providers of video players, such as Adobe, Apple, and Microsoft may have implemented the same ABR capabilities allowing their players to consume the similar HTTP-based video streams, there is no ubiquitous DRM solution that is freely available to enforce content protection upon the plethora of consumer-owned devices with which MSOs would like to target their IPTV services. As a result, MSOs may be forced to implement multiple DRM solutions to target multiple COAM platforms, and this results in video being encrypted with multiple DRM solutions—a duplication of video content being transported over a CDN, and a duplication of pre-encrypted content being stored on video servers.

Video content protection consists of three layers of protection: encryption of the content, distribution of licenses to players, and distribution of keys. Historically, the content encryption mechanisms have been proprietary to the vendors' DRM architectures. International standards have been created to harmonize the encryption algorithms so that the same encryption mechanism can be shared by multiple DRM architectures. One such standard gaining traction within the DASH community is the ISO MPEG DRM-interoperable Common Encryption (CENC) Standard [34].

An MSO providing IPTV services will usually be providing multiple tiers of video access to their subscribers for example, a basic video package (local video affiliates), "enhanced" video package (CNN, TBS, etc.), premium video packages (HBO, etc.), and access to individual video channels and video assets. Before requesting a video stream, a user has some form of user interface on their PC or tablet that provides the user with the video-watching choices available to the user. It is typical that an IPTV-consuming device would be challenged for user credentials to first determine which video choices to show in the navigational interface. There are standards and protocols for authorizing and authenticating users and devices, for example, Online Multimedia Authorization Protocol (OMAP) [35] and OAuth 2.0 Authentication [36].

After successfully authenticating a user and a device, the user is able to request a video stream (by requesting files represented by a URL). It is this request for a file that causes the video client to request a DRM license that will provide the DRM key allowing the video client to decrypt the associated DRM-protected video stream.

The preceding discussion in section 2.5.2 is a grossly simplified work-flow for digital rights management. The important point is that the IP video ecosystem is changing rapidly in an attempt to create common encryption, authentication, and DRM business rule management to allow MSOs to deliver protected content to a large collection of consuming devices without requiring the MSO to maintain silos of video content based upon the multitude of DRM solutions supported by consumer-owned and managed devices.

2.5.3 Advertising Insertion

Video content to be consumed by an IPTV service is acquired via satellite, off-air-receiver or terrestrially. That video is routed into an encoder that generates the associated manifest file and adaptive bit-rate encoded MPEG-4 content, and that content is then routed through a DRM encryptor for content encryption and is placed on the CDN's origin servers.

The video production and distribution financial eco-system is dependent on the generation of revenue from subscription services or from the insertion of advertising. Agreements between programmers and MSOs allow the MSO to insert ads during ad breaks in linear and on-demand content. This discussion is about the insertion of advertising performed by the MSO. The assumption is that any advertising insertion performed on behalf of the programmer/content originator has been performed by that programmer prior to delivering video content to the MSO.

The actual advertising asset is delivered to the MSO whether on tape, via satellite or over IP. Metadata about which network the ad can be inserted on, how often the ad can be inserted, and the window of time in which the ad can be inserted are also delivered with the asset. That ad asset is typically encoded into H.264 format and stored on an ad server. The ad server is simply another origin server on the CDN. Recall that the video programming is being encoded into multiple bit rates as part of the adaptive bit-rate encoding process. The ad assets must also be encoded at the bit rates and same video profiles as the entertainment content to avoid any client player anomalies.

For both on-demand and linear video, recall that each ABR video stream is delivered to a consumer as a unicast stream, unlike the traditional broadcast delivery of linear video to a traditional QAM-base TV. As such, the advertising inserted can be tailored specifically to the consuming device, allowing a different ad to be inserted into each video stream. This allows dynamic ad insertion resulting in accurate ad placement forecasting, user targeting, the ability to control the frequency of an ad's appearance, and accurate ad placement reporting. For most HLS-based ad insertion implementations, the video client is not involved in the ad placement decisioning process; all insertion of ads take place in the network, rather than at the video player itself.

For On-Demand video, an Ad Decisioning System determines which ads will be placed in each of the ad placement opportunities prior to the streaming of the on demand video asset to the consuming player. This allows the manifest file for the video to be generated once, containing the URLs for the entertainment asset chunks as well as the ad asset chunks.

For linear video, the manifest file representing the URLs for each of the entertainment asset chunks can be dynamically modified to insert the URLs representing ad asset chunks. This allows ads to be either inserted between segments of the entertainment asset or to replace preexisting ad content. This manifest manipulation is supported by a manifest manipulator that is a proxy for requests for manifest files made by a video client. In order for this manifest manipulation-based ad insertion to appear seamless to the end user, the chunks representing the entertainment asset and the ad assets must be conditioned to align the ad break boundaries with the entertainment asset chunk boundaries. The manifest file must also contain tags indicating where the advertising URLs can be inserted into the manifest file.

When linear MPEG2 video is provided to an MSO, if the MSO and the content provider have negotiated for the MSO to have the opportunity to insert ads into a video stream, then the video stream must contain markers indicating when an opportunity to insert ads exist. The format of these messages is well-defined by the SCTE-35 specifications, known as "Digital Program Insertion Cueing Message for Cable" [37].

When the MSO's video encoder encounters an SCTE-35 message indicating the opportunity to insert an ad, the encoder needs to determine which ad should be inserted and cause the manifest file associated with that video stream to be modified to include the video chunks associated with that advertisement.

There are a number of systems needed in order to seamlessly place ads into linear or on-demand video:

- The video encoders which have to recognize when an ad placement opportunity exists in a video.
- A Placement Opportunity Information System (POIS) that keeps track of the number of ads and the duration of those ads that can be inserted into a particular linear video stream or on-demand program.
- A Campaign Management System (CMS) in which ad campaigns are defined to indicate the timing window for a campaign, the time of day that an ad is to be seen, as well as defining the networks on which an ad is to be inserted.
- An Ad Decisioning System (ADS) that determines what ad to insert.
- A Manifest Manipulator that modifies a video program's manifest file to include URLs associated with the ad asset.
- A Content Information Service (CIS) that determines or verifies that ad assets are available for insertion on an Ad Server.
- Optionally, a Subscriber Information Service (SIS) that contains demographic information specific to a subscriber that can be used to determine which ad to deliver to a specific video player based upon information about the associated subscriber.

The providers of these systems have choices when it comes to the standards and protocols used to allow these systems to interoperate. The Society of Cable Telecommunications Engineers (SCTE) have a family of specifications known as SCTE-130 [38], and the Interactive Advertising Bureau (IAB) has a Video Ad Serving Template (VAST) [39], Video Multiple Ad Playlist (VMAP) [40], and Video Player-Ad Interface Definition (VPAID) [41] that can also serve a subset of these communications functions. In addition, CableLabs' Event Signaling and Management API specifications describe the interfaces for communications between encoders, packagers, and POISs [42].

1) The encoder receives video with SCTE indicators which implies ads can be inserted.

2) The encoder talks with w/ the POIS to see if this opportunity can be fulfilled. IE An Ad w/ the correct duration & other criteria exists.

3) The content may be segmented at ad insertion points (encoder learns this via the POIS). Content is sent to segmenter if true.

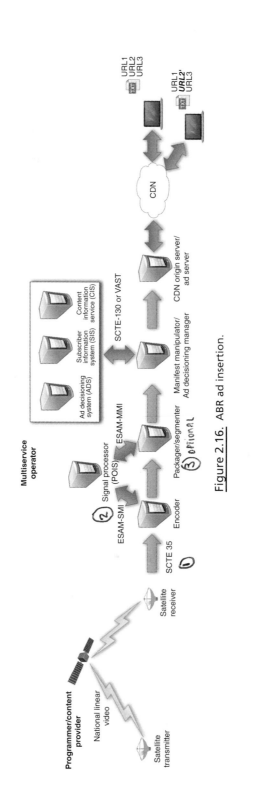

Figure 2.16. ABR ad insertion.

Figure 2.16 shows a simplified view of ABR-based ad insertion for linear video. In this example, linear video is received by an MSO over a satellite receiver. The video is provided by the programmer with SCTE-35 cues embedded within the MPEG-2 video transport streams indicating where the MSO's ad insertion opportunities exist. The encoder recognizes the SCTE-35 cue tones, exchanges messages with the POIS to determine whether the ad placement opportunity should be ignored or not and conditions the video stream appropriately. The packager is likewise informed by the POIS whether to segment the program stream on ad break boundaries.

When the subscriber requests to watch the linear video, the player requests the manifest file from the CDN. The Manifest Manipulator queries Ad Decisioning System, which in turn interacts with the CIS and SIS to determine which ads are appropriate for that video stream. In Figure 2.4, the first video player's manifest file retrieved directly from the origin server merely contains non-manipulated URLs. The second video player's manifest file fetched through the Manifest Manipulator is updated has been manipulated to include a URL 2, which represents the URL for an ad to be inserted for that client. Note that an ad segment would normally be represented by more than one URL chunk; this diagram has been simplified.

What we have described is the systems needed to replicate, over IP, the ad-viewing experience that people have long experienced on traditional cable systems. However, an IPTV ad-viewing experience can offer much more. There are opportunities for ad targeting, ad individualization, and interactive ads. Any of these topics could consume an entire book, and will not be addressed here, but the important technologies related to inserting targeted or interactive ads are the same as we have described for simple linear and on-demand ad insertion.

2.5.4 Emergency Alert Messaging

An MSO is required to provide emergency alert messaging to subscribers via their TVs. These messages warn of tornados, flash floods, and local and national emergencies. It is generally accepted that IPTV providers should provide equivalent warnings to IPTV clients.

An MSO maintains systems that receive emergency alert messages as both text and audio files along with metadata indicating to which geographic areas an emergency alert message pertains. The MSO typically monitors servers maintained by FEMA and state-specific server that support the Common Alerting Protocol (CAP) [43]. These servers provide the appropriate state and national emergency alert messages that MSOs are responsible for providing to QAM-based video subscribers.

For legacy QAM-based video systems, the MSO would know the location of the customers' STBs and would only receive emergency alert messages for the geographic area in which their customers' homes were located (i.e., Dallas, TX, or New York City).

With the advent of IPTV, the location at which the video user receives IPTV services can be very different from the user's billing address. An IPTV video-rendering device would need to register its location with the MSO, via the service address of the users' account, by "opting-in" to inform the MSO as to which geographic areas the user is interested in receiving EAS messages, or by the MSO attempting to discern the location of the device via geo-IP address location or self-reported GPS co-ordinates [44].

The simplest solution is to interrupt the viewed video stream and use the manifest manipulator to "redirect" all video manifest files to include URLs for the emergency audio message and to static video files that represent a message that an emergency alert is being broadcast.

More complicated solutions involve pushing emergency notifications to each device outside of the delivery of video streams. This solution becomes complicated for MSOs that provide video services to multitudes of different COAM devices for which the MSO has no responsibility for providing the actual client player on the device.

2.5.5 Closed Captioning and Accessibility

The Twenty-first Century Communications and Video Accessibility Act in the United States requires the delivery of closed captioning information to IP video-rendering devices. The format for closed captions is defined in CEA-608/CEA-708 [45] and must be provided by the content providers in the MPEG-2 transport streams that are the inputs to the MSOs' encoders.

The display and synchronization of closed captioning data at an IP client varies based on the ABR format used by the video player. The caption formats for Apple HLS is CEA-608, for Adobe HDS is onTextData, for Microsoft SmoothStreaming is "text tracks" and for MPEG-DASH is SMPTE-TT [46].

Consumer-owned and managed devices provide great flexibility for providing access to video programming for consumers with disabilities. COAM devices such as the Apple iPAD can be paired over Bluetooth with braille readers that a blind user can use to navigate the user interface for a video player. Likewise, these COAM devices typically have screen readers and the ability to easily magnify/zoom the user interface as well as compatibility with hearing aids. In addition, some COAM devices provide the ability to customize the inputs for the device for users with limited control of their hands or to help users with attention and sensory impairments.

2.5.6 Sports Blackouts

There are many providers of sports-related video programming who have contracts that restrict when and where certain sports videos can be viewed. For example, a particular baseball game may only be viewable within a certain distance from the live sports field if a certain percentage of the seats in the stadium are sold for that game. Sports are "blacked-out" in a geographic area when the content provider is contractually prohibited from delivering the video to potential viewers in that area.

Historically, the content provider could black out an event by directly controlling the satellite receiver physically located in the geographic area to be blacked out. With the trend of centralizing video acquisition, there is no tie between the location of a satellite receiver and the area where the video will be viewed.

To implement the centralized acquisition of video for potentially blackout content requires two things from the content providers: they must implement SCTE-35 Blackout Indicators and implement an Event Scheduler that supports the Event Scheduling Interface (ESNI) specification as defined by CableLabs [47]. The Blackout Indicators are signals embedded within a video stream that indicate the beginning and ending of a blackout event. The Event Scheduler is used to inform the MSO as to what alternative content should be inserted into a video stream when content is being blackout out for subscribers in a geographic area.

The MSO has to implement a POIS also known as an Alternate Content Decisioning System that will be responsible for making decisions about whether a particular user is to receive potentially blacked-out content, and if the user is to not receive the content, determine what alternative content to provide. The POIS decisions are determined by information provided by the content provider using the ESNI specifications and based on other information that only the MSO can determine, such as the physical location of the device requesting to view the video.

As of January 2014, the ESNI/POIS-based solution described earlier had not been reduced to practice in a production environment, but the authors assume that such a solution will be deployed as a long-term solution.

2.6 CDN ARCHITECTURE AND SERVICES

Content Distribution Networks (CDN) are created to solve a number of problems that are as old as the Internet itself: How can an entity deliver content, whether video or web pages as quickly as possible, with the least latency, at the lowest cost, to the widest quantity of consumers who are spread over a disparate geographic area?

We are choosing to use the terms "Content Distribution Network" and "Content Delivery Network" interchangeably and will abbreviate both terms as "CDN." A CDN is an interconnected network of servers used to deliver web assets to consuming devices in an efficient manner, taking advantage of caching of content to reduce or eliminate the retransmission of a single asset to multiple consumers.

In this chapter, we will describe an MSO's goals for deploying a CDN for video distribution over IP, the architecture of a CDN, services supplied by a CDN and areas for future CDN research. Private CDN components can be built or purchased by MSOs or can be run on behalf of MSOs by companies such as Alcatel Lucent, Arris, Concurrent, Huawei, Verivue, and others.

The goals of a CDN are to assure the cost-effective availability of IPTV video content, maintain or improve the quality of the IPTV viewing experience, and limit operational maintenance challenges including reporting on network utilization and usage (Figure 2.17).

Without a CDN, when accessing a webpage, web object, or streaming video via a URL, the request from the consuming device and the response from the web server that hosts the web object must traverse some number of network elements between the web server and the consuming device and that traversal takes some amount of time, typically measured in milliseconds which could extend to seconds.

The idea of the CDN is to place a copy of the requested web object onto an Edge Web Cache (EC) that is closer to a consuming device. The benefits of caching the content closer to the consuming device are twofold: As there are fewer network hops for the content to traverse, it should take less time to be received by the consumer and the

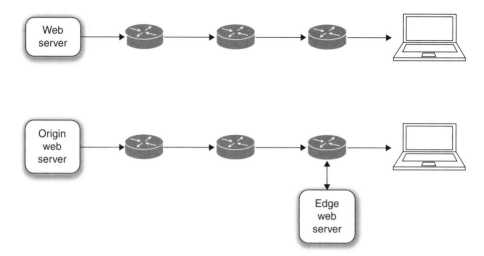

Figure 2.17. Delivery of a web asset without and with a caching server.

provider of the content can forgo paying for transport over the network hops between the origin server and the edge web server when subsequent consuming devices request the same content.

The first commercial uses of a CDN were to host web-site graphics on a CDN rather than on a web server that hosted the HTML that referenced the Web site graphics. The idea was to free the HTML-serving web server from the task of serving the largest, static elements comprising the web page. The HTML code associated with a website was merely modified to continue to use the domain name of the original web server, but the URL of the associated graphics resolved to an IP address hosted by the CDN provider (Figure 2.18).

CDNs are now used to efficiently distribute much more than static Web-site graphics; CDNs are now used to provide end-users with almost every type of file, including video media, streaming audio, or software updates. It should also be noted that a CDN not only can provide financial savings to the content owner (vs. hosting all of the content from a central set of web servers), but the CDN can also provide a form of network resiliency when abusers attempt to perform denial of service attacks against a web host.

A CDN's components can be placed into three categories: delivery, service, and storage. The service components deal with the following:

- Reporting, analytics, and billing;
- Provide a management console (including a means of managing the inventory of items being served by the CDN);
- Provide a "routing" appliance performing some form of content distribution and the dynamic selection of caches.

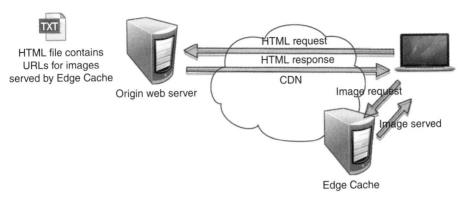

HTML file contains
URLs for images
served by Edge Cache

Origin web server

HTML request

HTML response

CDN

Image request

Image served

Edge Cache

Figure 2.18. Original use of CDN to cache large graphic files for websites.

The storage components deal with the following:

- Content ingestion;
- Provide publishing points from which content originates to be distributed through the CDN;
- Provide content storage and replicates that storage to ward against failure of storage devices.

The delivery components deal with the following:

- Content caching and delivery;
- Distributed deployment of content.

A CDN is typically comprises publishing points, storage nodes, origin servers, a hierarchy of caches that include intermediate caches and edge caches, request routers, monitoring systems, and accounting systems. The origin servers act as the sources of the web content on the MSO's network that will eventually be cached by the Edge Caches. When a user/ consumer of content requests content from a web server, a Request Router determines which Edge Cache should serve the content and the request is directed to an Edge Cache. If the Edge Cache does not have the requested piece of content, the Edge Cache can retrieve that missing content from another cache or from the origin servers. The monitoring systems judge the efficiency of content being distributed by the CDN, where "efficiency" refers to both minimizing time to deliver content and minimizing the financial costs of that delivery. Accounting systems keep track of the amount of data served by any one Edge Cache so that the MSO can generate reports about the popularity of content, for internal accounting purposes, and could be used to implement usage-based billing (Figure 2.19).

Commercial, third-party, CDNs are adding more and more "premium" services such as mobile device ad insertion, content management, and website cloaking; for the purposes of this book, we will not delve into these "add-ons," but rather concentrate on the use of CDNs to provide core IPTV video distribution services for MSOs. Note that

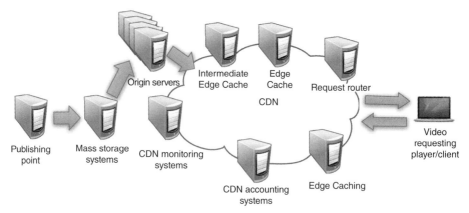

Figure 2.19. CDN component overview.

many of the CDN components widely in use are open source software projects, such as Edge Cache software based on Apache Traffic Server [48].

While CDNs were first created to store image files for websites, engineers at MSOs are most likely interested in building or acquiring a private CDN for distribute of IP-based video to customers on or off of the MSO's private high-speed data network. For this chapter, we will be describing the use of CDNs to distribute video assets, but the reader can infer the delivery of any type of file in the following CDN discussions. For those MSOs opting to offer video-on-demand (VOD) services over IP, the storage requirements imposed upon the Storage Nodes of a CDN can be quite large. As you will see in the subsequent chapter on IPTV, in order to support delivery of adaptive bit-rate video, the MSO will be encoding each video asset into multiple bit-rates, each of which would need to be stored on the storage nodes for delivery as a VOD asset. At a length of 1.5 h a movie encoded at three bit rates of 3 Mbps, 1.5 Mbps, and 700 Kbps would consume 3.4 GB for a single movie. The storage requirements will be even higher if the MSO chooses to store the movies in multiple encrypted formats to support multiple DRM solutions. As a result, CDN storage nodes can consist of storage appliances containing multiple terabytes or even petabytes of storage. Given the time and effort needed to pre-encode and store a large VOD library, an MSO should maintain multiple storage nodes at separate locations that are automatically synchronized, to ward against service outages were one storage location to become unavailable.

For those MSOs that are not offering VOD services over IP, but are only deploying linear IPTV offerings, the storage requirements for a storage node are much smaller than for a VOD offering as the very nature of linear video is such that linear video is acquired, encoded, and immediately distributed across the CDN without the need for long-term storage of the video content. Conversely, for those MSOs capturing linear video services to store those video assets on behalf of their subscribers by implementing a Network Digital Video Recorder (NDVR/Cloud DVR/Remote DVR), the storage requirements within the CDN are potentially much higher than the storage needed to merely support VOD [49].

For both VOD and linear video distribution across the CDN, the network throughput requirements of the origin servers and each caching server is typically in the range of 3.6 to 18 Gbps (meaning 4 1Gig-E interfaces to 2 10Gig-E interfaces).

It should be noted that the physical hardware upon which the caching, routing, and management functions are being performed have become common off-the-shelf (COTS) hardware as MSOs (and other CDN consumers) wish to avoid being locked into using a single vendor or set of proprietary hardware that may become unsupported or otherwise quickly obsolete.

Note that this section pertains to using a CDN to deliver video over IP networks to an IP-capable consuming device. It is also possible to use the storage and delivery capability of a CDN to deliver video content to legacy QAM-based VOD servers for delivery of VOD content to traditional QAM-based STBs. We will not address the technology necessary to perform that "bridging" of IP to QAM video worlds other than to refer the readers to this resource [50].

2.6.1 Content Publishing

The CDN can only delivery content if the CDN's origin servers are first populated with the content. There are many different application-based content publishing systems that are commercially available, such as the Platform™, LimeLight, or Alcatel Lucent's Velocix platforms. These publishing platforms and the associated CDN origin servers support standards (and some proprietary clients) to move large video files, whether via FTP, RSYNC, or a proprietary solution such as Aspera, now owned by IBM. The goals of a content publishing system are the following:

- Verify that video content has been placed onto the CDN origin servers, including verification of the bit-for-bit integrity of the file transfer.
- Provide notification of success or failure of content publication to an operator (via email, SNMP traps, or some other dashboard-like notification).
- Manage the authorization of content to be retrieved by a customer. For the delivery of on-demand video content, the video assets could be placed on the origin server weeks in advance of the date after which the video content are allowed to be viewed by the customer.

Given the thousands of video assets that may be made available to a subscriber in any calendar month, the content publication system needs to be able to support both the publication of individual video assets and for bulk publishing tasks.

2.6.2 Content Caching and Delivery

The Edge Caches of a CDN will support multiple mechanisms for delivering video assets (and other files): via simple file downloads, progressive download, on-demand streaming, and linear streaming. With the appropriate software licensing in place, the CDN will support several video delivery protocols: Adobe Flash, Microsoft Smooth

Streaming, Microsoft SilverLight, and HTTP. The subsequent IPTV chapter will describe the efforts to converge on a single video distribution model such as MPEG-DASH will would alleviate the need for a CDN to support (and license) multiple proprietary video delivery protocols.

Any IP-based device that is capable of accessing a wired or wireless network can be served video content by a CDN, whether the device is a mobile device (smartphone), SmartTV, STB, gaming console, PC, laptop or tablet. As of 2014, adoption of IPV6 is still not widespread, but there should be no technical barriers preventing a CDN from delivering content to either IPv4 or IPv6-capable devices [51].

Video files are segmented into smaller "chunks" and placed on the CDN origin servers. When a user requests a file, a Request Router, directs the file request to the nearest Edge Cache using DNS-based routing or a HTTP-redirect [52, 53]. If the Edge Cache has the chunks that comprise the file, then the Edge Cache begins serving content chunks to the client. For any chunks that the Edge Cache does not possess, the Edge Cache will request those chunks from other origin servers or from the storage nodes. The Edge Cache delivers that chunks to the client as soon as the chunks arrive at the Edge Cache; this allows content to be received by the client before the entire file is in the Edge Cache's possession. If network performance degrades between the Edge Cache and another one of its sources for video chunks, the Edge Cache will request chunks from other sources dynamically. The Edge Caches request file chunks as those chunks are needed; there is usually no need to pre-position content on the Edge Cache prior to the first client requesting the content.

The mechanisms for choosing from where an Edge Cache should request a missing chunk is complex, the source of many PhD dissertations, and is the "secret sauce" by which CDN vendors often attempt to differentiate themselves. Suffice it to say that the algorithms for selecting the cache from which to pull a missing chunk should be based upon cache availability and loading of that cache (CPU and memory utilization), upon network link latency and throughput, and of course, the location and availability of the missing chunks. There will be multiple network paths between CDN caches, and each network path may represent different real dollar costs to an MSO. The CDN management servers are responsible for maintaining the map of those costs if the MSO wishes to use network link costs in the calculation that the Edge Cache performs to determine whether to request missing chunks over a particular network link. While not specific to CDNs, one DNS-based mechanism for choosing an Edge Cache to service a request is the recursive DNS specification current in IETF draft form, known as EDNS-client-subnet [49].

The Edge Caches have a finite quantity of storage (whether memory or disk) so the cache cannot maintain copies of all content indefinitely. The caching algorithms used to determine when to replace content in the cache are varied: from least frequently use, least recently used, to the improbable to implement Belady's Algorithm [54].

Consumer behavior with regards to the consumption of video from the CDN varies based on time (of day, week, etc.), geography, content pricing model, and even the ease with which the content can be discovered in the video client. The more sophisticated Edge Cache content retention algorithms and chunk requesting algorithms will take into account actual performance data based on time, content type, network utilization, and reactive and predictive logic.

2.6.3 Transcoding vs. Storage

There remains an ongoing debate about whether it is better to store a video asset in every format the MSO customer devices require, or to store a video asset in a single format and then transcode or re-package that video into different containers or transport formats on-the-fly when the video is requested. While this debate will continue, my prediction is that the IPTV and consumer electronics industry will eventually converge upon a single MPEG-based video codec that will obviate the need to support multiple video codecs in the future [55].

2.6.4 Management and Reporting

The CDN management server can typically be used to associate delivery options to an asset or group of assets or for a consuming device of group of devices. This allows the MSO to throttle low-valued content to a certain throughput rate or to force low-valued content onto the "least-expensive" network routes. The inverse is also supportable. Note that the CDN management server can also be used in conjunction with the PacketCable Multimedia/COPS devices as described in Chapter 3, Section 3.8.2.2 in order to arrange for QoS prioritization for certain video streams over a DOCSIS CM network.

Recall that we described the CDN as being a "hierarchy of caches." The idea is that popularly requested content is going to reside on the Edge Caches because those are the caches that are closest to the clients requesting the content. By maintaining the popular content on the caches closest to the client, the MSO can reduce the quantity of content being transmitted from the storage and origin servers all the way across the network to the requesting clients. As the Edge Caches serve the majority of the content, the desire is for those Edge Caches to do that serving as quickly as possible. This is why it is typical for the Edge Caches to be hardware devices that are populated with a lot of memory so that the video files are maintained in memory and do not need to be read from a hard drive, resulting in faster transmission times. The secondary level of caches would be mostly hard drive-based (Figure 2.20).

The management servers are also responsible for real-time reporting on cache availability and network status, generation of usage reports, and for detecting and alerting operators of network security issues. The CDN management server will typically also interact with a content management system (CMS) such that the CMS can instruct the CDN management system to remove all video assets from the CDN associated with a particular content provider. In other words, the CDN management system would typically have a set of well-defined APIs that a CMS can call upon to control which content is allowed or disallowed on the CDN.

Upon deploying or operating a CDN, the MSO will be greatly interested in the answering the questions:

- How is my CDN being used?
- How many concurrent sessions are being supported?
- How many concurrent video sessions can my CDN support; what is the aggregate session usage?

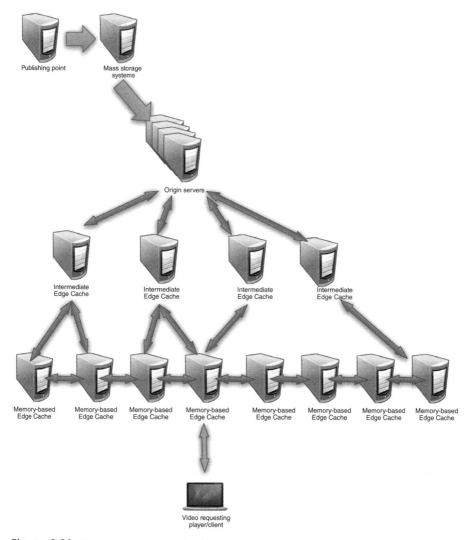

Figure 2.20. Hierarchical cache algorithms cause caches to pull content from the most efficient peer.

- What is the content demand curve, meaning the percentage of videos viewed versus the percentage of video titles available?
- What is the session usage by time of day/day of week/and so on?
- What is the session usage by content type and consuming device type?
- What is the health of each cache in the CDN?

These are all questions that can be answered by the reporting supported by the management server. Large video CDN deployments inevitably become "Big Data" projects in that the CDN is capable of generating Terabytes of usage data per month. The MSO can utilize this data to determine what types of video content are popular with different classes of device. This information will prove useful to determine where to invest both marketing and client software development dollars. One example of the mechanism for collecting health data from each cache is the implementation of the "stats over http" plugin created for the Apache Traffic Server [48]. "Splunk>" is a commercial analytics software package for parsing usage data created by caches that is often used for CDN deployments.

The analysis of the CDN usage is important as it will demonstrate which Edge Caches have the highest "miss" rate, meaning that those caches tend to have to seek out missing chunks from other caches at a high rate. This knowledge allows the MSO to tune those caches by increasing their storage capabilities or spot other patterns in usage in a geographic area. There are browser-based CDN performance measurement tools available, such as Gomez [56], Catchpoint [57], and WebPagetest [58, 59].

Most of this discussion about CDNs has been under the assumption that the MSO has implemented or is at least managing their own CDN. Such a private CDN is fine for serving content to customers on the MSO's high-speed data network, but does not provide network reachability to their customers who are accessing the MSO's IPTV services from another provider's IP network (i.e., while the customer is out of the home at say, a coffee shop). While the private CDN management servers will provide accounting and usage information about usage on the private CDN, there are whole sets of standards related to deriving similar accounting information from a peered "foreign" CDN [54]. This topic would require another book to cover, so the authors direct the reader to the IETF content distribution interworking drafts.

2.7 CCAP ARCHITECTURES AND SERVICES

The industry decision to centralize our data and video applications on a single platform was driven by strategic objectives with regard to services and total system cost criterion. In addition, the CCAP would provide the key transition technology essential for the orderly migration of existing video services to IP video. The transition to IP, a key goal of the CCAP business requirements, fosters lower cost and improved innovation across the entire service delivery framework.

One of the many goals for CCAP was to dramatically increase system capacity and density to enable MSOs the ability to protect and maximize existing critical infrastructure investment. Specifically, the industry sought the ability to deliver various access technologies from the same chassis in order to reduce space, power, and cooling requirements as much as possible in the chronically cramped headend and hub-sites. DOCSIS is likely to offer a useful life matching that of the HFC itself; around 20 years based on some estimates. While HFC is being gradually replaced with all fiber networks, or where new-build opportunities present themselves, transition to new access technologies will follow the migration to all-fiber networks. The CCAP platform will need to provide the flexibility necessary for linecards of differing access technologies to coexist with appropriate hardware and software integration.

It is possible, though not especially likely, that business partnerships may emerge among vendors in order to introduce capabilities the other partner lacks. It is more likely that vendors with broad experience across several technologies will be able to develop capabilities into their CCAP platforms more quickly, such as the demand for such capabilities as EPON linecards for their CCAP systems in situations in which the MSO cannot deploy a dedicated OLT for their business customers in a given service area.

Another important goal of the CCAP is to enable virtual node splitting and service group spanning. In order to fully realize this goal, the platform must provide downstream and upstream connections for each HFC node. This deployment mode can artificially increase CCAP deployment costs, as the number of ports that must be purchased will be significantly greater than the video and data traffic load requires in many cases. Another spoiler on the "easy node splitting" goal is the facts that (i) a laser may serve more than one node and (ii) large nodes may need to be field-split. Both of these realities will quickly put us back into the business of re-wiring our hubs. While virtual node splitting is an elegant concept, in practice it is unlikely to yield significant operational savings, and may in fact drive up initial capital costs and strand capital for ports that may remain underutilized for many years.

No single architectural decision in the past decade has offered as great a potential opportunity to shake up fundamental assumptions about the kind of business MSOs will be operating over the next 10 years as the CCAP. It isn't even the hardware itself that is as important as the changes occurring within the MSO's business models that drove convergence of disparate platforms. Customer's opting for "Internet-only" services from the MSO are already the fastest growing segment of the customer base industry-wide. This trend is likely to continue as new over-the-top (OTT) and MSO provided IP video services displace more traditional video products in addition to a growing list of compelling new IP based applications of every conceivable variety. While the "crystal ball" foretelling the future remains cloudy on some of the important details, it is clear that IP protocol will play a very significant role in the future of all applications and that this will necessitate significant changes across the entire industry. CCAP, as a tool to facilitate transition, will play a vital role in completing a smooth transition to IP video transport and an "all IP" service offering more generally, in lock-step with the ongoing evolution of the HFC plant and every increasing capacity needs per user.

The CCAP platform was broadly scoped to make it suitable for a wide variety of applications and adaptable to an equally large number of existing and planned system architectures. Understandably, this leads to significant debate within the community on the prioritization of the development of new system capabilities and the availability of those features. Additionally, an honest appraisal of the CCAP's potential over time demands ongoing discussion and how the unified platform fits within the MSO's organization and what might need to change operationally.

MSOs will rely on the CCAP to be able to change as the cable plant itself evolves, up to and including the point at which the plant is no longer HFC. The transition to all-fiber networks over the coming years and the evolution of access technologies in order for the MSO to remain competitive form an important cornerstone of CCAP's "future proofing" capabilities.

Historically, both MSOs and the system vendors have separated their video and data engineering teams, under separate leadership, using different development environments, and applying different test procedures. Additionally, these organizations have generally been staffed with personnel that have very different backgrounds and skillsets. However, the development of well-defined operational models appropriate to the CCAP demands a hybridization to achieve the desired outcome of systems and operations unification. Like the CCAP system itself, evolution of the operational models are a "work in progress" and represents important challenges the industry will face over the coming years as both the delivery platform and the manner of delivery fundamentally change.

From its earliest inception, the CCAP platform was envisioned to provide a central role in performing IP forwarding and streaming of legacy multiple program transport stream (MPTS) video. A typical forward spectrum plan consists of analog, digital channels, and HSD DOCSIS narrowcast channels. The spectrum allocation for digital channels can be further be categorized into two types of video services: broadcast and narrowcast. Broadcast is the transmission of programs or signals to the entire cable plant. Source signals such as analog and digital video are modulated onto frequencies for broadcasting across the cable plant unmodified and uninterrupted across the HFC network. Cable was built on this premise and some content will always be broadcast to the subscribers. Narrowcast is the "narrow" broadcasting of selected video contents or data signals to a subset of the cable plant. With this capability, the MSOs can target content delivery to subscribers based on their geographical characteristics within the cable plant or subset of subscribers within a cable plant. To further put this into perspective and context of video and data transports on a cable plant, an analog channel occupies a 6-MHz channel to enable the delivery of a single source or program (i.e., ABC, NBC, CBS, or FOX) whereas a digital channel can deliver on average 2–12 sources or programs on the same 6 MHz channel.

Putting it all together, CCAP unifies the CMTS, switching, routing, and QAM functions at the headend so all data, video, voice, and OTT services can be "switched" into IP mode as needed prior to its conversion to RF [60] or optical signals. Further, CCAP enables the MSOs the fluidity to configure delivery on either a geographic basis, a per-service basis, or on a service group basis. This converged capability combining with the richness of access technologies enables MSOs to manage the QAM channel as fungible 6 MHz resources. Intriguing possibilities exist for the CCAP system, including the incorporation of several unique access technologies into a single chassis for use in mixed residential and commercial deployments over both fiber optics and hybrid fiber coax (HFC) plant. Recently, distributed CCAP solutions have been proposed to satisfy specific business objectives, including situations in which the service area is served directly from the headend rather than from a localized hub site. Distributed architectures may also prove more suitable in providing tailored services to individual Multiple Dwelling Units (MDU) or may offer certain advantages to system capacity and port density beyond that which would be practical on an integrated CCAP. Alternatively, the integrated CCAP architecture truly takes advantage of the CCAP platform as a virtualized switch where the MSOs can deliver data and video services on a per-service or per-service group basis.

2.7.1 A Case for Continued Support of Modular Head-End Architecture

Predating the current CCAP system definition, the modular head-end architecture (MHA) was defined to separate the CMTS core from the edge QAM in order to increase system capacity, improve competition within the market, and decrease total cost per QAM channel [61]. The resulting standards provided for a universal edge QAM that could be used for either video or data services. A significant number of existing DOCSIS 3.0 CMTS deployments continue to rely on MHA, which defined the Modular CMTS (M-CMTS). Edge QAM technology evolution has accelerated in the years since D3.0 hardware first became available and as a result, the video engineering teams have continued to push for continual improvement in density and capabilities. Additionally, there has been an interest in improving the configuration management and monitoring of edge QAMs (via the CableLab's EQAM Provisioning and Management Interface standard), leading to new standards in this area as well [62]. Concurrently, there has been movement within the industry to migrate to a standard Edge Resource Management (ERM) protocol such as the Edge Resource Management Interface (ERMI) defined by CableLab's ERMI protocol standard [63].

We have very likely reached the "golden age" of edge QAMs, a point at which the technology is at or very near its zenith. This evolution has occurred much more quickly in the external edge QAM segment of the market then it has for integrated CMTS linecard QAMs where density remains comparatively low and cost per channel remains substantially higher. With many MSOs continuing to invest heavily in these new high-density external edge QAMs to expand their HD channel lineups, the need to continue to support these investments implicates a further need for these external edge QAMs to be supported in the incoming CCAP platform.

Replacing legacy low-density QAMs with a CCAP system, while possible, is difficult to cost-justify given the significant deltas in terms of cost and capability between current CCAP linecard offerings and competing edge QAM products. Though requiring MHA would indeed add to the already daunting list of requirements vendors must satisfy in producing their CCAP platforms, opportunities emerge that could fundamentally change the migration story for both video and legacy data services on the CCAP platform. Not all MSOs or vendors agree with this perspective and view the additional effort to add MHA support as a distraction from the development and ongoing support of the CCAP system as a fully integrated, "single box" solution. However, with large investments still being made in modern, high-density edge QAM chassis that may support as many as 156 channels per port, there will remain significant motivation among several MSOs to continue leveraging those investments for as long as possible. With the lower cost per port for an external edge QAM relative to a QAM on a DOCSIS linecard likely to be a factor for some time to come, convergence to a unified platform may in fact be hindered or delayed by the lack of MHA support in the CCAP without it. This in turn could lead to preserving the "status quo" of two independent systems for video and data until such time as the equation for the cost benefits changes (Figure 2.21).

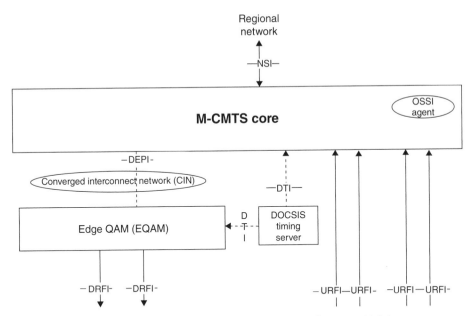

Figure 2.21. Modular CMTS network diagram from MULPIv3.1.

2.7.1.1 MHA Background

An "integrated" CCAP chassis includes all QAMs via onboard linecards and all control and electronics are therefore centralized. By comparison, the MHA architecture (see Cablelabs CM-TR-MHA-v02-081209 and CM-SP-M-OSSI-I08-081209 [64, 65]) relies on Gigabit Ethernet interfaces (now typically 10 Gbps Ethernet) to physically interconnect with the external QAM chassis using the DownstreamExternal PHY Interface (DEPI) [66] and the CableLabs DEPI protocol specification CM-SP-D EPI-I08-100611), a kind of software layer that operates as the control plane between CMTS core and Universal Edge QAM. DEPI provides a type of layer 2 "pseudowire" tunneling service (which itself was based on the L2VPN specification) between the CCAP core and the Universal Edge QAM (UEQAM), with the capabilities dependent on DEPI's configured mode of operation. DEPI supports two modes of operation with differing characteristics. Packet Stream Protocol (PSP) is one mode of operation and DOCSIS MPEG-TS (D-MPT) is the other mode (Figure 2.22), MHA Architecture.

MHA requires the CMTS to support both DEPI tunnel modes. At a high level, PSP provides more granular control of streams to promote improved round-trip latency at the expense of greater complexity and higher overhead. D-MPT is less complex and is used when differentiation of streams to control latency isn't needed or desired. In addition, an external Edge QAM must support not only the DEPI interface but also DOCSIS Timing Interface (see CableLabs DTI protocol specification CM-SP-DTI-I05-081209 [66]) in order to operate as a Universal Edge QAM which imposes certain limitations. Work is underway to augment and extend the specification for a new timing protocol suitable for remote PHY applications where a portion of the

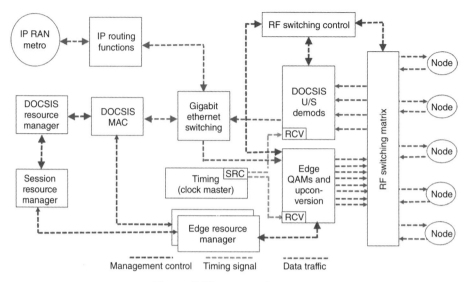

Figure 2.22. MHA architecture.

QAM resides within the HFC plant and potentially many hundreds of miles distant from the CMTS core. Current practice is for the QAM and CMTS core to be co-located within the same hub or headend, often in adjacent racks in large part due to the distance limitations imposed by the current timing protocol and more generally as a result of system packaging.

2.7.2 Mixed Access Technologies

At the present time, the majority of CCAPs provide only a single access technology via dedicated DOCSIS 3.0 access cards and support both operation over HFC and RFoG via a scheduler configuration switch. Very soon, more advanced D3.1 capabilities will be introduced through a combination of hardware and software which will offer considerably higher capacity than DOCSIS 3.0. Most CCAP systems will likely incorporate EPON support as well in the near future. In some cases, pluggable HFC access optics may be incorporated as well. This kind of flexibility will allow the CCAP to provide business and residential services from the same chassis on mixed HFC and all-fiber networks whenever it is necessary or desired to use a common platform rather than dedicated systems.

Considerations for hot sparing and available slot space for each access technology, chassis forwarding capacity, and the software environment necessary to support each access technology complicate system design in supporting mixed access technologies, but may be necessary in hub sites where rack space is severely limited, or a transition plan is in effect to migrate a population of users from one technology to another. The migration use case is perhaps the most compelling, though it will be many years in most service areas before both the physical plant and the access technologies are likely to see significant change.

At the present time, the most immediate "mixed access technology" deployment scenario may be via Gigabit Ethernet and DOCSIS linecards in situations where there are empty slots in the CCAP chassis. The GigE linecard could be reconfigured for use in direct interconnect of business customers in situations where a more dedicated Ethernet switching router cannot be installed in the hub site due to space limitations and a single customer, or a small number of customers, would not be able to justify separate equipment even if there was sufficient rack space, power, and cooling for separate equipment.

2.7.3 RFoG and CCAP

HFC is being gradually replaced by all-fiber networks whenever the cost to substantially upgrade the coax portion of the plant approaches that of fiber to the premises on both an initial cost and long-term operational basis. However, one of the primary values of RFoG is that it allows the already deployed access technology to remain intact even after the fiber optic network is extended. RFoG allows the existing DOCSIS and video QAM infrastructure to continue to be utilized over fiber optics with the addition of an RF-over-Glass Optical Network Unit (R-ONU) installed onto the side of the home or business similar to the manner in which a Network Interface Device (NID) is attached to a building to act as the demarcation point. The R-ONU converts the optical signals back into electrical signals, allowing an existing STB, CM, or MTA to be used as an alternative to switching CPE equipment to EPON ONUs and SIP voice service terminal equipment.

Even though many of the details of the RFoG system are transparent to the existing CMTS or CCAP, the system software must support the ability to limit the number of simultaneous transmitters to one in order to prevent optical domain collision of the burst transmissions, a phenomenon called Optical Beat Interference (OBI). Unlike previous versions of DOCSIS, DOCSIS 3.1 requires a scheduler configuration switch that permits only a single transmitter at any one time for optimal support of RFoG architectures. Previous versions of DOCSIS support RFoG via product requirements that may be unique to a single MSO or software load and was not formalized into the DOCSIS specifications directly until DOCSIS 3.1.

2.7.3.1 RFoG Background

The RFoG system is specified to simultaneously interoperate with 1 G EPON, 10 Gig EPON, and GPON. This is accomplished by using different wavelengths ("colors" of light to separate the networks) for each technology. Fiber, optical passives, and all construction necessary to enable PON services are identical, allowing future addition of EPON services to a home after an initial RFoG installation and can operate either simultaneously with RFoG or as a CPE replacement, on a home-by-home basis. Once EPON equipment and passive optical filters are added at the hub, EPON ONUs can be installed anywhere in an RFoG network.

Any existing service offered by the MSO over HFC can be delivered over RFoG. In an HFC architecture with 500 homes passed per node (500 HHP), the headend and/or hub are connected to the node via fiber optics, and from the node electrical RF signaling transmits to the serving area using coaxial cable for the remaining distance of a mile or so. Unlike the HFC where a node serves many individual housing or office units, the RFoG architecture consists of 32, or as many as 64 (depending on available optical budget) devices operating as "mini-nodes" that terminate the fiber connection at the customer premises location rather than within the plant as is common practice in HFC.

RFoG also can provide the unique capability to operate both as a mid-split (to the modem) and a legacy split (to settops) within the same home at the same time. Unlike HFC conversions to mid-split, this allows RFoG to scale to very high upstream tier rates without the need to retire set-tops that may require low or fixed data carriers. Only homes needing higher upstream tiers would utilize upgraded RFoG ONUs, while other homes on the network could continue to utilize legacy-split RFoG ONUs.

RFoG was designed to provide support for DOCSIS and video QAM over fiber optics, with the standard developed by the SCTE's IPS Working Group 5 resulting in the SCTE 174 (2010) standard. RFoG was designed to allow both transition to EPON and to provide a means for DOCSIS/QAM to coexist with EPON during the migration period, using the same physical fiber resources by allocating different wavelengths for each access technology. The enabling hardware on the customer premises side is an R-ONU, referred to as a "micro node" according to the standard, a device that supports the RFoG access network located behind a passive optical splitter. When the network is upgraded, the RFoG elements can be retained, while the MSO deploys the necessary equipment, such as EPON OLTs and ONUs, for a full EPON roll out. Because both RFoG and HFC systems can concurrently operate out of the same headend or hub site on the same physical fiber, RFoG can be used to increase capacity for the RF services even when fiber count is very low at the feeder line.

The smaller service group size of 64 HHP and fewer does add additional cost to the deployment in terms of labor, materials and electronics. However, the benefits of the smaller service groups include a dramatic increase in effective capacity available per subscriber as well as the inherent benefits of an all-passive optical network into the home or business. Additionally, once the fiber foundation has been laid, migration to even higher bit rate access technologies such as EPON can be accomplished while continuing to leverage RFoG for DOCSIS and video QAM even within the same service group, into the same home or business.

2.7.3.2 Mixed RFoG and EPON

The same PON network can support both the RFoG and EPON by using different optical wavelengths over a wave-division multiplexing (WDM) system. The combination of the DWM optical distribution network (ODN) and PON provides a powerful tool for managing fiber resources that are costly or impractical to increase in fiber count and aids with the process of migration. For example, RFoG uses 1310 nm in the

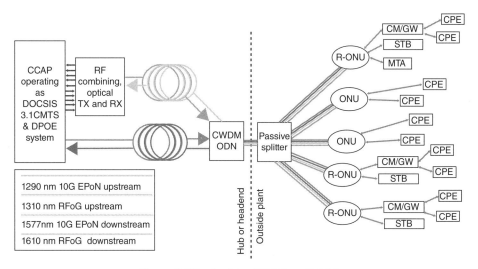

Figure 2.23. Combined EPON and RFOG.

return path and 1550, or 1610 nm the forward path. 10 Gbit EPON uses 1577 nm in the forward path and 1290 nm in the return path, avoiding collision of wavelengths within the WDM system. Some MSOs already use EPON and RFoG together in order to offer both video services and data services such as VPN and very high-speed Internet to commercial customers that have a fiber drop to their service address. More recently, interest in using RFoG for residential services has increased as a possible solution to bridge the gap and allow a more gradual migration from DOCSIS and video QAM services to EPON under specific circumstances. Combined EPON and RFOG are depicted in Figure 2.23.

A detailed description of EPON and DPoE can be found in Section 4.3.1, DOCSIS provisioning of EPON [67].

2.7.4 Distributed Architectures

There has been some degree of interest in architectures which locate all or portions of the electronics normally found in an integrated CCAP chassis into distributed chassis components located on the HFC plant, in closer proximity to the customer service address in order to decrease attenuation and attenuation that could permit the use of very high modulation orders with operating margin to spare. The solutions proposed thus far can be separated into two broad categories of solutions that have gained certain popularity: Solutions for MDU applications; and solutions that extend rather than replace the current "integrated" architectures (Figure 2.24). The problem space that has evolved from the earliest discussions initiated several years ago focused primarily on ways to get around technical limitations in current hardware designs, or better suited to specific HFC architectures that were not well served from a centralized

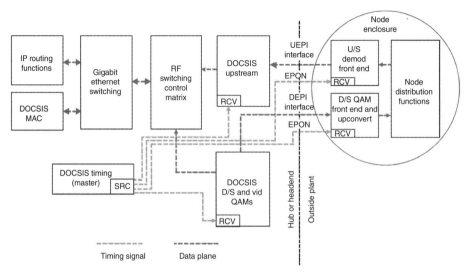

Figure 2.24. Distributed CCAP architecture: Remote PHY.

CCAP located in a distant hub or head-end site. Some of the challenges that led to the currently proposed solutions include the following:

- Limited space in the hub and headend that necessitates other approaches to increase density beyond what is possible within the limitations of current or planned CCAP chassis and linecard design;
- HFC architectures that are characterized by long distances from the headend, such as when there is no intermediary hub site between headend and customer drop (collapsed hub, home-run to head-end "blast and split"-type architectures);
- Technological limits on capacity and/or port density that may be improved upon with a distributed approach;
- MDU applications that differ markedly from what is needed for general-purpose residential deployments;
- Opportunities for improved cost efficiencies if smaller upstart vendors are allowed to compete for this new equipment category, in theory.

Despite certain advantages that may be realized with some of the more novel distributed architecture solutions, the change in "packaging" incurs certain drawbacks by virtue of the optimization of the solution to fit well in certain situations and thus more poorly in others. New problems can be created by these distributed architectures that must be managed and their impacts well understood well in advance of any large-scale deployment. This includes having a very good understanding of total costs, such as costs that have been "shifted," but not necessarily eliminated in the system.

The "Distributed CCAP" concept was ultimately decided on by the MSOs to be an electronics packaging decision driven primarily by purchase orders and MSO-specific requirements rather than demanding a public specification development. However, certain aspects of the architecture have recently been identified, which will require protocol development in order to guarantee interoperability. Specific decisions had to be made to insure this interoperability, particularly in situations where most of the QAM remains in the CCAP chassis core while the low-level transmission components are split and relocated onto the plant.

2.7.4.1 Distributed CCAP's Potential Challenges

Some of the challenges that the various distributed architectures introduce, based on the type of solution proposed, include the following:

- Within certain distributed architectures that locate a sizeable percentage of the logic of the system in the distal end of the distributed architecture, there are difficulties in dividing the DOCSIS MAC from the PHY layer, and locating only those portions of the MAC that are needed in the distributed chassis. The protocol that is required between the centralized portion and the remote portion along with attendant details such as timing requirements remain a work in progress. When the CCAP is operating exclusively with non-DOCSIS access technologies, this may become less problematic given differences in MAC and PHY between DOCSIS and protocols such as EPON.

- Full function "mini-CCAPs" that are located on the plant that uses a generic, high-capacity Ethernet switch router in the headend in order to aggregate the distributed CCAPs to increase system capacity create new problems for handling flow-level dynamic QoS without necessitating changes to the aggregation switch that necessarily making what was a "generic switch" DOCSIS aware. Static configuration eliminates the problem and removes the control plane encumbrances. Additionally, a full system such as this generally is higher in cost tends to increase management and increases operational complexity in ways that make it unsuitable for general-purpose deployments.

- Most distributed architectures dramatically alter how well-understood traffic engineering principles are acted upon and may require very different financial modeling to perform "just in time" capacity increases, except in cases where only the PHY portion of the system is distributed and the overall design of the system remains largely unchanged.

- Selection of remote PHY-only solution, one which leaves all higher level functionality in a centralized chassis, may be the most prudent method to expand capacity and/or port density of the system beyond roughly 200 SG's per chassis. It is therefore possible to improve upon density concerns stemming from more conventional architectures that are limited by the number of physical coaxial ports that can be mounted onto a backplane, without substantially compromising the quality of the cabling, and do so without deviating significantly from the current model.

- Potential for increased complexity in management of the CCAP system, with the mandate that software abstraction layers will make it possible to view and manage the distributed system as though it is a single chassis via as-yet-to-be-defined management interfaces.

- Potential for significant changes to operational model in order to service, repair and/or replace components that are considered part of the CCAP data system, which is generally managed and controlled by different personnel than those who manage the legacy CMTS and new CCAP systems today that usually staff secure, access controlled, well-cooled and power-protected hub sites and headends.

The development of solutions tailored to specific applications, rather than a more "general-purpose" notion of distributed architecture will likely continue to evolve. It is expected that products will be available in coming years to satisfy the unique use cases defined for specific HFC architectures and MDU applications. However, for the foreseeable future, it appears quite likely that integrated CCAPs, located in climate controlled facilities, will remain favored for the majority of deployments. Cisco, Arris, Casa, Gainspeed, and Harmonic all have prepared detailed plans for distributed CCAP solutions to meet MSO unique needs.

2.7.5 CCAP Migration and Deployment Considerations

There are four distinct phases of migration that the CCAP will go through coinciding with feature introductions and business decisions within the MSO. The first of these phases is perhaps the most obvious: Migration from the existing CMTS platforms to the new CCAP platforms, which we refer to throughout this section as "initial deployment." The second phase involves migrating video transport and control to the CCAP titled "convergence," a potentially lengthy phase given shifting business needs and project dependencies. The third phase involves migrating to new access technologies within the common CCAP chassis and enhancing existing video transport capabilities that we've titled "enhancement," though the exact capabilities that will be the subject of enhancement are the subject of ongoing debate. The fourth phase is characterized by decommissioning of legacy functionality no longer deemed necessary, a final phase that could signal the terminus for legacy video or DOCSIS depending on assumptions such as CPE useful life, and the useful life of the HFC that equipment is tied to. This phase is referred to simply as "maturation," the final stage at which the CCAP itself may have reached the end of the line.

At each major phase, software and hardware changes are all but guaranteed. For example, rich IP multicast controls and resource management are an example of enhanced video capabilities that will occur sometime after D3.1 capabilities are integrated based on current prioritization of feature of research and development. Transition from CMTS to CCAP and migration from disparate video transport systems to the converged platform necessitate changes to the testing of new software capabilities and support for new hardware, with all the attendant operational shifts, training, and toolset changes that must occur in advance. It may be a very exciting time in the first few years after initial deployment.

2.7.5.1 DOCSIS 3.0 M-CMTS Migration to CCAP with MHA Support

A scenario common to many MSOs that deployed DOCSIS 3.0 equipment when it first became available will often make use of the MHA (M-CMTS) system design and might have four or more external UEQAM chassis connected to the system. If this system is to be replaced *in situ* with minimal disruption, the MSO has some decisions to make. On the one hand, if the UEQAMs are of higher density than the integrated QAMs available on the first generation CCAP linecards, then the MSO may actually lose density. If the UEQAMs are of an older generation with relatively low channel density per port, then the change is more easily justified and use of the internal QAMs is limited only by software capabilities to support video transport.

If video and/or data services are being transported by very recently acquired "full spectrum" (158 channels per port) external QAM chassis and the new CCAP's onboard linecards only support 64 CEA channels per port, then an interconnect between the CCAP and edge QAMs should not only preserve the data QAM investment, but allow the video QAMs to be migrated onto the CCAP as well. For example, the CCAP was supplied with two 8-port 10 Gbps Ethernet aggregation cards that act as the Core Interconnect Network (CIN) between the CCAP and two sets of external edge QAMs: Four each of data UEQAMs and four additional video QAMs that were software upgraded to become UEQAMs for use with the DEPI control plane and DTI timing clock source (most Edge QAMs built in the past several years have a DTI timing interface whether they use them or not, a software change adds DEPI to allow interconnection to an M-CMTS over Gbps Ethernet) [66, 68].

The higher capacity provided by the new CCAP chassis allows for more capacity per slot and therefore more GigE interfaces per card. As a result, the new CCAP can interconnect a larger number of external edge QAMs within the same chassis footprint and salvage the investment in external QAMs for several more years as successive product generations in CCAP linecard generations either fill the gap in terms of legacy 6-MHz channel QAM channel density, or the need for legacy QAM support becomes moot with increasing focus on IP deliver of content over the native wideband OFDM channels of DOCSIS 3.1, the final version of DOCSIS.

Other considerations apply to the migration as well, such as the physical dimensions of the system. While it would be convenient to assume that you can remove the CMTS from the rack and install the new CCAP directly in its place, that's seldom plausible and the downtime that may result could go into many hours to re-cable. Even connector types may differ, such as an ordinary F connector versus high-density miniature coax (MCX) and microminiature coax (MMCX), which may demand adapters (and new points of failure) and/or different coaxial cabling types between the system and the splitting/combining gear and/or the optical transmission and optical receiver gear.

2.7.6 Displacement or Augmentation: Transition Planning

When DOCSIS 3.1 linecards are introduced, the new cards can be deployed by displacing the DOCSIS 3.0 linecards and network combining to the CCAP changed to reflect the service groups activated with D3.1. Note that D3.1 is backward compatible with earlier generations of DOCSIS and continues to support D3.0 CPE transparently. At a certain point in the future (perhaps more than 15 years), enough of the video services

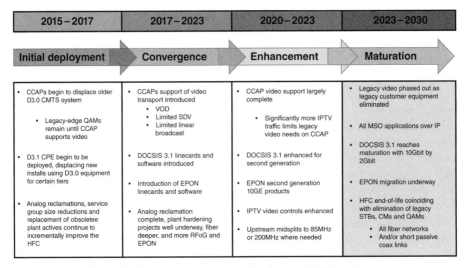

2015–2017	2017–2023	2020–2023	2023–2030
Initial deployment	Convergence	Enhancement	Maturation

• CCAPs begin to displace older D3.0 CMTS system
 	
• Legacy-edge QAMs remain until CCAP supports video
 	
• D3.1 CPE begin to be deployed, displacing new installs using D3.0 equipment for certain tiers
 	
• Analog reclamations, service group size reductions and replacement of obsoletee plant actives continue to incrementally improve the HFC
	• CCAPs support of video transport introduced
	  • VOD
	  • Limited SDV
	  • Limited linear broadcast
	
	• DOCSIS 3.1 linecards and software introduced
	
	• Introduction of EPON linecards and software
	
	• Analog reclamation complete, plant hardening projects well underway, fiber deeper, and more RFoG and EPON
		• CCAP video support largely complete
		  • Significantly more IPTV traffic limits legacy video needs on CCAP
		
		• DOCSIS 3.1 enhanced for second generation
		
		• EPON second generation 10GE products
		
		• IPTV video controls enhanced
		
		• Upstream midsplits to 85MHz or 200MHz where needed
			• Legacy video phased out as legacy customer equipment eliminated
			
			• All MSO applications over IP
			
			• DOCSIS 3.1 reaches maturation with 10Gbit by 2Gbit
			
			• EPON migration underway
			
			• HFC end-of-life coinciding with elimination of legacy STBs, CMs and QAMs
			  • All fiber networks
			  • And/or short passive coax links

Figure 2.25. CCAP migration, technology and HFC evolution.

may have been migrated to IP and enough of the STB CPE's retired that the QAM resources may be reallocated or discontinued entirely. Specific service groups, such as those feeding MDUs or brand new housing developments, may not have the same legacy concerns. CCAP migration, technology and HFC evolution are depicted in Figure 2.25.

In the scenarios given in the following text, examples are provided for deploying the CCAP in an environment with significant investments in external edge QAMs, the migration to all-IP transport of video services and the migration from DOCSIS to EPON over a hypothetical 15-year time period. The scenarios for an integrated CCAP without the need or desire to support external edge QAMs are in most respect simpler but do not offer the same opportunities as a result.

2.7.6.1 Phase 1: Initial Deployment
This scenario assumes an early DOCSIS 3.0 M-CMTS system that has been attached to new high-density UEQAMs. The video QAM, resource, and session controllers remain separate.

1a. New CCAP supporting MHA deployed to replace existing M-CMTS *in situ:*
   - Connect existing data UEQAMs to CCAP 10GE aggregation cards, test, and validate;
   - Adapt cabling as necessary between CCAP and RX/TX optical transmission equipment or between RF combining and CCAP;
   - DOCSIS downstream continues to operate over DEPI to UEQAMs as with previous M-CMTS;
   - DOCSIS upstream supplied by integrated linecards in the CCAP identical to the previous M-CMTS.

1b. New CCAP that does not support MHA deployed to replace existing M-CMTS *in situ:*

- Re-cable CCAP back panel to combining network, displacing existing UEQAMs via the onboard QAMs;
- Adapt cabling as necessary between CCAP and RX/TX optical transmission equipment or between RF combining and CCAP;

Note that video QAMs and control systems remain separate during initial deployment to reduce complexity and allow more time for development, integration, and testing of video transport capabilities.

2.7.6.2 Phase 2: Convergence

Basic video transport capabilities have been integrated into the CCAP, allowing convergence.

2a. CCAP basic video convergence capabilities enabled, CCAP supports MHA:

- Install DOCSIS 3.1 linecards and software upgrades;
- Software upgrade of existing high-density video QAMs to UEQAM (DEPI stack, DTI, or DTP timing interface) to allow these QAMs to be used with the CCAP for video transport;
- Connect all UEQAMs to CCAP 10GE aggregation cards;
- Enable video transport capabilities.

2b. CCAP basic video convergence capabilities enabled, CCAP does not support MHA:

- Install DOCSIS 3.1 linecards and software upgrades;
- Enable video transport capabilities.

At this phase, video capabilities may remain relatively basic across most CCAP products, such as support for narrowband services such as VOD, with limited SDV capabilities. Some enhanced capabilities may appear such as support for linear broadcast and support for edge encryption that some MSOs prefer. The MSO has the choice of using the new D3.1 linecards to operate new and legacy DOCSIS services over, or relegate the legacy DOCSIS and video services to external Edge QAMs based on a "best-fit" approach whenever MHA support exists in the CCAP.

2.7.6.3 Phase 3: Enhancements

EPON linecards are introduced for all-fiber networks, while video transport capabilities have continued to be improved upon. Second-generation DOCSIS 3.1 products are introduced.

- CCAP enhancements;
- Install EPON linecards and software upgrades where EPON is needed for commercial or rare residential opportunities;
- Install second-generation D3.1 linecards offering higher capacity than the first generation provided for multigigabit services;
- DOCSIS 3.1 deeply deployed and mature in this phases timeframe, with coax still serving the majority of endpoints (85% of total);

- Enable enhanced video services, including linear broadcast, full participation in SDV control plane and enhancements to management interfaces for legacy video transport;
- IPTV services enhancements including more mature management interfaces and controls;
- Possible introduction of software-defined networks (SDNs) to configuration management of the CCAP system.

D3.1 linecards will ultimately replace D3.0 linecards. EPON linecard might be introduced at this point though slot space, sparing strategies and the remaining useful life of the HFC suggest EPON will not play a significant role in residential connects for the time period associated with this phase. Increased capacity needs driven by customer demand and migration of video content streaming to IP may demand substantial upgrades to the chassis and second generation linecards by the end of phase 2 or the beginning of phase 3 to satisfy capacity needs. This timeframe would coincide with roughly 7 years after initial deployment.

2.7.6.4 Phase 4: Maturation—Further Enhancement and Decommissioning of Legacy Capabilities

DOCSIS will be a very mature technology by this stage and DOCSIS 3.1 will have reached its design limits. While still very important in terms of CCAP's capabilities, DOCSIS will increasingly be displaced by other access technologies such as EPON in the parts of the network that have been rebuilt as all-fiber when and where financially justifiable to do so.

- Further enhancement of IP-centric capabilities, decommissioning of legacy video, and data capabilities;
  - Onboard linecards or distributed PHY solutions provide full service over D3.1 with legacy support for D3.0-style bonding groups for large installed base of DOCSIS 3.0 CPE;
  - EPON linecards provide for business service and some small percentage of residential customers over newly rebuilt PON networks.
    - Decommissioning of legacy capabilities has begun.
    - Mature external UEQAMs no longer needed on some service groups due to elimination of associated legacy STBs.
    - MHA may no longer be needed if capacity in CCAP system is adequate.
    - Legacy voice services no longer needed, all voice services are migrated to SIP.
    - Replace DOCSIS linecards with EPON linecards as service groups are rebuilt and CPE equipment replaced.

Continuing evolution toward fiber networking and all-IP transport of services has eliminated the need for much of the legacy capabilities, though significant investments will remain until all equipment has been reclaimed or replaced and the HFC itself is end of life. Many of the legacy capabilities will remain in software given complexity in total removal, though hardware resources and linecards necessary to support certain legacy capabilities will have reached a point where they may simply no longer be needed and their slot space and board real estate may be reclaimed.

2.7.6.5 Challenges to CCAP Migration: Development and Planning

In most organizations, Edge QAMs approved by the MSOs internal quality assurance (QA) team for use with the CMTS core are often different makes and models than those used by the video transport systems and operate very different software stacks. Additionally, testing of integrated QAMs within the CMTS platform has generally not been performed by video engineering that has traditionally had a parallel video transport and control system. Testing methodologies will need to be changed in order for video engineering requirements to find their way into the CCAP. This includes test procedures, test environments, and qualification procedures for both basic and advanced video capabilities that must now be merged in order for video transport capabilities to be properly qualified and tested prior to deployment.

One important change to internal QA testing will be the merger of test suites and validation for protocols not generally tested against by the individual QA groups. For example, the video QAM testing usually does not include extensive testing or validation of DEPI capabilities when external edge QAMs are being evaluated and quite naturally video quality assurance teams do not generally perform testing of QAMs integrated on the linecards of a CMTS. Conversely, data QAM testing does not generally validate if the EDIS or ERMI resource manager software interface is implemented correctly or behaves appropriately, nor that any required encryption suite (such as PowerKey) correctly exchanges keys and conforms with the requirements. These considerations are not insurmountable, but they are new and demand changes to how most organizations operate and careful coordination between teams.

2.8 WI-FI ARCHITECTURE AND SERVICES

Wi-Fi is becoming a network of choice for both SPs and consumers alike. The past few years have seen renewed interest in large-scale Wi-Fi deployment from SPs.

There are a number of factors influencing this phenomenon. For example, more than ever consumer electronics manufactures are shipping products with Wi-Fi. Even the products that support connectivity using 3GPP and 3GPP2 specifications include support for Wi-Fi. At the same time, consumer preferences are evolving—more and more consumers are demanding access to communication services anywhere and anytime.

Additionally, over the past decade, Wi-Fi performance has improved exponentially. The latest Wi-Fi standard—802.11ac—promises support for speeds greater than 1 Gbps. Many Wi-Fi products, including 802.11n, support multiple input–multiple output (MIMO): transmit beamforming (TxBF) and operations in 5 GHz spectrum. These technologies promise greater reliability and even better performance.

In this section, we review some of the recent technological advancements in the Wi-Fi domain. These technologies lead to improvements in Wi-Fi performance, resulting in a network that is highly reliable, manageable and amenable to monetization. Both wired and wireless SPs are considering offering broadband wireless services to their customers through extensive deployment of Wi-Fi networks. Wired SPs consider Wi-Fi networks as the last leg to their broadband customers giving them the opportunity to use their subscribed broadband services while enjoying the freedom of mobility. Wireless SPs use their Wi-Fi networks (typically in the form of public hotspots) as a complementary wireless technology to offload part of their traffic demand and reduce the congestion on the much costlier licensed spectrum.

2.8.1 New Air Interfaces

We first review the most recent Wi-Fi air interface technology in 5 GHz unlicensed band, 802.11ac. We, then, briefly describe the current effort in IEEE 802.11 working group to define new enhancements for high-efficiency WLAN.

802.11ac: Building on the success of 802.11n, 802.11ac introduces enhancements in multiple fronts including multi-antenna techniques, modulation scheme, and channel bandwidth.

TxBF allows an access point (AP) to concentrate its signal energy at the client location. AP does this by encoding a single stream of information across multiple antennas and carefully controlling the phase of the transmitted signal from each antenna. This results in better SNR and throughput at the client and can reduce interference across the network. 802.11ac mandates explicit TxBF as the only acceptable form of beamforming. In explicit beamforming, a wireless client estimates the channel information and feeds it back to the AP to help it calculate the correct beamforming vector. Explicit beamforming has higher gain compared to its implicit counterpart, which relies on AP to estimate the channel. The channel estimated at AP is not generally equivalent to the downlink channel from AP to the client and as such implicit beamforming is always inferior in performance compared to explicit beamforming. Note that both explicit and implicit beamforming are defined in 802.11n, while multiple explicit beamforming methods are defined in the standard. 802.11ac standardizes a single explicit beamforming based on the null data packet (NDP) approach to improve interoperability [69].

802.11ac also increases the maximum number of spatial streams from AP to the clients to 8 from 4 in 802.11n. Spatial stream employs the concept of spatial division multiplexing, which allows an AP to send multiple streams of data simultaneously to a client using multiple antennas. This results in higher throughput at the client. The maximum number of supported spatial streams at the client, however, remains at 4. The eight spatial streams from AP can be achieved only when 802.11ac uses multiuser MIMO (MU-MIMO), which is also a new multi-antenna technique introduced in 802.11ac. The MIMO technology in 802.11n operates only in single-user mode, that is, between the AP and a single client at a time. MU-MIMO, on the contrary, allows AP to simultaneously transmit multiple spatial streams to multiple clients provided that the clients are separated enough such that they do not interfere with each other. An AP with eight antennas, therefore, can use all of its degrees of freedom and transmit eight spatial streams, dividing the spatial streams to multiple clients.

Other enhancements in 802.11ac include the use of higher channel bandwidth and 256 QAM modulation scheme. 802.11ac introduces new channel bandwidths of 80 and

160 MHz. The 160-MHz channel bandwidth option can be either in contiguous mode or in noncontiguous (80 + 80 MHz) mode. The first generation 802.11ac products support only 20-, 40-, and 80-MHz channel bandwidth. As of the writing of this book, the FCC spectrum rules do not allow for a 160-MHz channel. Channel bandwidth of 80 MHz + 80 MHz and 160 MHz are expected in the second-generation 802.11ac products. The FCC is also expected to change some ruling to allow up to 1000 mW transmission in the UNII-1 band, which will allow devices to use channel bandwidth of 80 MHz + 80 MHz.

Using the techniques mentioned earlier, the 802.11ac PHY layer data rates significantly increase. It is estimated that the wave-1 802.11ac products (3 spatial streams, 80 MHz channel bandwidth and explicit transmit beamforming) results in theoretical PHY rate of 1.3 Gbps. Wave-2 802.11ac products (4 spatial streams, 160 MHz channel bandwidth, and MU-MIMO) are estimated to provide up to 3.47 Gbps PHY rate.

High-efficiency WLAN (HEW) IEEE 802.11 working group (WG) has recently formed a study group (SG) on high-efficiency WLAN. While previous 802.11 standards have focused on improving the peak data rates, HEW is expected to introduce new technologies that enhance the real-world throughput achieved by wireless customers. Instead of peak throughput, other metrics including: average per-client throughput, fifth percentile per client throughput and area throughput are considered for optimization. In addition, dense deployment environments with a mix of clients/APs and traffic types including outdoor situations will be considered.

2.8.2 Hotspot 2.0

Hotspot 2.0 is the name of a program in Wi-Fi alliance (WFA) that aims to enhance Wi-Fi user experience while roaming. The interoperability certification program in WFA is called Passpoint, and the devices that pass such interoperability testing are called Passpoint certified.

As of late, the release 2 of Hotspot 2.0 has been published by WFA. The focus of Hotspot 2.0 release 1 was on enabling mobile devices to discover the APs that have roaming agreements with their home networks and securely connect without any user intervention. Hotspot 2.0 release 2 focuses on additional functionalities including online signup and over the air credential provisioning.

Hotspot 2.0 relies on IEEE 802.11u, 802.1X authentication with selected EAP methods and 802.11i security. The last three functionalities are part of the WPA-2 enterprise certification program in Wi-Fi alliance. IEEE 802.11u protocol enables mobile devices to communicate with APs pre-association and understand the capabilities of the network. 802.11u uses two protocols: generic advertisement service (GAS) and the access network query protocol (ANQP) to enable the pre-association communications with the AP.

2.8.3 Handoff Algorithms

Handoff algorithms are critical to support multimedia applications over Wi-Fi. As a mobile client moves from the coverage area of one AP to an adjacent AP, the signal strength received from the first associated AP drops and the client needs to change its association. This transition, however, is time consuming and the resulting delay can be disruptive to the ongoing session. With new QoS and security features, this handoff delay can be even more significant as additional management frames related to these features need to be exchanged during this transition period. A number of solutions have been proposed to reduce the handoff delay.

IEEE 802.11r (Fast BSS Transition) protocol allows a wireless client to fully authenticate only with the first AP in the domain and use shorter association procedure with the next APs in the same domain. The domain is defined as a group of APs that support fast BSS transition protocol and are connected over the distribution system (DS). The first AP that the client authenticates to will cache its PMK (pairwise master key) and use it to derive session keys for other APs. A similar approach called opportunistic key caching (OKC) is used to reduce handoff delay between multiple APs in a network where those APs are under common administrative control.

2.8.4 Radio Resource Management/Self-Organizing Networks for Wi-Fi

Radio resource management/self-organizing network (RRM/SON) is a useful approach for interference mitigation and performance improvement, especially in dense Wi-Fi deployments.

Wi-Fi networks may be large in scale, comprising hundreds of thousands or millions of operator managed APs. Cable MSOs across the world are deploying Wi-Fi hotspots to meet the demands of their customers. Self-organizing methods are required for the efficient management of the Wi-Fi resources with large numbers of APs. Wi-Fi SON approaches can include techniques supported by each AP for immediate response to air interface conditions. Wi-Fi SON approaches can also include placing centralized SON servers in the cloud or network that provides a high-level management of specific parameters based on a wider view of the Wi-Fi network that may be available to individual APs. The goal of the RRM/SON is to provide operators with a centralized Wi-Fi SON control based on a wide view of the Wi-Fi access network, which consists of wireless controllers as well as standalone APs from different vendors.

2.8.5 Home Wi-Fi Networks

Wi-Fi is ubiquitous in houses with broadband connection. The demands on home networks have increased over time. In the past, customers would connect a few computers directly to a CM using Ethernet or USB. Then customers wanted the convenience of receiving wireless data at their laptop so they could read E-mail or surf the web. Customers achieved this by going to a retail outlet and purchasing a Wi-Fi AP, which they would plug into the CM. The placement of the AP in the home would provide a level of service and coverage that the customer would accept because they were the ones who purchased and placed the unit. Now cable MSOs are taking on the responsibility of providing and installing the AP. This raises the customer's expectation as to how well the Wi-Fi works in their home. An additional complication is that customers want to stream videos to their smart phones, laptops and tablet computers. The following is a list of key challenges that operators manage when using Wi-Fi for home networks:

- Interference management;
- Wi-Fi coverage;
- Connection and link reliability;
- Connection and performance management.

Wi-Fi performance can be affected by a number of factors, including construction material of the house, location of the AP, and interference from other Wi-Fi networks using the same/adjacent/alternate channel as well as non-Wi-Fi sources such as 5 GHz cordless phones. An optimum placement of the AP in the customer house makes a big difference in Wi-Fi signal coverage and performance. APs used in residential deployments normally come equipped with omnidirectional antennas with minimal to no directional antenna gain. For equal coverage in each direction, AP should be placed in a central location in the house. In a multistoried house, the AP should be placed on the middle floor. Wi-Fi signals attenuate differently through different material. Furniture and accessories built with metal can cause significant attenuation. Operators should avoid placing Wi-Fi devices directly behind computers, monitors, TVs or other metal obstructions.

Operator should also choose a Wi-Fi channel to avoid co-channel, adjacent channel, and alternate channel interference. If such choice is not available, to minimize the impact of adjacent and alternate channel interference, the AP should be at least 20 feet away (assuming open air) from the other nearby APs.

Sources for interference are many and dynamic in nature. To manage interference, operators should consider the following additional tools:

- Support for automatic channel selection (ACS) at boot up and during operation;
- Support for dual band–dual concurrent (DBDC) with support for multiband steering;
- Support for RRM and SONs.

2.8.6 Community Wi-Fi

Many cable operators are deploying Wi-Fi hotspots to meet the demands of their customers. While deploying dedicated public Wi-Fi hotspots is a great way to provide public Wi-Fi services to customers in public areas such as malls, stadiums, parks, and so on, they may not be the most economical tool to provide Wi-Fi services in residential neighborhood and sparsely populated areas.

Cable operators are complementing their public Wi-Fi hotspot deployments by using existing Wi-Fi assets (e.g., residential Wi-Fi Gateways) to provide public Wi-Fi services to customers in residential neighborhoods. This allows operators to use an existing residential broadband connection as backhaul for both public and private Wi-Fi services. The private Wi-Fi is for exclusive use by the "broadband customer" paying for the broadband services for the residence. The public Wi-Fi is for the operator to provide services to other customers.

This allows operators to leverage unused capacity on existing residential Wi-Fi infrastructure to offer Wi-Fi network access to their on-the-go subscribers. The operators can also use this excess capacity to offer services to retail and roaming—partner operator's—subscribers.

The residential subscribers accessing the network from inside their homes have prioritized access to the Wi-Fi resources. The residential Wi-Fi infrastructure is configured in a manner that allows an independent access channel. This independent access channel provides service quality, safety, and privacy for both residential and visiting customers. Roaming and on-the-go subscribers are only allowed to use the Wi-Fi network capacity that is not currently used by the subscriber at home (Figure 2.26).

In order to offer community Wi-Fi services, an operator enables residential Wi-Fi GW with two Wi-Fi SSIDs—Private SSID and Public SSID (Table 2.1).

The following is a list of few key attribute of community Wi-Fi network that operators manage:

- Network selection;
- Traffic priority;
- Privacy and security;
- Traffic separation;
- RRM.

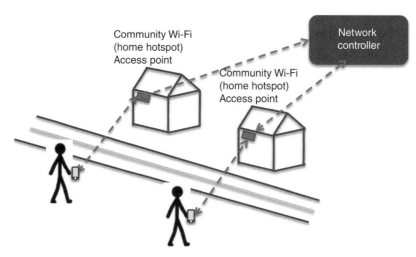

Figure 2.26. Community Wi-Fi home hotspots.

TABLE 2.1.   Community Wi-Fi Service Parameters

	Private SSID	**Public SSID**
Purpose	Private SSID on the residential GW is for exclusive use by the "broadband customer" paying for the broadband services for that residence.	Public SSID on the residential GW is for the operator to provide services to other customers (e.g., broadband customer not in range of their private SSID).
SSID configuration	Customer configured SSID	Operator managed/configured SSID
Access control	Residential customer controls access on private SSID	Operator controls access on public SSID
Traffic separation	Customers on the public SSID on the same residential GW are not allowed to communicate directly with the devices on the private SSID.	Two users on the public SSID are not allowed to communicate with each other directly. Additionally, users on the public SSID cannot directly communicate with the devices on the private SSID.
Security	Customer may or may not enable authentication/encryption.	All users of the community Wi-Fi services will need to network authenticated.
Services	Operator use the private SSID to offer services (e.g. voice, video) to the broadband customer.	Operators may offer different services to different customers on the public SSID.
QoS on Wi-Fi air interface	Different priority to traffic on the private SSID than the traffic on public SSID	Different QoS profiles to different customers or group of customers public SSID
Backhaul	Since both private and public SSID are hosted on a single residential GW, the same physical network is used to backhaul traffic on both private and public SSID. Different treatment (e.g., QoS, forwarding) of traffic from public and private SSID. Different treatment (e.g., QoS, forwarding) of traffic from public and private SSID	Since both private and public SSID are hosted on a single residential GW, the same physical network is used to backhaul traffic on both private and public SSID. Different treatment (e.g. QoS, forwarding) of traffic from public and private SSID. Different treatment (e.g. QoS, forwarding) of traffic from public and private SSID
Accounting	No need for per-user accounting on private SSID.	Per-user accounting is a requirement

2.8.7 Public Wi-Fi Hotspot

In addition to residential Wi-Fi deployments, cable operators are deploying outdoor Wi-Fi in high traffic areas such as public places, stadiums, and malls to provide outdoor Wi-Fi service to their customers. As an example, at the time of this writing, Cablevision, a cable operator on the US East Coast, has deployed thousands of outdoor Wi-Fi APs, which is used to provide outdoor Wi-Fi access to their residential customers.

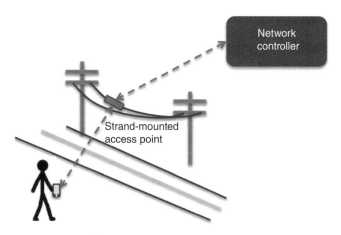

Figure 2.27. Public Wi-Fi hotspot.

Public Wi-Fi hotspots provide an excellent extension to residential Wi-Fi network and are becoming very popular with customers (Figure 2.27).

2.8.8 Cable Wi-Fi

MSOs are deploying Wi-Fi APs within their footprints in order to provide their customers with value added service outside and potentially inside their home. This service can be used on all the family's devices, including smart phones (e.g., iPhone) and can significantly reduce the cell provider's data charges for customers.

To leverage that investment, some MSOs such as Bright House Networks, Cox Communications, Optimum, Time Warner Cable, and Comcast's XFINITY now allow, in selected markets, each other's high-speed data customers to use thousands of Wi-Fi APs broadcasting the CableWiFi® wireless network name.

Subscribers gain access to a visited network by using their home operator subscription. All visited networks support subscriber access via a web portal sign in with username and password for roamers as a minimum requirement. Cable operators have the option to deploy additional methods for network access with operator configured secure clients, for example, with WPA. The visited network proxies credentials to the home network, where the home network then performs subscriber authentication and device authorization. Upon successful initial authentication, the subscriber device can be redirected to a home network hosted welcome page if the home network provides the page uniform resource locator (URL) to the visited network. The service is intended to support roamers in a visited network for a predetermined length of time such that the roamer is not forced to sign in repeatedly during the configured time period, even when moving between visited network APs. A roamer session may end due to session or idle time outs. The time values are set (i) by a RADIUS attribute sent from the home network to the visited network, (ii) visited network local policy if a home network attribute is not received, or (iii) visited network policy that may override larger values received by the home network. The application and enforcement for these time outs are set between two network operators as part of their roaming agreement [70].

Each of the providers has created methods for their subscribers to locate CableWiFi hotspots via their Web sites and mobile applications. Many of these hotspot locator implementations were created and are maintained by third-party SPs. The Wireless Service Locator (WSL) tool was created in 2010 by CableLabs to enable AP location information exchange between cable operators and third-party SPs. The tool provides a simple way to implement Wi-Fi roaming partner relationships between operators and their roaming partners, as well as a method to appropriately filter and distribute AP information to selected third parties. Operators can also define groups of APs in order to implement business rules around which APs should be shared and with whom (Figure 2.28).

Recently, WSL 2.0 was introduced in order to provide operators with additional data exchange methods and formats to align with industry standards such as the WBA WRIX-L 1.2 [71]. The new functionality enables web service methods to enable near real time data updates.

WSL is a CableLabs-hosted web application. It serves as a central database for the footprint data of the wireless APs of participating operators and wireless access providers. Its purpose is to allow operator's to share AP location data with each other as they launch wireless services for their customers and work together through roaming agreements (Figure 2.29).

Mapping providers or other associates can use data from the WSL. The operators use WSL to authorize mapping SPs, and define which APs and broadcast network names are shared with individual mapping providers. These mapping providers typically build mobile and web applications that allow the operator's subscribers to locate the operator's hotspots, their partner's hotspots and metadata about those APs. Providing end users with the ability to locate wireless APs adds value to the end user's subscription and can drive business at the AP locations.

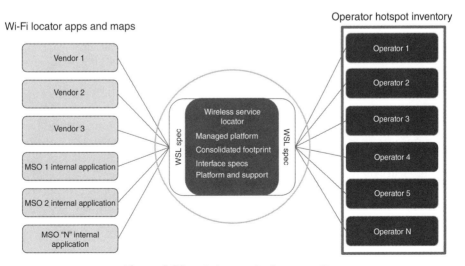

Figure 2.28. Wireless service locator, WSL.

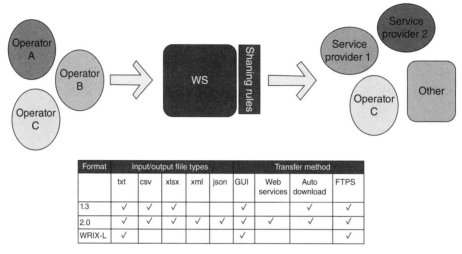

Format	Input/output flile types					Transfer method			
	txt	csv	xlsx	xml	json	GUI	Web services	Auto download	FTPS
1.3	✓	✓	✓			✓		✓	✓
2.0	✓	✓	✓	✓	✓	✓	✓	✓	✓
WRIX-L	✓					✓			✓

Figure 2.29. WSL database for sharing access point location data.

2.9 CONCLUSION

In this chapter, we have described the key residential network architecture and services of MSOs.

DOCSIS delivers high-speed data with high bandwidth and QoS guarantees to various traffic types on the HFC network. This made DOCSIS the technology of choice for MSOs, and a suitable vehicle toward all IP migration.

For residential voice services, MSOs have successfully applied an incremental approach, initially delivering traditional residential POTS to analog phones using the PacketCable 1.5 architecture, and then deploying the more powerful IMS-based PacketCable 2.0 architecture to deliver more complex communications services to a more diverse set of endpoint devices. This evolutionary strategy has enabled MSOs to grow their residential voice market share. Currently, MSOs collectively provide residential voice service to more than 38 million customers in North America alone.

The 3GPP IMS architecture and the comprehensive suite of IMS interface specifications provide a highly available and scalable IP-based service delivery platform for real-time multimedia services. IMS started life as a mobile-centric solution, but has since been embraced and updated by SPs that deploy other access technologies, including cable, such that it has become the dominant IP-based communications services delivery platform in today's network.

One of the key architectural features of IMS is to separate the application layer from application-agnostic core functions such as routing, security, and basic session establishment. This arrangement simplifies the development of new applications, since they can leverage these common core functions.

IMS continues to evolve to keep abreast of new technologies. IMS Release 13 incorporates WebRTC into the IMS architecture. Many IMS vendors have developed a virtualized version of their IMS-core solution. AS vendors are beginning to provide open APIs that enable SPs and other third-party developers to easily create and deploy new and innovative services.

MSOs have deployed private CDNs for distribution of IPTV services, whether linear, on-demand, or both. The CDNs consist of caching devices that efficiently cache and deliver IP video assets without requiring proprietary software or plugins on the subscribers' devices to do so. The delivery of content is only one part of the CDN: the CDN also contains components needed to ingest video, report on the health of the network components, and provide feedback to the MSO on the popularity of content, and the devices that consume that content. In a commercial environment where the types and capabilities of consumer-owned video-rendering devices is increasing rapidly, a well-executed CDN can arm the MSO with the knowledge necessary to determine which content and which devices merit expenditures to support. A CDN implementation is the only means for creating a widely deployed IPTV service. The old means of setting up 100 s of purpose-built video streaming servers is no longer sufficient to service the ever-increasing demands of an IPTV-hungry world.

IPTV video services can be delivered today to a wide variety of leased and consumer-owned IP devices over content delivery networks in increasingly non-vendor-proprietary means. These IPTV video services include linear/live and on-demand video and replicate and improve upon the traditional QAM-based video viewing experience. The MSOs maintain the ability to monetize the video services, and provide the equivalent of features required via regulation on the equivalent QAM-based video services.

New areas of research can be performed to provide efficiencies for IPTV delivery: introduction of IP multicast to reduce bandwidth usage, alternatives to TCP for transport protocols, and open standards for digital rights management across consumer-owned devices.

Initially, MSOs intended to only converge video and data convergence via CCAP; ultimately, CCAP will include the convergence to IP and from coaxial networks to fiber networks. As MSOs continue to evolve the CCAP concept and the actual CCAP hardware matures, significant changes are expected to ripple throughout the industry. It is generally acknowledged and even accepted at this point that the cable industry is migrating to IP as the primary means of transporting services and to achieve spectrum efficiency. Further, that all new applications will be available only over IP. It remains to be seen if full integration of legacy transport capabilities into the CCAP will occur quickly enough to provide a seamless migration from our existing disparate video and data systems, or if the CCAP is used more as a springboard to deliver IP video with the focus shifting to improving these sorts of capabilities instead.

All MSOs may not migrate to IP delivered content at the same time. Some MSOs will lead while others will trail by several years based on their own unique market needs and the competitive landscape as appraised at any given time. CCAP vendors will be forced to reconcile those differences, including any conflicts that emerge as to what is needed and when with MSOs wanting more legacy functionality at a time when most of the vendor's resources may be committed to integrating new access technologies and IP video intelligence to meet the demands of the other MSOs. The shaping of the CCAP system will demand much from the MSOs, including hitherto unheard of levels of internal and external debate, consensus building, and broadly similar visions of where the industry is headed. These are guaranteed to be very interesting times, with the pace of change driven largely by some of the biggest decisions that the industry has faced in its history with impacts felt almost immediately. What does it mean to be a cable company in the twenty-first century? How do we get content to more screens while preserving existing investments for as long as possible? Are we an ISP with content, or a content provider that is also an ISP? What part will customer demands for greater mobility and content diversity play in our decision making? The answers to these questions are critical to the success and longevity of the CCAP. While spoken in hushed words in private rooms, fiber will play an equally important role in the future of cable and, therefore, of the CCAP itself. If successful, the CCAP may well survive the retirement of all of the traditional cable transport systems including the HFC plant itself

Complementary to their wireline services, MSOs offer Wi-Fi services to their residential customers. Wi-Fi was originally conceived as a best-effort non-manageable wireless network operating in unlicensed bands with no specific QoS guarantees. Recent advancements in 802.11 technologies have led to faster and more secure Wi-Fi air interface. In addition, new interference and network management solutions, both at the air interface and the core, have become available over the past few years. Today's Wi-Fi networks performance is more robust than ever and the quality of experience of Wi-Fi users has substantially improved. Both wireless and wired SPs have investigated heavily on Wi-Fi turning it into a carrier grade, scalable, and controlled wireless network. With the ongoing efforts in standardization forums like IEEE, WFA, and WBA to improve the quality of Wi-Fi services, it is expected that operators will more and more invest in their Wi-Fi networks. Operators can achieve this, both as a value-add to their customers and as new revenue opportunities, by deploying large-scale Wi-Fi networks.

REFERENCES

1. PacketCable IMS Delta Specifications, Session Initiation Protocol (SIP) and Session Description Protocol (SDP); Stage 3 Specification 3GPP TS 24.229, PKT-SP-24.229-C01-140314, March 14, 2014, Cable Television Laboratories, Inc.
2. PacketCable Residential SIP Telephony Feature Specification, PKT-SP-RSTF-C01-140314, March 14, 2014, Cable Television Laboratories, Inc.
3. IETF RFC 3261 (June 2002): "SIP: Session Initiation Protocol."
4. PacketCable CMS to CMS Signaling Specification, PKT-SP-CMSS1.5-I07-120412, April 12, 2012, Cable Television Laboratories, Inc.
5. IETF RFC 3261 (June 2002): "SIP: Session Initiation Protocol."
6. IETF RFC 4566 (June 2006): "SDP: Session Description Protocol."
7. PacketCable 1.5 Network-Based Call Signaling Protocol Specification, PKT-SP-NCS1.5-I04-120412, April 12, 2012, Cable Television Laboratories, Inc.
8. PacketCable 1.5 PSTN Gateway Call Signaling Protocol Specification, PKT-SP-TGCP1.5-I04-120412, April 12, 2012, Cable Television Laboratories, Inc.

9. IETF RFC 3551 (July 2003): "RTP Profile for Audio and Video Conferences with Minimal Control."

10. IETF RFC 4566 (June 2006): "SDP: Session Description Protocol."

11. IETF RFC 3264 (June 2002): "An Offer/Answer Model with Session Description Protocol (SDP)."

12. IETF RFC 6665 (July 2012): "SIP-Specific Event Notification."

13. IETF RFC 3680 (March 2004): "A Session Initiation Protocol (SIP) Event Package for Registrations."

14. IETF RFC 3856 (August 2004): "A Presence Event Package for the Session Initiation Protocol (SIP)."

15. IETF RFC 3842 (August 2004): "A Message Summary and Message Waiting Indication Event Package for the Session Initiation Protocol (SIP)."

16. IETF RFC 3550 (July 2003): "RTP: A Transport Protocol for Real-Time Applications."

17. 3GPP TS 23.228: "IP multimedia subsystem; Stage 2."

18. 3GPP TS 24.229: "IP multimedia call control protocol based on Session Initiation Protocol (SIP) and Session Description Protocol (SDP); Stage 3."

19. IETF RFC 3264 (June 2002): "An Offer/Answer Model with Session Description Protocol (SDP)."

20. IETF RFC 3550 (July 2003): "RTP: A Transport Protocol for Real-Time Applications."

21. IETF RFC 3327 (December 2002): "SIP Extension Header Field for Registering Non-Adjacent Contacts".

22. EF RFC 3608 (October 2003): "SIP Extension Header Field for Service Route Discovery During Registration".

23. PacketCable IMS Delta Specifications; Session Initiation Protocol (SIP) and Session Description Protocol (SDP); Stage 3 Specification 3GPP TS 24.229.

24. CableLabs. (2012) Content Encoding Profiles 3.0 Specifications OC-SP-CEP3.0-I04-121210. CableLabs.

25. Adobe. Configure Adobe HTTP Dynamic Streaming and HTTP Live Streaming, http://help.adobe.com/en_US/flashmediaserver/devguide/WSeb6b7485f9649bf23d103e5512e08 f3a338-8000.html.

26. Pantos, R. (2013). HTTP Live Streaming draft-pantos-http-live-streaming-12, http://tools.ietf.org/html/draft-pantos-http-live-streaming-12.

27. Zambelli, A. (2009). IIS Smooth Streaming Technical Overview, Microsoft Corporation.

28. 3GPP. (2013). Overview of 3GPP Release 9 V0.3.0. 3GPP.

29. Open IPTV Forum. (September 7, 2010). OIPF Release 2 Specification HTTP Adaptive Streaming [V2.0].

30. ISO. ISO/IEC 23009-1: 2012, "Information technology—Dynamic adaptive streaming over HTTP (DASH)—Part 1: Media presentation and segment formats."

31. Microsoft. (2008). Microsoft Play Ready Content Access Technology White Paper. Microsoft.

32. Google. (2013). Wide Vine DRM.

33. Adobe. (2012). Adobe Access 4 Datasheet. Adobe.

34. ISO. ISO/IEC 23001-7: 2011, "Information technology—MPEG systems technologies—Part 7: Common encryption in ISO base media file format files."

35. Open Authentication Technology Committee. (2012). Online Multimedia Authorization Protocol. OATC.

36. Hardt, D. E. (2012). The OAuth 2.0 Authorization Framework. IETF.

37. Society of Cable Telecommunications Engineers. (2007). Digital Program Insertion Cueing Message for Cable. SCTE.

38. Society of Cable Telecommunications Engineers. (2011). ANSI/SCTE 130-1 2011 Digital Program Insertion—Advertising Systems Interfaces. SCTE.

39. iab. (2009). Digital Video Ad Serving Template (VAST) Version 2.0. iab.

40. iab. (2012). Video Multiple Ad Playlist (VMAP) Version 1.0. iab.

41. iab. (2012). Digital Video Player-Ad Interface Definition (VPAID) 2.0. iab.

42. CableLabs. (2013). Real-time Event Signaling and Management API. CableLabs.

43. FCC. (2009). CFR-2009-Title47-Vol1-part 11. Federal Communications Commission.

44. OASIS. (2010). Common Alerting Protocol Version 1.2. Organization for the Advancement of Structured Information Standards.

45. ANSI/CEA. (2013). Digital Television (DTV) Closed Captioning ANSI/CEA-708E.

46. SMPTE. (2010). Timed Text Format (SMPTE-TT). SMPTE.

47. CableLabs. (2013). Event Scheduling and Notification Interface OC-SP-ESNI-I02-131001. CableLabs.

48. Apache. (2013). *Apache Traffic Server.* Apache.

49. *Cartoon Network LP, LLLP vs. CSC Holdings, Inc.* (2007).

50. CableLabs. (2008). Data-Over-Cable Service Interface Specification Modular Head end Architecture CM-TR-MHA-V02-081209. CableLabs.

51. IETF. (1998). *Internet Protocol, Version 6 (IPv6).* The Internet Society.

52. IETF. (2013). *Request Routing Redirection Interface for CDN Interconnection.* IETF.

53. IETF. (2000). *Hyper Text Caching Protocol (HTCP 0.0).* The Internet Society.

54. Belady, L. (1966). *A study of replacement algorithms for virtual storage computers.* IBM Systems Journal, Volume 5, Issue 2, pp. 78–101.

55. Fischer, Y. (2012). *Comparing Just-in-Time Packaging with CDN Storage for VoD and nDVR Applications, Proceedings of the Canadian SCTE, March 2012.* (C. SCTE, Ed.), Exton, PA.

56. *Gomez Performance Network.* (2013). (Compuware), http://www.compuware.com/en_us/application-performance-management/solutions/apmaas/overview.html

57. Catchpoint. (2013, Dec). *catchpoint.com.* Retrieved on 12/13/2013, from catchpoint.

58. Webpagetest.org. (2013). *Webpagetest.org.* Retrieved Dec 2013, from Webpagetest: Webpagetest.org

59. IETF. (2002). *Content Distribution Interworking (CDI) AAA Requirements draft-ietf-cdi-aaa-reqs-01.* IETF

60. Hardin, G. (2007). *Narrowcast Services—Unifying Architectures.* NCTA.

61. CableLabs. (2008). Data-Over-Cable Service Interface Specification Modular Headend Architecture CM-TR-MHA-V02-081209. CableLabs.

62. CableLabs. (2008). Edge QAM Provisioning and Management Interface Specification CM-SP-EQAM-PMI-01-081209. CableLabs.

63. CableLabs. (2011). Edge Resource Management Interface CM-SP-ERMI-I04-110623. CableLabs.

64. Cablelabs. (2008). Modular Headend Architecture Technical Report CM-TR-MHA-v02-081209. CableLabs.

65. CableLabs. (2008). M-CMTS Operations Support System Interface Specification CM-SP-M-OSSI-I08-081209. CableLabs.

66. CableLabs. (2010). Downstream External PHY Interface CM-SP-DEPI-I08-100611. CableLabs.

67. CableLabs.(2013).DPoEIPNetworkElementRequirementsDPoE-SP-IPNEv1.0-I06-130808. CableLabs.

68. CableLabs. (2008). DOCSIS Timing Interface CM-SP-DTI-I05-081209. CableLabs.

69. IEEE 802.11ac: Enhancements for Very High Throughput for Operation in Bands below 6 GHz, Dec. 2013.

70. Wi-Fi Roaming Architecture and Interfaces Specification, http://www.cablelabs.com/wp-content/uploads/specdocs/WR-SP-WiFi-ROAM-I03-140311.pdf

71. Wireless Roaming Intermediary eXchange (WRIX—i, l, d & f), http://www.wballiance.com/resource-center/specifications/

3

OPERATIONS, ADMINISTRATION, MAINTENANCE, PROVISIONING, AND TROUBLESHOOTING FOR RESIDENTIAL SERVICES

Brian Hedstrom, Stuart Hoggan, Sergio Gambro
and Mehmet Toy

3.1 INTRODUCTION

Operations, administration, maintenance, provisioning, and troubleshooting (OAMPT) capabilities, which add troubleshooting functionality to the traditional OAMP, are necessary to order and maintain the residential services described in Chapter 2. In addition to OAMPT, services need to be billed to subscribers.

Service order process begins with sales. Once the service order is in, the provisioning process installs equipment at customer premises, sets up equipment at customer premises and central offices, sets up user accounts, and creates new circuits. The provisioning processes will also include checklists that need to be strictly adhered to and signed off, and integration and commissioning processes that will involve sign-off to other parts of the business life cycle.

Operations encompass automatic monitoring of the environment, detecting and determining faults, and alerting network administration. Network administration collects performance statistics, accounts data for the purpose of billing, plans capacity based on usage data, maintains system reliability, administers network security, and maintains the service database for periodic billing.

Cable Networks, Services, and Management, First Edition. Edited by Mehmet Toy.
© 2015 The Institute of Electrical and Electronics Engineers, Inc. Published 2015 by John Wiley & Sons, Inc.

Maintenance involves routine equipment checks, upgrading software and hardware, fixes, new feature enablement, backup and restore, standard network equipment configuration changes as a result of policy or design, and monitoring facilities.

When there is a failure, failed components are identified by diagnostic tools and troubleshooting. Troubleshooting procedures involve knowledge databases, guides, and process to cover the role of network operations engineers from initial diagnostics to advanced troubleshooting.

OAMPT for residential services are challenging mainly due to the involvement of a very large number of customer premises equipment (CPE), facilities, and connections (i.e., in the order of millions). As a result, automation for each OAMPT function is crucial for multiple system operators (MSOs). Autoprovisioning of CPE and connections, administration of network security via Data over Cable Service Interface Specification (DOCSIS) protocol and back-office systems are developed to resolve these scalability issues.

In the following sections, we will discuss OAMPT for residential services described in Chapter 2.

3.2 OPERATIONAL SYSTEMS AND MANAGEMENT ARCHITECTURES

The DOCSIS service and device management approach use the set of management categories defined by the International Telecommunication Union (ITU) Recommendation M.3400 [1], referred to as the FCAPS model, to organize the requirements for the configuration and management of the CMTS/CCAP and CM devices. Telecommunications operators, including MSOs, commonly use this model to manage large networks of devices. The FCAPS model is represented by the individual management categories of fault, configuration, accounting, performance, and security management categories.

Fault management is a proactive and on-demand network management function that allows nonstandard/abnormal operation on the network to be detected, diagnosed, and corrected. A typical use case at the network layer involves network elements detecting service-impacting abnormalities; when detected, an autonomous event (often referred to as an alarm notification) is sent to the network operations center (NOC) to alert the MSO of a possible fault condition in the network affecting a customer's service. Once the MSO receives the event notification, further troubleshooting and diagnostics can be performed by the MSO to correct the fault condition and restore the service to proper operation. Similar use cases apply at the service level.

Configuration management provides a set of network management functions that enables system configuration building and instantiating, installation and system turnup, network and device provisioning, autodiscovery, backup and restore, software download, status, and control (e.g., checking or changing the service state of an interface). DOCSIS configuration management is typically performed at the network layer (e.g., device provisioning at the CMTS/CCAP and CM).

Accounting management is a network management function that allows MSOs to measure the use of network services by subscribers for the purposes of cost estimation and subscriber billing. Subscriber Accounting Management Interface Specification (SAMIS), as defined in DOCSIS [2], is an example of an implemented accounting management function.

Performance management is a proactive and on-demand network management function. The ITU Recommendation M.3400 [1] defines its role as gathering and analyzing "statistical data for the purpose of monitoring and correcting the behavior and effectiveness of the network, NEs, or other equipment and to aid in planning, provisioning, maintenance and the measurement of quality." A Performance management network layer and service-level use case might include the NOC performing periodic (15 min, for example) collections of quality-of-service (QoS) [3] measurements from network elements to perform monitoring and identification of any potential performance issues that may be occurring with the service being monitored. With the historical data that has been collected, trending analysis can be performed to identify issues that may be related to certain times of day or other corollary events. The MSO can run reports on the data to identify suspect problems in service quality, or the NOC application can be provisioned, so that when certain performance thresholds are violated, the MSO is automatically notified that a potential service quality problem may be pending. Significant intelligence can be integrated into the NOC application to automate the ability to detect the possible degradation of a customer's service quality, and take actions to correct the condition.

Security management provides for the management of network and operator security, as well as providing an umbrella of security for the telecommunications management network functions. Security management functions include authentication, access control, data confidentiality, data integrity, event detection, and reporting. A security management use case might include providing authentication and data confidentiality when transferring a software image file that contains the entire software load for the device to a network element [4].

3.2.1 Cable Modem Management

Figure 3.1 illustrates the cable modem (CM) management architecture from the MSO back-office interface perspective. The CM and CMTS reside within the network layer where services are provided to end subscribers and various metrics are collected about

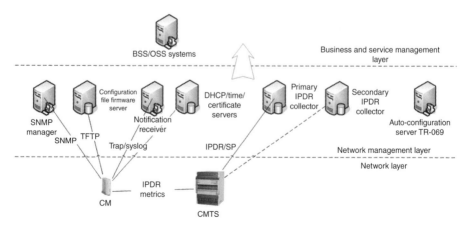

Figure 3.1. Cable modem management architecture.

TABLE 3.1. CM Management Servers

Management Server	Function
SNMP manager	Performs SNMP configuration and queries against a CM's SNMP agent.
Configuration file server	Transferring configuration files, via TFTP or optionally HTTP to the CM upon reinitialization.
Firmware file server	Transferring firmware images, according to the Secure software download mechanism, to a CM
Notification receiver	Receives autonomous SNMP notifications and syslog messages from a CM.
DHCP server	Assigning an IPv4 and/or IPv6 address as well as other DHCP parameters in order for the CM to obtain it's configuration file and register on the network
Time server	Provides a CM with current time of day, ToD.
TR-069 server	A CM does not communicate directly with the TR-069 server. Rather, if the CM is an eDOCSIS device and includes an eSAFE that supports the TR-069 protocol, the eDOCSIS device will communicate with the TR-069 server.

network and service performance, among other things. Various management servers reside in the network management layer within the MSO back-office to provision, monitor, and administer the CM within the network layer. The management servers are listed in Table 3.1.

Finally, the business and service management layer is where higher level MSO business processes (e.g., order management, billing, customer care, customer relationship, and SLA management) are implemented via BSS/OSS systems. These BSS/OSS systems utilize the data and information from the network management layer, which interrogate data from the network layer.

3.2.2 CMTS/CCAP Management

Figure 3.2 illustrates the CMTS/CCAP management architecture from the MSO back-office interface perspective. The CMTS or CCAP [5] resides within the network layer where services are provided to end subscribers and various metrics are collected about network and service performance, among other things. Various management servers reside in the network management layer within the MSO back-office to provision, monitor, and administer the CM within the network layer. The management servers are listed in Table 3.2.

The business and service management layer is where higher level MSO business processes are implemented via BSS/OSS systems. For example, the subscriber usage metrics collected via the IPDR/SP protocol at the IPDR collector is sent northbound to the BSS systems for billing and mediation purposes.

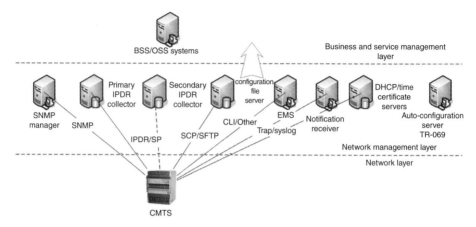

Figure 3.2. CMTS/CCAP management architecture.

TABLE 3.2.   CMTS/CCAP Management Servers

Management Server	Function
SNMP manager	Performs SNMP queries against a CMTS/CCAP SNMP Agent. Note: In DOCSIS 3.0 and earlier generations, SNMP was used for configuration.
Configuration file server	Transferring XML configuration files, via SCP/SFTP to the CMTS/CCAP for configuration management purposes.
DHCP server	Assigning an IPv4 and/or IPv6 address as well as other DHCP parameters.
Notification receiver	Receives autonomous SNMP notifications and syslog messages from a CMTS/CCAP.
Certificate revocation server	Provides information and status for security certificates.
Time server	Provides a CM with current time of day, ToD.
IPDR (IP detail record) collectors	Bulk data statistics, such as usage metrics, are collected via the IPDR/SP (streaming protocol) protocol using primary and secondary IPDR Collectors.

3.2.3 DOCSIS 3.1 Service and Network Management Overview

The major service and network management features introduced in the DOCSIS 3.1 protocol include configuration, monitoring, and reporting on the feature set:

- DOCSIS light sleep mode (DLS) [6];
- Backup primary channels;
- Active queue management (AQM);
- Proactive network maintenance (PNM) [7].

The DOCSIS service and device management approach used for DOCSIS 3.1 remains the same as DOCSIS 3.0 for the CM management model. With respect to DOCSIS 3.1 management of the CMTS, the management model is aligned with the CCAP management approach, where configuration management using XML-based configuration files is used. In summary, the DOCSIS 3.1 management converges the CCAP and CMTS management models.

3.3 SERVICE ORDERS

Throughout the preceding chapters, you have read about the complex nature of the services delivered by MSO's today. Many times, you will read about the "BSS/OSS" layer; and while the network layer is a complex environment, the BSS/OSS layer is equally complex in its own right.

The service order is in effect the translation of the offer the end consumer purchased, into the various commands that initiate the activation or change in service in the network in order to deliver that service to the customer. Just as the network has undergone a complex transformation thanks to changes in technology, so have the service ordering platforms. In order to properly explain the service order, we need to explain some of the common terminology that comes with it.

3.3.1 Order Entry/Sales Channels

Considering that cable TV started in the early 1950s in the United States, there were only a limited means for taking a customer order, by telephone, or by door-to-door sales. We've come a very long way from then so you'll be surprised to hear that those two sales channels are still operating today and provide a great deal of the sales volume for an MSO. With the growth in Internet connections in the home, operators have invested heavily in letting customers place orders themselves via the Web. Additionally, operators have partnered with retailers and e-tailers as resellers of their services and that has given customers many different choices for acquiring new services from an MSO. While adding all these sales channels may have expanded how the customer can purchase services, it has also generated complexity in enforcing business policies in a consistent manner through all those channels. Also purchasing a new service is something very different than changing your existing services.

MSOs have to deal with increased competition not only from Satellite providers but also Telco providers (like Verizon and AT&T in the United States) as well as overbuilders and increasingly from cord cutters and over the top providers. In order to keep customers while maintaining or increasing revenue, they have bundled more and more services into the same offers to customers. Basic video services as opposed to digital or premium channels like HBO are all levels of video services that are can be bundled together to provide a more compelling service to a customer.

Determining the right offers and bundles is a complex process that is a continual process that mostly involves finance and marketing organizations of the operator. Ultimately, the service provider's offers are built around a series of products, services, installation choices, and devices that bundled together the customer purchases usually at an initial discount price. This offer is then distributed to the sales channels to sell (not all offers available everywhere). Each of those offers then has restrictions on whether it can be sold to new or existing customers, to customers with a certain credit worthiness, or with specific qualifications, all of which will increase the complexity of the order process.

Now that the offer is defined and available to the sales channel to sell, we can talk about where that ordering step actually happens. In order to understand the service order properly, MSO history needs to be understood.

When cable TV started in the United States, it was the 1950s, and Information technology was not even a term most executives considered a department of their company. They had a problem, they needed to take orders for their service, and they needed to bill customers for those services. A number of companies came to the rescue and created software applications that would basically be "cable in a box." These systems, affectionately known as "the billers" were constantly enhanced to do more and more of the back office functions needed for a company to survive. They were the customer management system (i.e., the system that determines the services a customer is authorized to receive). They became the CPE inventory system in order to understand which customers had which device. After all, CPE was a major capital expense for these companies and given that they charged for each box, they were also profit generators. These billers also became work order managers and workforce management tools, in order to schedule employees to go to a customer home and install or disconnect service. And here is where these systems initiate the service order.

The service order is basically a breakdown of the offer the customer has chosen into its various parts. All systems use names like service codes or rate codes to determine the translation of a particular part of the offer into something that will be charged to a customer. The main function of the service order is to break down the offer that was purchased into the various service components that are being requested. Different parts of the service order will be transmitted to downstream provisioning systems based on the services ordered and the status of a customer. For example, adding HBO to an existing customer who already has a digital STB installed does not require the intervention of a technician, so the service order can trigger a provisioning hit on the video interface that in effect is a command to the Conditional Access (CA) System of the operator.

Increasingly, customers have choices for every service a provider offers and with this increased competition is an increased volume of change of service orders. While in the past, an operator could simply remove all the services off an account and re-add all the newly authorized services, (basically a disconnect and a re-connect), this has some nasty repercussion on a customer account. In today's services, a disconnect of your services means that you may lose some of the features that are unique to you, like your email address, your phone number, or potentially the content you've recorded on your DVR, all of which are very poor customer experiences; therefore, it's very important to translate the service order into the intention of what the customer wanted to happen. An existing customer who is only removing HBO services from their account needs to only receive a video hit to de-authorize the single-premium channel. Due to the fact that many of the systems involved in the service order were developed during the times when intention was irrelevant, the service provider space has developed systems and tools to derive intention out of the service order. Ultimately, those intentions are sent through vendor-specific interfaces to the downstream provisioning platforms that will be discussed in Section 3.4.

Increasingly, the service provider network is being utilized to deliver more complex services that are considered over the top; and as such, they are not provisioned within the service provider network but rather through a partner. Similar to service orders sent to proprietary Customer Account System (CAS), each partner tends to have a proprietary interface for their particular business need. These particular integrations can be performed at the service order or the provisioning component of the service provider's back-office.

3.4 PROVISIONING

The provisioning system of the BSS/OSS stack is a critical component of any service provider's back-office. Without this critical component, none of the products and services sold to a customer will be activated, or once activated will not be shut down. Given the history mentioned in Section 3.3, the provisioning system originated in the "billers" as it was the only system that maintained the subscriber information and could interact with the Video CA system. With the advent of broadband Internet service, there was the need to interface to the DOCSIS DHCP server to authorize the customers particular CM with a specific level of service. By now, the engineering and IT organizations of the service providers had acquired enough experience to purchase software packages from specialized vendors and customize them for their specific needs.

Most service providers leverage a proprietary interface from the "billers" and then extrapolate the device information to be pushed into a specific Lightweight Directory Access Protocol (LDAP) data structure containing critical modem information that is used by the DOCSIS provisioning process to determine the correct boot file for that particular subscriber.

3.4.1 CM Provisioning and Software Updates

Provisioning of the CM [4, 8] is primarily performed using the configuration file download process. Binary configuration files are constructed using the TLV definitions from the DOCSIS specifications. The CM is required to support the TFTP protocol for downloading the configuration files. The configuration file process for the CM is shown in Figure 3.3.

Software updates to the CM are performed using the secure software download (SSD) mechanism defined by the DOCSIS security specification [9]. A "monolithic" software image containing the CM firmware as well as any embedded DOCSIS (eDOCSIS) device firmware (e.g., eMTA) is loaded onto a firmware file server in the MSO's back-office for secure download to the CM via TFTP when the CM receives the appropriate firmware upgrade trigger [4]. The SSD process is illustrated in Figure 3.4.

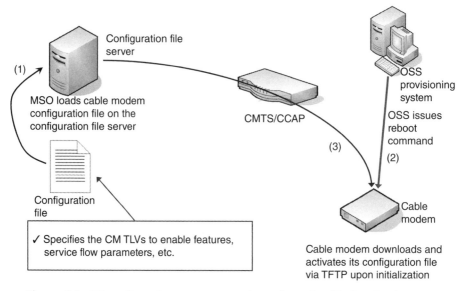

Figure 3.3. CM configuration management via configuration file download process.

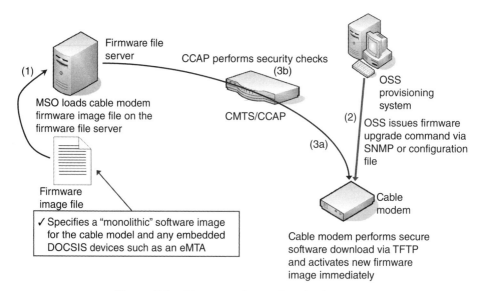

Figure 3.4. CM secure software download process.

3.5 FAULT MANAGEMENT

Fault management is a proactive and on-demand network management function. The ITU Recommendation M.3400 [10] defines it as "a set of functions which enables the detection, isolation and correction of abnormal operation of the telecommunication network and its environment." A typical use case involves network elements detecting service-impacting abnormalities; when detected, an autonomous event (often referred to as an alarm notification) is sent to the NOC to alert the MSO of a possible fault condition in the network affecting a customer service. Once the MSO receives the event notification, further troubleshooting and diagnostic testing can be performed on-demand by the MSO to correct the fault condition and restore the service to proper operation. Fault management includes the following function sets:

- Alarm surveillance,
- Fault localization,
- Fault correction, and
- Testing.

3.5.1 Alarm Surveillance

The DOCSIS protocol provides an event mechanism to support autonomous events to be triggered and sent by CM and CMTS/CCAP devices to event monitoring applications [2, 5]. The protocol defines DOCSIS-specific events as well as standard events defined by the IETF. When certain conditions occur as defined within the protocol specifications, the associated events are triggered. The event mechanism can be configured to send events via SNMP notifications, syslog messages, or log the events into the device's local log. The local log supports the ability of on-demand polling for event and alarm information.

DOCSIS events are defined with several severity levels, including informational, notice, warning, error, and critical to convey the impact of the raised event/alarm to the customer, individual service, node, CMTS, and so on. This can provide a level of priority in addressing faults since one event might be service impacting to one customer while another fault might be service impacting to many customers.

3.5.2 Fault Localization

Before the MSO performs a truck roll to assess the physical plant, there are a number of remote monitoring and testing functions within the CMTS/CCAP and CM, which can be performed on-demand to test the plant conditions to isolate faults. There are DOCSIS-specific spectrum analysis measurement and proactive network maintenance measurement tools that can be utilized as specified by the DOCSIS protocol. There are layer-3 trace route and ping tests to trace packet flows. In addition, there are many other DOCSIS protocol and IETF standard metrics for monitoring the health of the network and testing the network links.

3.5.3 Fault Correction

Once the fault has been localized and identified with a root cause, the correction can be put in place. Correcting the fault might be as easy as remotely updating a parameter in the CM configuration file, rebooting a network element due to a software glitch, or might be more complex such as encountering a service affecting software defect that requires the device vendor to issue a new software release. If a plant hardware device failed, a truck roll may be necessary to replace the failed component in the network. If a customer CPE device failed, the customer may be required to swap their CM in for a new one.

3.5.4 Testing

Testing was discussed under the Section 3.5.2. Other forms of testing include bandwidth throughput testing to determine the level of service received. There are many forms of testing not necessarily tied to localizing faults. Such forms include assessing the overall health of the service, the devices providing access to the service, and the overall network delivering those services. There are also some levels of testing beyond the MSO network into the home network, between the CM and the CPE devices that sit behind the CM, for example, testing the Wi-Fi network providing high-speed data to wireless devices in the home network.

3.6 PERFORMANCE MANAGEMENT

Performance management is a proactive and on-demand network management function. The ITU Recommendation [1] defines its role as gathering and analyzing "statistical data for the purpose of monitoring and correcting the behavior and effectiveness of the network, NEs, or other equipment and to aid in planning, provisioning, maintenance and the measurement of quality." A performance management network layer and service-level use case might include the NOC performing periodic (15 min, for example) collections of QoS measurements from network elements to perform monitoring and identification of any potential performance issues that may be occurring with the service being monitored (residential high-speed data service, for example).

With the historical data that has been collected, trending analysis can be performed to identify issues that may be related to certain times of day or other corollary events. The MSO can run reports on the data to identify suspect problems in service quality, or the NOC application can be provisioned, so that when certain performance thresholds are violated, the MSO is automatically notified that a potential service quality problem may be pending. Significant intelligence can be integrated into the NOC application to automate the ability to detect the possible degradation of a customer's service quality, and take actions to correct the condition.

At the CATV MAC and PHY layers, performance management focuses on the monitoring of the effectiveness of cable plant segmentation and rates of upstream traffic and collisions. Instrumentation is provided in the form of the standard interface statistics and service queue statistics. For example, this includes collecting statistics of parameters such as number of frames lost at the MAC layer and number of codeword errors at the PHY layer. Additional metrics include variables to track

PHY state such as signal-to-noise ratios, transmit and receive power levels, propagation delays, microreflections, in channel response, and sync loss. There are also counters to track MAC state, such as collisions and excessive retries for requests, immediate data transmits, and initial ranging requests. These monitoring functions are used to determine the health of the HFC network and whether the offered QoS to the subscriber is met. The quality of signal at the PHY layer is an indication of plant conditions [5].

At the LLC layer, the performance management focus is on bridge traffic management. If the CMTS and CCAP implement transparent bridging, they support various statistics defined by the IETF Bridge MIB.

The CMTS and CCAP also include diagnostic log capabilities, which provide early detection of CM and cable plant problems.

DOCSIS 3.0 performance management requires an efficient mechanism for collecting large data sets as described earlier. The identified data sets are as follows:

- The CMTS/CCAP resident CM status information;
- Additional granularity of QoS statistics for bonded and nonbonded channels to aid in network capacity planning and dimensioning;
- Enhanced signal quality monitoring for granular plant status;
- Minimizing redundant information collection associated with differing services provided by the CMTS (statistics for PacketCable voice may incorporate large data sets for DOCSIS PHY and MAC);
- Support for CM and CMTS/CCAP host resource statistics, such as memory and CPU utilization.

The DOCSIS 3.1 performance management requirements include extending the DOCSIS 3.0 efficient mechanism for collecting large data sets, which includes

- Proactive network maintenance for monitoring signal quality [11], and
- Status and statistics for OFDM downstream and OFDMA upstream channels [12].

3.6.1 CMTS and CCAP Performance Metrics Collection

For the DOCSIS 3.0 and 3.1 identified performance metric data sets implemented in the CMTS and CCAP, both SNMP and IPDR/SP protocols are available for bulk data collection of the data sets. MSO business policies and back-office application architectures dictate which protocol is utilized for collecting the various performance metrics. Figure 3.5 highlights the IPDR/SP network model for bulk data collection of PM metrics to a performance management application in the MSO back-office. The CMTS/CCAP is the "IPDR recorder" performing the measurements of the PM metrics (e.g., recording the information to be sent to the back office application). IPDR data records are streamed over the IPDR/SP protocol to a collector system. The data records conform to an IPDR/SP service definition that is published as part of the

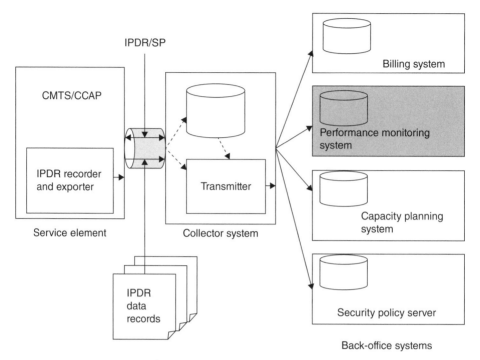

Figure 3.5. IPDR/SP network model.

DOCSIS OSSI specification. The collector system then extracts the IPDR data records and can reformat the data into any business-to-business application interface the back-office is expecting.

3.6.2 CM Performance Metric Collection

For the DOCSIS 3.0 and 3.1 identified performance metric data sets implemented in the CM, the SNMP protocol is used for collecting these metrics.

The CM performance metric sets include the following:

- RF MAC domain metrics;
- SC-QAM and OFDM downstream channel metrics;
- SC-QAM and OFDMA upstream channel metrics;
- Energy management 1x1 mode statistics;
- Service flow statistics;
- DOCSIS path verify (DPV) statistics;
- Spectrum analysis statistics;
- Signal quality metrics;
- QoS statistics;
- Downstream Service Identifier (DSID) statistics.

3.7 BILLING SYSTEMS AND FORMATS

The billing systems were probably the first software systems used by service providers. While much of the analogue services sold to customers in the 1960s were transmitted via dedicated hardware, the billing systems were software platforms that evolved out of necessity to create accurate and timely bills for services as well as to assist the business in growing as their needs changed.

The name "billing" is a misnomer because it indicates just one of the many functions that has been built into these software applications. The systems cover nearly every aspect of the operations of a modern MSO. The product catalogue, the CPE inventory, the order entry system, the service order manager, the taxation engine, and the biller, as well as the workforce manager (determining the number of availability of installation technicians) and the serviceable home database of the operator (which premises are serviceable for what services), are some of the main components but are by no means an exhaustive list of the functions the biller performs today (Figure 3.6).

There is hardly a single aspect of the MSO's business today that does not involve some portion of the "biller." Operators have been working for years to separate each function into discrete components with defined interfaces so that they can implement best-of-breed systems rather than having to be tied to a single service provider. An

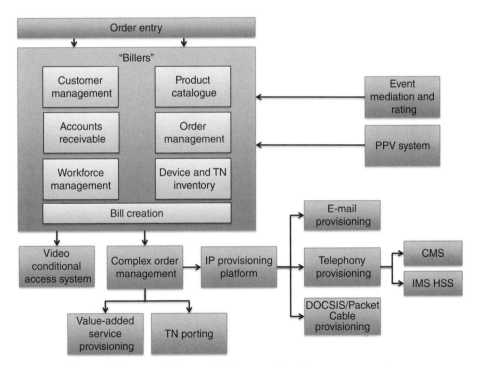

Figure 3.6. Simplified view of business and service management layer.

example is the implementation of the telephone number (TN) inventory system. Some operators have utilized the system provided by the "biller," while others have utilized external TN inventories. Some operators have multiple "billing" providers in order to maintain leverage and competition between the vendors. In all cases, the billing systems providers have developed a deep knowledge of the MSO business and will continue to play a large part of the MSO's back-office for years to come.

3.7.1 Downstream Interfaces

In a typical IT architecture, the billing systems seem to be at the top of the architecture with the ordering process taking place at the beginning of every action. Downstream of the billing system, there are the various subsystems like the video CA platform (Cisco and Motorola CAS systems being the most prevalent in the United States) or the DOCSIS provisioning platform. The format of the interface for the video CA system is proprietary to each of the above mentioned vendors, and each billing provider has had to develop and certify their system with that platform. Service providers have developed many applications to augment or extract functions from the billers in order to meet business needs and they now certify these applications with the billers. A typical example of downstream system is the complex order management platform that is used to break down service orders into suborders to systems like the IP provisioning system used to activate high-speed data and digital voice services. Vendors have published specific APIs that many providers have wrapped in Web services and exposed to their back-office systems.

3.7.2 Upstream Interfaces

While the biller may be at the top of the architecture drawing from an IT viewpoint, it is by no means upstream of everything. A few examples of systems that upload information to the biller on a regular basis are the pay-per-view system and the event mediation and rating platforms that provide summarized billable events. Each of these systems has integrated with each biller according to vendor-specific interfaces. While the billing vendors had wanted to provide a single solution for all services, their offerings in the way of IPDR or CDR (call detail record) mediation and rating did not meet the requirements of the MSO community, so many of them installed separate platforms to perform that function. In the case of the event mediation and rating platform, it would generally consolidate all charges of the same taxable type into a single entry to be applied to the customer account. All details of the individual transactions would remain in the mediation and rating platform. This allows the service provider to minimize the amount of data uploaded into the billers as well as providing cost savings. Generally, the uploading of large quantities of data takes place through secure FTP of files that are uploaded through specific batch jobs developed by the billing vendors.

When a service provider wants to build an interface to a biller, there is typically a certification process in order to utilize the biller's proprietary interface. These are typically projects that the service providers IT or engineering organization undertake once and rarely have to modify after implementation. Modifying or certifying to the vendor specific interface is an expensive proposition, so service providers avoid it as much as possible.

3.7.3 Billing

The core function of the "billing" system is to generate an accurate bill that is also usually the most contentious part of a customer experience with an operator. It will not surprise the reader to learn that the bill is a highly analyzed document that is very carefully crafted with the input from many departments/functions within the service provider. Marketing and customer care are obvious choices, but also government affairs/ regulatory, legal, tax, and finance all participate in the design of the bill with IT. The design of the billing cycles is carefully planned with accounting and customer care to ensure a constant stream of cash coming into the company and yet optimizing the customer experience for those that want to pay their bill on the 18th of each month. Each customer is assigned to one of these billing cycles; and once a month, their bill is calculated. Charges are usually broken into (i) monthly recurring charges, (ii) one-time charges, (iii) taxes and fees, (iv) discounts, and (v) credits. When a customer pays their bill, their account is "credited" and when they purchase a video-on-demand movie, their account is debited. The bill is calculated at the time their billing cycle is run. This is a batch process that generally occurs multiple times per month depending on the particular setup the service provider has chosen.

3.7.4 Product Catalog

As mentioned earlier, the billing systems have, at their core, a product catalogue that is the definition of the services, products and prices for each. The complexity of the product catalogue should not be underestimated. The product catalogue is a proprietary component of each platform and each vendor utilizes specific terminology that has been engrained in all users of the system, thereby making the platform "sticky."

The product catalogue generally breaks down the offer into many services, each with it's own code. That particular service will usually generate revenue that needs to be accounted for within the MSO's chart of accounts. It will also generate fees to be placed on a customers' bill. It will carry a specific description to be placed on the bill. The charges will carry a specific configuration to ensure they are placed within a specific section of the bill and are summed up with specific other types of charges.

The service will have specific taxation properties; and when the service is ordered by a customer, it will trigger specific "provisioning tasks" that will send information to downstream systems like the video CAS platform or the complex order management platform. All of these are parameters that need to be determined and configured within the billing product catalogue making the offer design process critically important and very complex.

3.8 SECURITY

When providing content and services over a network to subscribers, it is important to apply proper security. There are many threats that exist that can negatively impact a subscriber's experience and service provider's business. These threats include spoofing, tampering, and information disclosure.

Servers and devices can be spoofed or cloned by malicious entities. This is usually done to either steal service or obtain subscriber identity information. When a device is cloned, it looks like the device of a paying subscriber and allows the hacker to steal service. Spoofing servers can cause the device or subscriber to connect to an unauthorized server that requests confidential information from them (including username/passwords) that would enable hackers to steal services from not only their cable service provider but also other service providers the subscriber has accounts with. Authentication is the method used to verify connection to valid servers and devices, and helps prevent spoofing.

When data are sent across a network, it can be modified by an unauthorized entity that exists between the sender and the receiver. This can cause problems with provisioning and delivery of content and services. Subscribers would not get what they requested or the quality may be degraded. Provisioning data could have errors resulting in subscriber overbilling or underbilling. Message integrity checking is the method used to make sure that data are not tampered with as it is sent over a network.

Hackers can exist on any part of a network: at the end points or located somewhere in between. This gives them access to the data sent over the network. Whether it's from the cable service provider or from the subscriber they will be able to receive or monitor the traffic. This can lead to theft of service, identity theft, and a number of other bad things. It is important to protect service content and subscriber information when sent over the network. Encryption is the method used to prevent information disclosure.

All of the residential network architectures discussed in Chapter 2 have security functions and controls to help prevent the threats listed earlier and more. The following sections explain what security mechanism exist for each architecture and recommendations on how to manage them.

3.8.1 Security in DOCSIS 3.0/3.1

DOCSIS 3.0/3.1 security features focus on preventing theft of service and information disclosure (loss of privacy). The main features are device authentication and traffic encryption (Baseline Privacy Plus or BP+), secure software download, and secure provisioning. DOCSIS 3.0 security details and requirements can be found in the DOCSIS 3.0 security specification [13], MULPI [11], and OSSI [12] specifications. SNMP MIB requirements are found in RFC 4131 [14] and the DOCSIS 3.0 OSSI specification.

3.8.1.1 Device Authentication and Traffic Encryption
Device authentication and traffic encryption are provided by a feature called Baseline Privacy Interface Plus (BPI+). It supports CM authentication, key exchange, and traffic encryption. BPI+ starts when a CM connects and registers on the DOCSIS network. X.509 digital certificates are used to authenticate the CM. A unique device certificate and private key are installed in each CM along with the manufacturer certification authority (CA) certificate that issued/signed the device certificate. The DOCSIS root CA certificate is installed in the CMTS and acts as a trust anchor for validation purposes. CableLabs manages the DOCSIS certificate public key infrastructure (PKI) shown below for the cable industry (Figure 3.7).

Figure 3.7. DOCSIS certificate PKI hierarchy.

The CMTS authenticates each CM before allowing it to access the network. Authentication begins with the CM sending its device certificate and manufacturer. CA certificate to the CMTS. The CMTS validates these certificates by verifying that they chain to a trusted root CA certificate. Validation includes other checks such as certificate expiration, revocation status, and MAC address verification.

There are number of SNMP MIBs for managing the DOCSIS certificate validation process (Table 3.3).

Once a CM's device certificate has been validated, it is considered trusted and can be used to exchange keys. The protocol used for key exchange is called Baseline Privacy Key Management (BPKM). The CMTS generates an authorization key and encrypts it with the device certificate public key. The encrypted authorization key is then sent to the CM. The CM decrypts the authorization key using the private key associated with the device certificate public key.

The authorization key is used by both the CMTS and CM to derive the key encryption key (KEK). The KEK is used to encrypt the traffic encryption key (TEK) that is used to encrypt the actual traffic. Authorization and TEK state machines are used to manage renewal functions for these keys. SNMP MIBS and configuration file TLVs are used to manage DOCSIS key exchange (Tables 3.4 and 3.5).

Encrypted traffic is assigned a security association identifier (SAID). Each encrypted DOCSIS MAC frame includes an SAID in its header. There are three different types of SAIDS: primary, static, and dynamic. A primary SAID is used for bidirectional traffic between the CM and CMTS. Encrypted multicast traffic from the CMTS to a group of CMs is supported using static or dynamic SAIDs. Static SAIDs are established during CM registration and dynamic SAIDs are established as needed.

In previous versions of DOCSIS security specifications, BPI + started after a CM completed registration. DOCSIS 3.0 security added the capability to have BPI + start before registration called early authentication and encryption (EAE). EAE secures provisioning (DHCP messaging and configuration file download) and CM registration traffic. The SNMP MIB used to configure EAE is named docsIf3MdCfgEarly AuthEncrCtrl.

TABLE 3.3. SNMP MIBS for Certificate Management [14]

Name	CM or CMTS MIB	Description
docsBpi2CmDeviceCertTable	CM	This table describes the Baseline Privacy Plus device certificates for each CM MAC interface.
docsBpi2CmtsDefaultSelf SignedManufCertTrust	CMTS	Sets trust status of self-signed manufacturer CA certificates.
docsBpi2CmtsCheckCertValidity Periods	CMTS	Enables/disables certificate validity period checking.
docsBpi2CmtsProvisionedCmCertTable	CMTS	Used to override CM device certificate validation status.
docsBpi2CmtsCACertTable	CMTS	Table of known CA certificates. Supports trust status configuration.

TABLE 3.4. SNMP MIBs for Managing Key Exchange [14]

Name	CM or CMTS MIB	Description
docsBpi2CmBaseTable	CM	This table describes the basic and authorization-relatedBaseline Privacy Plus attributes of each CM MAC interface.
docsBpi2CmTEKTable	CM	Table describes the attributes of each traffic encryption key (TEK) association at the CM.
docsBpi2CmtsAuthTable	CMTS	Table describes the attributes of each authorization association at the CM.
docsBpi2CmtsTEKTable	CMTS	Table describes the attributes of each TEK association at the CMTS.

TABLE 3.5. Configuration File TLV Encodings for Managing Key Exchange [13]

Name	Description
Authorize wait timeout	Specifies the retransmission interval, in seconds, of authorization request messages when in the authorize wait state.
Reauthorize wait timeout	Specifies the retransmission interval, in seconds, of authorization request messages when in the ReAuthorize wait state.
Authorization grace time	Specifies the grace period for reauthorization, in seconds.
Operational wait timeout	Specifies the retransmission interval, in seconds, of key requests when in the operational wait state.
Rekey wait timeout	Specifies the retransmission interval, in seconds, of key requests when in the Rekey wait state.
TEK grace time	Specifies grace period, in seconds, for re-keying the TEK.
Authorize reject wait timeout	Specifies how long, in seconds, a CM waits in the authorize reject wait state after receiving an authorization reject.

It is possible for a CM to inform the CMTS that it does not support BPI+. This may be because it is an older CM (DOCSIS 1.0) or it only partially supports BPI+. When this happens, the CMTS allows the CM to register and become operational without having to complete device authentication. This can be used by hackers to bypass BPI+device authentication and steal service via MAC address cloning.

DOCSIS 3.0 security has a feature called BPI+Enforce where different levels of BPI+enforcement can be applied by the CMTS when a CM registers. The strongest level requires that all CMs successfully complete BPI+before network access is allowed. The SNMP MIB used to configure BPI+Enforce settings is docsIf3MdCfgBpi2EnforceCtrl.

3.8.1.2 Secure Software Download

The integrity of software images downloaded to the CM for update is verified using the DOCSIS SSD mechanism. A digital code verification certificate (CVC) is used to digitally sign the image. This signature is validated before the CM replaces the old software image with the new one.

The format of the digital signature is PKCS#7, and it contains the CVC and the signing time. These items are used by the CM during the validation process. The digital signature is appended to the software image file. When the image file is downloaded to the CM, it verifies that the CVC was issued/signed by the trusted DOCSIS root CA public key, that the organization name in the CVC matches the organization of the manufacturer of the CM, that the CVC has the correct extensions, and it also checks the validity start time of the CVC and the signing time. The image file must be signed by the manufacture and may be co-signed by the cable operator. CableLabs manages a certificate PKI for issuing CVCs to CM manufacturers and cable operators.

DOCSIS SSD has a built in method for revoking CVCs and signed images. The CM maintains the start time of the CVC validity period and the signing time of the last image update. These values are called time-varying control (TVC) values. When a signature of an image is validated, the CM checks that the CVC validity start time and the signing time are equal to or greater than the TVC values. If they are greater, the CM updates the TVC values after successful completion of the other validation checks. If they are less than the TVC values, the image file is rejected. This method automatically revokes the old CVC when a new CVC is used and revokes images with signatures having older signatures. Cable operators can control the signing time (keep it the same) on images to make it easier to switch back and forth between image download versions if needed.

Downloading an image involves first enabling SSD and then triggering the download. SSD is enabled by including the CVC in the configuration file. Once the virtual CM (vCM) has validated the configuration file CVC, it is ready for triggering. Triggering the download is accomplished using configuration file TLV encoding settings or SNMP MIB sets.

TLV encoding 21 is used set the server URL and TLV encoding 9 is used to set the filename. If the filename is different than what the vCM already has, it will trigger a TFTP download of the file from the server. When using SNMP, the docsDevSwServer MIB is used to specify the server URL, docsDevSwFilename MIB is used for the filename, and docsDevSwAdminStatus MIB is used to start the download. It is possible to disable the configuration file TLV encoding trigger when using SNMP.

3.8.1.3 Secure Provisioning

A DOCSIS CM is provisioned by sending its configuration file to the CM, which it then forwards to the CMTS. Since the CM is untrusted, it is important that security controls are in place to verify service settings. Otherwise, it would be easy for hackers to change those settings and steal service. DOCSIS security has a number of methods to protect provisioning. The configuration file contains a message integrity check (MIC) value called the extended MIC that enables the CMTS to verify that the service settings are those provided by the provisioning system and have not been tampered with.

A DOCSIS 3.0 CMTS also has the capability of only allowing configuration file downloads from authorized servers. It can also learn the settings in a configuration file as it is downloaded and verify that they match what the CM requests during registration. CMTS configuration file learning is a very effective method for preventing hackers from modifying CM service settings.

To help prevent unauthorized use of IP addresses, DOCSIS 3.0 security has a feature called source address verification (SAV). SAV only allows traffic to be forwarded from CMs that are using IP address assigned by the cable operator. It supports dynamic IP addresses assigned using a DHCP server or static IP address ranges that have been allocated to the subscriber (Table 3.6).

3.8.1.4 CM Cloning

One of the most common DOCSIS attacks is to clone the MAC address of an authorized CM and steal service. All hackers have to do is snoop downstream traffic for active MAC addresses and copy them to other CMs. A good way to mitigate CM cloning is to enforce certificate device authentication using BPI+. Once a cable operator has upgraded their network to DOCSIS 1.1 CMs or higher, they can use the BPI+Enforce feature to make sure that all CMs complete device authentication.

TABLE 3.6.   SNMP MIBs for Managing SAV [12]

Name	CM or CMTS MIB	Description
docsIf3MdCfgSrcAddrVerifEnabled	CMTS	Enables/disables SAV for the MAC domain.
docsSecCmtsSavControl CmAuthEnable	CMTS	Enables/disables SAV static IP address policy settings.
docsSecSavCmAuthTable	CMTS	Provides a read-only set of SAV policies based on configuration file settings.
docsSecSavCfgListTable	CMTS	Defines the CMTS configured subnet prefix extension.
docsSecSavStaticListTable	CMTS	Defines a subnet prefix extension based on CM statically signaled subnet prefixes to the CMTS.
docsSecCmtsCmSavStatsSavDiscards	CMTS	Provides the information about number of dropped upstream packets due to SAV failure.

Another effective method is to bind/associate the CM's MAC address with the CMTS domain it registers on. If that MAC address ever shows up on a different CMTS domain, the provisioning system can block the CM from registering and becoming operational.

There are two types of clones: MAC address clones and sometimes what is called "perfect" clones. MAC address clones are created when the hacker clones only the MAC address. "Perfect" clones occur when the hacker clones the MAC address, the certificate, and the private key. Clones are usually registered on different CMTS MAC domains. Analysis of the CMTS docsBpi2CmtsAuthCmBpiVersion MIB value for each MAC domain can be done to identify the type of clone and which CMTS MAC domain has the original/good CM vs. the cloned CM for a particular MAC address.

DOCSIS 1.0 MAC address clones can be identified when all the docsBpi2Cmts-AuthCmBpiVersion MIB values are "bpi." It is difficult to identify a DOCSIS 1.0 MAC address clone from the original CM.

DOCSIS 1.1 or higher MAC address clones can be identified when one of the docsBpi2CmtsAuthCmBpiVersion MIB values are "BPI+" and the rest are "bpi." The CMTS MAC domain having a value of "BPI+" has the original/good CM registered and the other CMTS MAC domains have the clones.

"Perfect" clones can be identified when all the docsBpi2CmtsAuthCmBpiVersion MIB values are "BPI+." It is difficult to identify a "perfect" clone from the original CM, but it is recommended to shut them all down since a "perfect" clone requires collusion from all users.

Allowing self-signed certificates can also be an issue. There is a configuration setting on the CMTS that allows it to accept self-signed certificates (docsBpi2CmtsDefaultSelf-SignedManufCertTrust). Self-signed certificates are not issued by a trusted authority; they can be created by anybody including hackers. Some cable operators enable self-signed certificates to support special CMs or test equipment that do not support compliant DOCSIS certificates. Enabling self-signed certificates essentially bypasses CM authentication and cloning protection provided by BPI+ and could make CMs look like "perfect" clones. The CMTS MIB docsBpi2CmtsDefaultSelfSignedManufCertTrust should be set to "untrusted" to prevent hackers from using self-signed certificates to clone modems.

3.8.2 PacketCable Security

PacketCable security has two architectures: PacketCable 1.5 and PacketCable 2.0. PacketCable 1.5 security uses Kerberos key management and PacketCable 2.0 security is based on IMS SIP security.

3.8.2.1 PacketCable 1.5

PacketCable 1.5 security [15] utilizes digital certificates, SNMPv3, IPsec, and Kerberos-based key management to secure various interfaces in a PacketCable network. An overview of the PacketCable 1.5 security architecture with operator guidance can be found in [16]. Security is needed between the MTA and various other interfaces in the network (Table 3.7).

TABLE 3.7. PacketCable 1.5 Security Interfaces

Interface	Security Mechanism Used
MTA–Key distribution center (KDC)	KERBEROS/PKINIT.
MTA–Provisioning server	SNMPv3 security, Kerberized key management.
MTA–Call management server	IPsec ESP, Kerberized Key management.
MTA–TFTP server	No protocol security. Configuration file authenticated via a hash.
MTA–MTA or MTA–MG	Encryption and authentication of the voice packets (RTP and RTCPRTCP).

An MTA has to authenticate itself to a service provider's network to establish services. This authentication mechanism is briefly outlined:

An MTA needs a Kerberos ticket for access to PacketCable services (i.e., secure provisioning and secure call signaling). To obtain this ticket, the MTA sends a request to a key distribution center (KDC) in the PacketCable network. The MTA in this request authenticates itself by using digital certificates. The manufacturer embeds these certificates in the MTA.

The KDC is a part of the PacketCable service provider's network and grants Kerberos tickets after authenticating MTA requests. The KDC itself has certificates supplied by the service provider. Once a KDC authenticates a valid MTA request, it sends the MTA a Kerberos ticket and an encrypted session key.

Now that an MTA has a ticket, it can request services from an application server (provisioning server or CMS). The MTA sends this ticket and additional key management data to an application server. The application server verifies the Kerberos client ticket. It sends back to the MTA additional key management data. After this exchange, the keys for communication are established.

Further sections examine the MTA provisioning process, its call flows, and security issues in the call flows in more detail.

3.8.2.1.1 MTA Provisioning Process. Device provisioning occurs when an MTA device is inserted into the network. Provisioning involves the MTA making itself visible to the network, obtaining its IP configuration, and downloading its configuration data. One of three MTA provisioning flows can be used: secure, basic, and hybrid. This section provides an overview to device provisioning using the secure flow and the steps involved in the process. Basic and hybrid flows do not require SNMPv3 or some KDC functions. For a detailed look at device provisioning, refer to the PacketCable MTA device provisioning specification [17] (Tables 3.8 and 3.9).

3.8.2.1.2 Provisioning Steps and Security Issues. Outlined in this section are the provisioning step descriptions and security issues a network operator should consider. The security issues have been outlined in accordance with the MTA provisioning steps.

a. MTA-1–MTA-4. MTA–DHCP interface

To perform Kerberized key management, the MTA must first locate the KDC. The MTA retrieves the provisioning realm name from DHCP and then uses a DNS look up to find the KDCFQDN based on the realm name.

TABLE 3.8. Provisioning Flows [17]

Flows	CM / MTA	CMTS	DOCSIS DHCP	DOCSIS TFTP	DOCSIS ToD	Prov Server	PKT DHCP	PKT DNS	PKT TFTP	MSO KDC	CMS	Telephony Provider KDC	SYSLOG
Start with DOCSIS 1.1 Initialization / Registration													
CM-1	DHCP Broadcast Discover (Option Code 60 w/MTA device identifier)												
CM-2	DHCP Offer (Option Code 122 w/Telephony Service Provider's DHCP server address)												
CM-3	DHCP Request (device ID, e.g., MAC Address)												
CM-4	DHCP ACK (CM IP, ftp srv addr, CM Configuration filename)												
CM-5	DOCSIS 1.1 CM config file request												
CM-6	DOCSIS 1.1 config file												
CM-7	ToD Request												
CM-8	ToD Response												
CM-9	CM registration with CMTS												
CM-10	CMTS Registration ACK												
Complete DOCSIS 1.1 Initialization / Registration													
MTA-1	DHCP Broadcast Discover (Option Code 60 w/ MTA device identifier)												
MTA-2	DHCP Offer (Option Code 122 w/name of provision realm)												
MTA-3	DHCP Request												
MTA-4	DHCP ACK												
MTA-5	DNS Request												
MTA-6	DNS SRV (KDC host name associated with the provisioning REALM)												
MTA-7	DNS Request												
MTA-8	DNS Response (KDC IP Address)												
MTA-9	AS Request												
MTA-9a	MTA FQDN Request												
MTA-9b	MTA FQDN Reply												
MTA-10	AS Reply												
MTA-11	TGS Request												
MTA-12	TGS Reply												
MTA-13	AP Request												
MTA-14	AP Reply												
MTA-15	SNMP Inform												
MTA-16	SNMP Get Request(s) for MTA device capabilities (optional / iterative)												
MTA-17	SNMP Get Response(s) containing MTA device capabilities (optional / iterative)												
MTA-18	MTA config file												
MTA-19	SNMP Set with URL encoded file download access method (TFTP or HTTP) and filename												
MTA-20	Resolve TFTP server FQDN												
MTA-21	TFTP server IP address												
MTA-22	Telephony config file request												
MTA-23	Telephony config file												
MTA-24	MTA send telephony service provider SYSLOG a notification of provisioning completed												
MTA-25	Notify completion of telephony provisioning (MTA MAC address, ESN, pass/fail)												
SEC-1	DNS Request												
SEC-2	DNS SRV (KDC host name associated with the telephony REALM)												
SEC-3	DNS Request												
SEC-4	DNS Response (MSO KDC IP Address)												
SEC-5	AS Request (PKINIT) (MTA Device Cert, MTA Manufacturer Cert, MTA FQDN, Prov CMS ID)												
SEC-5a	MTA FQDN Request												
SEC-5b	MTA FQDN Reply												
SEC-6	AS Reply (PKINIT) (TGT with MTA service provide FQDN)												
SEC-7	TGS Request (CMS Kerberos ticket)												
SEC-8	TGS Reply (CMS Kerberos ticket)												
SEC-9	AP Request												
SEC-10	AP Reply												

TABLE 3.9.   Description of Provisioning Flows [17]

Flow Name	Flow Description
CM1–CM10	These are the steps required to provision DOCSIS cable modem. Please refer to Ref. [15] for more information.
MTA1	DHCP broadcast discover: The MTA sends a broadcast DHCPD is cover message. This message includes option code 60 (Vendor Specific Option) in the format "pktc1.0:xxxxxx."
MTA2	DHCP offer: The MTA accepts the DHCP offer from the primary or secondary DHCP. The DHCP offer must include the provision in grealm.
MTA3	The MTA sends a DHCP request broadcast message to accept the DHCP offer.
MTA4	DHCP Ack: The DHCP server sends the MTA a DHCPACK message, which contains the IPv4 of the MTA and also contains the FQDN of the MTA.
MTA5	DNS Srv Request: The MTA requests the MSOKDC host name for the Kerberos realm.
MTA6	DNS Srv Reply: Returns the MSOKDC host name associated with the provisioning realm
MTA7	DNS Request: The MTA now requests the IP Address of the MSOKDC.
MTA8	DNS Reply: The DNS server returns the IP Address of the MSOKDC.
MTA9	AS Request: The AS request message is sent to the MSO KDC to request a Kerberos ticket.
MTA10	AS Reply: The AS reply message is received from the MSO KDC containing the Kerberos ticket. Note: The KDC must map the MTAMAC address to the FQDN before sending the AS reply.
MTA11	TGS Request: If MTA obtained TGT in MTA10, the TGS request message is sent to the MSO KDC.
MTA12	TGS Reply: The TGS reply message is received from the MSO KDC.
MTA13	AP Request: The AP request message is sent to the provisioning server to request the keying information for SNMPv3.
MTA14	AP Reply: The AP reply message is received from the provisioning server containing the keying information for SNMPv3. Note: The SNMPv3 keys must be established before the next step using the information in the AP reply.
MTA15	SNMP Inform: The MTA sends the provisioning server SNMP entity an SNMPv3 Inform requesting enrollment.
MTA16	SNMP Get Request (Optional): The provisioning server SNMP entity sends the MTA one or more SNMPv3 Get requests to obtain any needed MTA capability information. The provisioning application may use a GETBulk request to obtain several pieces of information in a single message.
MTA17	SNMP Get Response: (Iterative) MTA sends the provisioning server SNMP entity a Get Response for each GetRequest.
MTA18	PacketCable does not define this protocol. The provisioning server may use the information from MTA16 and MTA17 to determine the contents of the MTA configuration file. Mechanisms for sending, storing and, possibly, creating the configuration file are outlined in MTA19.
MTA19	SNMP Set: The provisioning server sends an SNMP Set message to the MTA containing the URL-encoded file access method and file name, the hash of the configuration file, and the encryption key (if the configuration file is encrypted).

TABLE 3.9. (cont'd)

Flow Name	Flow Description
MTA20	DNS request: If the URL-encoded access method contains an FQDN instead of an IPv4 address, the MTA has to use the service provider network's DNS server to resolve the FQDN into an IPv4 address of either the TFTP server or the HTTP server.
MTA21	DNS Reply: DNS response: DNS server returns the IP address against MTA20 DNS request.
MTA22	TFTP/HTTP Get Request: The MTA sends the TFTP server a TFTP Get Request to request the specified configuration data file. In the case of file download using the HTTP access method, the MTA sends the HTTP server a request for the specified configuration data file.
MTA23	TFTP/HTTP Get Response: The TFTP Server sends the MTA a TFTP Response containing the requested file. In the case of file download using the HTTP access method, the HTTP server sends the MTA a response containing the requested file. The hash of the configuration file is calculated and compared to the value received in step MTA19. If encrypted, the configuration file has to be decrypted.
MTA24	Syslog Notification: The MTA sends the voice service provider's syslog (identified in the configuration data file) a "provisioning complete" notification. This notification will include the pass–fail result of the provisioning operation.
MTA25	SNMP Inform: The MTA sends the provisioning SNMP entity an SNMP INFORM containing "provisioning complete" notification. Note: At this stage, the MTA device provisioning data is sufficient to provide any minimal services as determined by the service provider (e.g., 611 and 911).

Threats: It is possible by faking DHCP messages that an MTA is pointed to a wrong provisioning server and a wrong KDC. Although such an attack would not affect PacketCable servers, it can lead to a denial-of-service attack against the subscriber.

b. MTA-5–MTA-8. MTA–DNS server interface

The MTA requests the MSOKDC host name for the Kerberos realm in its DNS server request. The DNS server returns the MSOKDC host name associated with the provisioning realm. The DNS request and reply are then used for resolving the KDC name to its IP address.

Threats: DNS servers are susceptible to attack and are quite often the first target of malicious users. Securing the DNS server is recommended to avoid denial of service.

c. MTA-9 AS request

The AS request is generated by the MTA and is sent to the KDC, using the information it obtained during the previous DHCP and DNS message exchanges, to request a Kerberos ticket for provisioning services.

d. MTA-9a–MTA-9b. MTA FQDN request/reply

The MTA authenticates itself with the MTA device certificate in the AS request, where the certificate contains the MTA MAC address, but not its FQDN.

In order to authenticate the MTA principal name (containing the FQDN), the KDC has to map the MTA MAC address (from the MTA device certificate) to the MTAFQDN, in order to verify the principal name in the AS request.

e. MTA-10. AS reply

The AS reply message is received from the MSOKDC and contains a Kerberos ticket.

f. MTA-11–MTA-12. TGS request/reply

This is an optional step. TGS request/reply is used for performance improvements, but it may not be supported by all MTAs. Even if an MTA supports this, it may be disabled via a DHCP suboption (see Section 6.1). In the case where a client obtained a TGT, that TGT is then used in the TGS Request/TGS Reply exchange to obtain a specific application server (provisioning server or CMS) ticket.

g. MTA-13. AP request

The AP request is generated by the MTA and is sent to the provisioning server using the information that it obtained during the previous DHCP, DNS, and AS reply message exchanges.

h. MTA-14. AP reply

The AP reply message is received from the provisioning server containing the keying information for SNMPv3.

i. MTA-15–MTA-19. SNMP messages

SNMP-based network management within PacketCable security supports SNMPv3. In PacketCable, the provisioning server includes the SNMP manager function. The provisioning server uses an SNMP set to provide the MTA parameters needed to find, authenticate, and decrypt its configuration file. The SNMP messages exchanged in these steps happen automatically and do not require any special configuration.

The provisioning servers ends an SNMP set message to the MTA containing the URL-encoded file access method and filename, the hash of the configuration file, and the encryption key (if the configuration file is encrypted).

j. MTA-20–MTA-21. DNS request/response to resolve TFTP/HTTP server IP address

MTA-20 is a DNS request used by the MTA to resolve the FQDN provided in the SNMP set in MTA-15–MTA-19.

k. MTA-22–MTA-23 telephony configuration file request, response

The MTA sends the TFTP server a TFTPGet request to request the specified configuration data file. The TFTP server in responses ends the MTA a TFTP response containing the requested file.

In the case of file download using the optional HTTP access method, the MTA sends the HTTP server a request for the specified configuration data file. The HTTP server sends the MTA a response containing the requested file.

The following MIB objects must be configured in the configuration file sent to the MTA in order for an MTA to properly provision (Table 3.10):

TABLE 3.10. Configuration File MIB Objects [18]

Object Name	Description/Value to be Set
PktcMtaDevEnabled	The MTA admin status of this device, where true (1) means the voice feature is enabled and False (2) indicates that it is disabled.
pktcMtaDevRealmTable	Contains information about each realm. At a minimum, each entry must contain the Realm name and the Organization Name (see later).
pktcMtaDevRealm	Realm name
pktcMtaDevRealmOrgName	The value of the X.500 organization Name attribute in the subject name of the service provider certificate. The configuration file needs to contain the correct organization name in this MIB object.
pktcMtaDevCmsTable	The pktcMtaDevCms table shows the IPsec key management policy relating to a particular CMS. The table is indexed with pktcMtaDevCmsFQDN. The following two items must be configured at a minimum.
pktcMtaDevCmsFqdn	The FQDN of the CMS.
pktcMtaDevCmsKerbRealmName	The Kerberos realm name of the associated CMS.
pktcNcsEndPntConfigCallAgentId	This object contains a string indicating the call agent name (ex:ca@abc.def.com). The call agent name after the character "@"must be a fully qualified domain name and must have a corresponding pktcMtaDevCmsFqdn entry in the pktcMtaDevCms table. A MTA may have multiple end points. There must be a Call Agent Id assigned to each active endpoint. There will be multiple Call Agent Id entries if there are multiple CMS.

After receiving the MTA configuration file, an MTA validates the following:

- "pktcMtaDevRealmName" MIB object of the realm table must be the same as the realm name supplied to the MTA in DHCP option 122, suboption 6.
- "pktcMtaDevRealmOrgName" MIB object of the realm table must be the same as the "organization name" attribute in the service provider CA certificate provided by the KDC in the AS reply (MTA–10. AS reply).

The TFTP server needs to be secured against denial-of-service attacks.

l. MTA-24. MTA to SYSLOG notification

The MTA sends the telephony service provider's SYSLOG (identified in the configuration data file) a "provisioning complete" notification. This notification will include the pass–fail result of the provisioning operation.

m. MTA-25 MTA provisioning complete notification

The MTA sends the provisioning server an SNMP inform containing a "provisioning complete" notification. At this point, the MTA device provisioning process is complete. The MTA now has provisioned its telephony endpoint(s).

3.8.2.1.3 Securing Call Signaling Flows. The secure signaling call flows occur after MTA-1–MTA-25. Several tables in the MTAMIB are used to control security flows SEC-1 through SEC-11. The CMS table (pktcMtaDevCmsTable) and the realm tables (pktcMtaDevRealmTable) are used for managing the MTA security signaling.

The pktcMtaDevCms table shows the IPsec key management policy relating to a particular CMS. For each CMS specified in the pktcMtaDevCmsTable table with pktc-MtaDevCmsIpsecCtrl set to true and assigned to a provisioned MTA endpoint, the MTA must perform the following security flows after the provisioning process and prior to any NCS message exchange. If there are CMS entries with pktcMtaDevCmsIpsecCtrl set to true, but the CMS is not assigned to an endpoint, the MTA may perform the security flows on start up, or it may perform the security flows on-demand, for example, if an endpoint is transferred to a new CMS for which it has not established SAs. MTA behavior in this instance is up to vendor implementation. For each CMS specified in pktcMtaDevCms Table with pktcMtaDevCmsIpsecCtrl set to false, the MTA does not perform the following flows and sends and receives NCS messages without IPsec (i.e., NCS packets are sent in the "clear") (Table 3.11).

TABLE 3.11.   Post-MTA Provisioning Security Flows [18]

Sec Flow	Flow Description
SEC-1	DNSSRV Request: The MTA requests the telephony KDC host name for the Kerberos realm.
SEC-2	DNSSRV Reply: Returns the Telephony KDC host name associated with the provisioning REALM. If the KDC's IP address isincludedintheReply,proceedtoSEC-5.
SEC-3	DNS Request: The MTA now requests the IP address of the telephony KDC.
SEC-4	DNS Reply: The DNS Server returns the IP Address of the telephony KDC.
SEC-5	AS Request: For each different CMS assigned to voice communications endpoints, the MTA requests a TGT or a Kerberos ticket for the CMS by sending a PKINIT request message to the KDC. This request contains the MTA Device Certificate and the MTA FQDN.
SEC-5a	MTA FQDN Request: The KDC requests the MTA's FQDN from the provisioning server.
SEC-5b	MTAFQDN Reply: The provisioning server replies to the KDC request with the MTA's FQDN. AS Reply: The KDC sends the MTA a PKINIT reply message containing the requested Kerberos ticket.
SEC-7	TGS Request: In the case where the MTA obtained a TGT in SEC-6, it now obtains the Kerberos ticket for the TGS request message.
SEC-8	TGS Reply: Response to TGS request containing the requested CMS Kerberos ticket.
SEC-9	AP Request: The MTA requests a pair of IPsecs implex Security Associations (inbound and outbound) with the assigned CMS by sending the assigned CMS an AP request message containing the CMS Kerberos ticket.
SEC-10	AP Reply: The CMS establishes the Security Associations and then sends an AP reply message with the corresponding IPsec parameters. The MTA derives IPsec keys from the subkey in the AP Reply and establishes IPsec SAs.

a. SEC-1–SEC-4.

b. As discussed earlier, the security flows begin just after provisioning and the MTA now has to establish SAs with the call management server. Sec-1–Sec-4 are DNS request/replies where the IP address of the telephony KDC is resolved. SEC-5 AS request.

    The AS request is generated by the MTA and sent to the KDC, using the information that it contained in the configuration file and obtained during SEC-1–SEC-4.

c. SEC-5a, SEC-5b. MTA FQDN request, reply

The KDC requests the MTA's FQDN from the provisioning server. The provisioning server replies to the KDC request with the MTA's FQDN.

d. SEC-5a, SEC-5b. MTA FQDN request, reply

The KDC requests the MTA's FQDN from the provisioning server. The provisioning server replies to the KDC request with the MTA's FQDN.

e. SEC-6. AS reply

The AS reply message is received from the MSO KDC and contains a Kerberos ticket.

f. SEC-7, SEC-8. TGS request/reply

TGS request: This is an optional step. Even if an MTA supports this, it may be disabled via a DHCP suboption. In the case where the MTA obtained a TGT in SEC-6, it now obtains the Kerberos ticket for the TGS request message.

g. SEC-9. AP request

The MTA requests a pair of IPsec SAs (inbound and outbound) with the assigned CMS by sending the assigned CMS an AP request message containing the CMS Kerberos ticket.

h. Sec-10. AP reply

The CMS establishes the SAs and then sends an AP reply message with the corresponding IPsec parameters. The MTA derives IPsec keys from the subkey in the AP reply and establishes IPsec SAs.

At this point, the MTA is fully provisioned and able to communicate securely with the provisioning server and the CMS.

3.8.2.1.4 Post-Provisioning Process. MTAs still need to maintain tickets and SAs after the provisioning process. This is because Kerberos tickets and SAs have a finite lifetime. For obtaining new tickets, Kerberos AP request/reply exchanges occur periodically to insure that there are valid security parameters between the client and an application server. In order to perform a successful exchange, the client must ensure that it has a valid Kerberos ticket before issuing an AP request. The following two parameters are associated with an MTA requesting a new Kerberos ticket:

Ticket Expiration Time (TicketEXP): A Kerberos ticket has a finite lifetime, indicated by when it was issued (initial authentication time or start time in a Kerberos ticket) and when it expires (end time in the ticket). For the purposes of this document, TicketEXP indicates the number of seconds that a ticket is valid for (end time–initial authentication time or start time).

Ticket Grace Period (TicketGP): This indicates to the client when it should request a new ticket. Grace periods can be configured through the MIB. An MTA requests a new ticket at the time: TicketEXP–TicketGP.

The following MIB objects are used to set grace periods:

- PktcMtaDevRealmPkinit grace period is used to control when a new application server ticket is requested using a PKINIT exchange. This includes TGTs for MTAs that support them. The minimum value for this object is 15 min and the default value is 30 min.
- PktcMtaDevRealmTgs grace period is used to control when a new application server ticket is requested using a TGS request/reply exchange. This is only applicable to MTAs that support TGTs. The minimum value for this object is 1 min and the default value is 10 min.

MTAs that store tickets in persistent storage (e.g., NVRAM) are able to reuse the same tickets after a reboot. This eliminates the need to request a new ticket from the KDC whenever are boot occurs. Instead, the frequency at which an MTA requests a new ticket will generally be constant, regardless of how many times the MTA is rebooted during the ticket lifetime (the only time the frequency would vary is if the MTA ticket expires while the MTA is without power). Storing tickets in NVRAM, then, provides a huge performance advantage because the KDC can be deployed without having to worry about peak load during a restart avalanche (as is the case with DHCP servers, for example). Instead, the maximum load a KDC would need to sustain is the average load. MTAs are required to store the ticket for the provisioning server, and up to several tickets for CMS. For example, if an MTA has two physical endpoints, it must have the ability to store a minimum of three CMS tickets. This will allow for each end point to be controlled by a different CMS, and handle some basic load balancing through redirection mechanisms that a CMS may employ. Note that Kerberos tickets are not to be issued for a period of time that is longer than 7 days.

An SA will be reestablished between the MTA and an application server periodically. The reestablishment can be initiated by either the MTA or the application server based on configuration.

There are two types of messages that the CMS can use for signaling the MTA to send back an AP request so as to reestablish security parameters:

- Wake Up–An application server sends this message when it initiates a new key management exchange. For instance, if there is an incoming call to an endpoint and an SA needs to be reestablished a wake-up message is used by the application Server.
- Rekey—The rekey message replaces the wake-up message and provides better performance, whenever an application server wants to trigger the establishment of a security parameter with a specified client. The use of the rekey message

eliminates the need for the AP reply message, thus reducing the key management overhead to a single roundtrip (the two messages are rekey and AP request). Rekey can only be used by a CMS, and requires that the client and application server share a valid server authentication key. In the case that no such key exists (e.g., the ticket and key have expired), the wake-up message must be used.

Rekey was originally developed to support call agent clustering. If multiple call agents inside a cluster are actually different hosts, then there needs to be a secure interface for sharing the server authentication key that is used in rekey messages. This interface is not specified by PacketCable and would be up to vendor implementation.

In either case, once this AP request is received (as a result of a wake-up or rekey) and verified by the application server, the server has to also establish the security parameters (in the case of a wake-up flow, the application server will also send back an AP reply).

- The MTA will always ensure there are valid SAs if the reestablish flag is set to True. This means that the CMS will not normally need to send wake-up or rekey messages to initiate SA establishment. The only time a CMS would need to issue a wake-up or rekey under these circumstances is in the case of an error (e.g., the MTA is not responding, even though there are active SAs).

Setting the MIB objects that have been defined can configure key management policies. The MIB object pktcMtaDevCms table contains the key management policies for each CMS in the associated realm. Each CMS has a pktcMtaDevCmsEntry, which contains key management parameters for an MTA–CMS interface. The MIB objects that are associated with reestablishment of security parameters can be grouped into the following:

- Solicited timeouts—These timeout parameters are applied only when the CMS-initiated key management with a wake-up or rekey message.
- Unsolicited timeouts—These timeout parameters are applied only when the MTA-initiated key management.
- Max retries—These parameters indicate the number of retries before an entity (MTA or application server) give up attempting to establish an SA.

SNMPv3 security parameters do not expire, so generally the MTA will not need to reestablish SNMPv3 security parameters. In the unlikely event that an MTA somehow loses SNMPv3 security parameters, it will simply initiate a new AP request/reply exchange. The SNMPv3 manager may lose its security parameters as well (due to are boot, for example). In this case, the SNMPv3 manager may force the MTA to reestablish SNMPv3 security parameters using a wake-up message. The MIB object associated with this condition is pktcMtaDevProv solicited key timeout. Although SNMPv3 security parameters do not expire, an operator may wish to schedule updates to SNMPv3 keys on a periodic basis. In this case, the provisioning would need to support the ability to periodically sent wake-up messages based on configuration.

3.8.2.1.5 Kerberos. Kerberos for service key management and KDC database synchronizations are described in the following. The service key that is shared between a KDC and an application Server, to encrypt/decrypt service tickets is a versioned key. Application servers are required to support multiple service keys at the same time. When the service key is changed, the application server must retain the old key for a period of time that is at least as long as the ticket lifetime used when issuing service tickets (i.e., up to 7 days). A service key may be changed due to routine key refresh, or because it was compromised. In the case of a compromised key change, the application server is required to mark a specific key, and all keys before it, as compromised, and must reject any tickets that were encrypted with a compromised key.

Currently all service keys are pre-shared using an out-of-band mechanism between the KDC and the device providing service. In the future, PacketCable may support a method that does not require these keys to be pre-shared.

For KDC database synchronization, each PacketCable KDC has a database that contains service keys for different application servers (one service key for each server). In the event that multiple KDCs are used within a single REALM, all the KDC databases need to be synchronized. This is important for two reasons:

1. **Service Key Management**—Each KDC needs to be updated with the latest service key when the key is changed (especially in the case of a key compromise).
2. **Automated Key Change**—Although PacketCable does not define an automated means for application servers to change their service keys on KDCs, some vendors are implementing this functionality. In such a scenario, the application server will update its service key with a single KDC. It is the responsibility of the KDC to then ensure that change is replicated across all KDCs.

Database synchronization is out-of-scope for PacketCable 1.5 security. Database synchronization is an important issue, however, and must be addressed in deployments consisting of multiple KDCs.

3.8.2.1.6 PacketCable PKI. PacketCable 1.5 security uses PKI as a means for authenticating and securing communications in a PacketCable network. CableLabs manages the maintenance and certificate issuance of the PacketCable PKI. The entities in the PacketCable PKI are the following: MTA root CA: This is the root CA that signs and issues certificates for MTA manufacturing CAs. The root CA does not issue certificates to end entities.

- MTA manufacturing CAs: The MTA manufacturer CA is the next rung in the PKI hierarchy. The certificates for these CAs are signed and issued by the root CA. MTA manufacturing CAs issue certificates to manufacturing end-entity devices.
- MTA manufacturing end entity: In PacketCable, the manufacturing end entity is an MTA device that authenticates itself by means of the public key certificate that is issued to it by its manufacturer CA.

- Service provider root CA: This is the root CA that signs and issues certificates for service provider CAs. The root CA does not issue certificates to end entities.
- Service provider CAs: The service provider CA is the next rung in the PKI hierarchy. The certificates for these CAs are signed and issued by the root CA. Service provider CAs issue certificates to local CAs and server end-entity devices.
- Local system CAs: The local system CA is the next level in the PKI hierarchy and is optional. The certificates for these CAs are signed and issued by the service provider CA. Local system CAs issue certificates to server end-entity devices.
- Server end entity: In PacketCable, the server end entity can be any of the PacketCable servers, which authenticates itself by means of the public key certificate that is issued to it by its service provider CA or local system CA (Table 3.12).

3.8.2.1.7 PacketCable Security Relevant MIB Objects. There are a number of SNMP MIB objects used to manage PacketCable security functions. Table 3.13 provides a list of these MIB objects with a brief description.

3.8.2.2 PacketCable 2.0

The PacketCable 2.0 standard is based on the IMS as defined by the 3rd Generation Partnership Project (3GPP). 3GPP is a collaboration agreement between various standards bodies. The scope of 3GPP is to produce technical specifications and technical reports for GSM and 3G mobile system networks.

Within the overall PacketCable 2.0 goal to leverage existing industry standards whenever possible, there is a specific objective to align with the IMS architecture and specifications being developed by 3GPP. Specifically, the PacketCable 2.0 architecture reuses many of the basic IMS functional entities and interfaces.

TABLE 3.12. Packet Cable PKI Certificates

Certificate Name	Description
MTA root CA certificate	Self-signed certificate, Issued by CableLabs. Established trust point.
MTA manufacturing CA certificate	Signed by root CA. Must be obtained by MTA device manufacturers. Cannot sign another CA certificate.
MTA certificate	Signed by MTA manufacturing CA. End entity certificate.
Service provider root CA certificate	Self-signed certificate, Issued by CableLabs. Established trust point for authenticating PacketCable servers.
Service provider CA certificate	Signed by root CA. Must be obtained for server.
Local system CA certificate	Signed by service provider CA. Optional: may be needed for server.
Server certificate	End entity certificate signed by service provider CA or local system CA.

TABLE 3.13. PacketCable 1.5 Security MIBs [18]

MIB Name	Description
pktcMtaDevMacAddress	The telephony MAC address for this device.
pktcMtaDevFQDN	The fully qualified domain name for this MTA.
pktcMtaDevEnabled	The MTA Admin Status of this device, where true (1) means the voice feature is enabled and false (2) indicates that it is disabled.
pktcMtaDevProvisioningTimer	This object enables setting the duration of the provisioning timeout timer. The timer covers the provisioning sequence from step MTA-1 to step MTA-23. The value is in minutes and setting the timer to 0 disables this timer. The default value for the timer is 10.
pktcMtaDevSnmpEntity	The FQDN of the SNMPv3 entity of the provisioning server to which the MTA has to communicate in order to receive the access method, location and the name of the configuration file during MTA provisioning.
pktcMtaDevRealmTable	Contains per Kerberos realm security parameters.
pktcMtaDevRealmEntry	List of security parameters for a single Kerberos realm.
pktcMtaDevRealmName	Kerberos realm name
pktcMtaDevRealmPkinitGracePeriod	For the purposes of the key management with an application server (CMS or provisioning server), the MTA obtains a new Kerberos ticket (with a PKINIT exchange) this many minutes before the old ticket expires. The minimum allowable value is 15 min. The default is 30 min.
pktcMtaDevRealmTgsGracePeriod	When the MTA implementation uses TGS Request/TGS Reply Kerberos messages for the purpose of the key management with an application server (CMS or provisioning server), the MTA obtains a new service ticket for the application server (with a TGS Request) this many minutes before the old ticket expires. The minimum allowable value is 1 min. The default is 10 min.
pktcMtaDevRealmOrgName	The value of the X.500 organization name attribute in the subject name of the service provider certificate.
pktcMtaDevRealmUnsolicitedKeyMaxTimeout	This timeout applies only when the MTA-initiated key management. The maximum timeout is the value that may not be exceeded in the exponential back-off algorithm.
pktcMtaDevRealmUnsolicitedKeyNomTimeout	Defines the starting value of the timeout for the AS Request/Reply back-off and Retry mechanism with exponential timeout.

pktcMtaDevRealmUnsolicitedKeyMeanDev	This is measurement of the mean deviation for the round trip delay timings.
pktcMtaDevRealmUnsolicitedKeyMaxRetries	This is the maximum number of retries before the MTA gives up attempting to establish a security association.
pktcMtaDevCmsTable	Contains per CMS key management policy
pktcMtaDevCmsEntry	List of key management parameters for a single MTA–CMS interface.
pktcMtaDevCmsFqdn	The fully qualified domain name of the CMS.
pktcMtaDevCmsKerbRealmName	The Kerberos realm name of the associated CMS.
pktcMtaDevCmsMaxClockSkew	This is the maximum allowable clock skew between the MTA and CMS
pktcMtaDevCmsSolicitedKeyTimeout	This timeout applies only when the CMS-initiated key management (with a wake-up or rekey message). It is the period during which the MTA will save a nonce (inside the sequence number field)from the sent out AP Request and wait for the matching AP Reply from the CMS.
pktcMtaDevCmsUnsolicitedKeyMaxTimeout	This timeout applies only when the MTA-initiated key management. The maximum timeout is the value that may not be exceeded in the exponential back-off algorithm.
pktcMtaDevCmsUnsolicitedKeyNomTimeout	Defines the starting value of the timeout for the AP Request/Reply back-off and Retry mechanism with exponential timeout for CMS. If provided, MTA-configuration file content overrides this value.
pktcMtaDevCmsUnsolicitedKeyMeanDev	This is the measurement of the mean deviation for the round trip delay timings.
pktcMtaDevCmsUnsolicitedKeyMaxRetries	This is the maximum number of retries before the MTA gives up attempting to establish a Security Association.
pktcMtaDevCmsIpsecCtrl	The value of "true (1)" indicates that IPSEC and IPSEC Key Management are used to communicate with the CMS. The value of "false (2)" indicates that IPSEC signaling security is disabled for both the IPSEC Key Management and IPSEC protocol (for the specific CMS).
pktcMtaDevProvKerbRealmName	The name of the associated Provisioning Kerberos Realm acquired during MTA4 (DHCP Ack). This is used as an index into the pktcMtaDevRealm Table
pktcMtaDevManufacturerCertificate	The MTA manufacturer's X.509 public key certificate, called MTA manufacturer certificate. It is issued to each MTA manufacturer and is installed into each MTA either in the factory or with a code download.
pktcMtaDevTelephonyRootCertificate	IP telephony root X.509 public key certificate stored in the MTA nonvolatile memory and updateable with a code download. This certificate is used to validate the initial AS Reply from the KDC received during the MTA initialization
pktcMtaDevCertificate	MTA'sX.509 public key certificate issued by the manufacturer and installed into the embedded-MTA in

(Continued)

TABLE 3.13. (cont'd)

MIB Name	Description
	the factory. This certificate, called MTA device certificate, contains the MTA's MAC address.
pktcMtaDevProvConfigHash	Hash of the contents of the configuration file, calculated and sent to the MTA prior to sending the configuration file.
pktcMtaDevProvConfigKey	Key used to encrypt/decrypt the configuration file, sent to the MTA prior to sending the configuration file. If the privacy algorithm is null, the length is 0.
pktcMtaDevProvUnsolicitedKeyMaxTimeout	This timeout applies to MTA-initiated AP request/reply key management exchange with provisioning server. The maximum timeout is the value, which may not be exceeded in the exponential back-off algorithm.
pktcMtaDevProvUnsolicitedKeyNomTimeout	Defines the starting value of the timeout for the AP request/reply back-off and retry mechanism with exponential timeout.
pktcMtaDevProvUnsolicitedKeyMeanDev	This is the mean deviation for the round trip delay timings
pktcMtaDevProvUnsolicitedKeyMaxRetries	This is the maximum number of retries before the MTA gives up attempting to establish an SNMPv3 security association with provisioning server.
pktcMtaDevProvSolicitedKeyTimeout	This timeout applies only when the provisioning server-initiated key management (with a wake-up message) for SNMPv3. It is the period during which the MTA will save a nonce (inside the sequence number field) from the sent out AP request and wait for the matching AP reply from the provisioning server.
pktcMtaDevRealmPkinit grace period	For the purposes of the key management with an application server (CMS or provisioning server), the MTA obtains a new Kerberos ticket (with a PKINIT exchange) this many minutes before the old ticket expires. The minimum allow able value is 15 min. The default is 30 min.
pktcMtaDevResetKrb tickets	This object is used to force an MTA to delete Kerberos tickets that it may have stored in NVRAM. It can be used to force deletion of the provisioning server ticket, all CMS tickets, or both the provisioning server ticket and all CMS tickets.

3.8.2.2.1 PacketCable 2.0 Reference Architecture. The PacketCable 2.0 architecture is based on the IMS architecture [19] (see Section 2.6.1), with the addition of some incremental extensions to support cable networks. These extensions include the use of additional or alternate components, as well as enhancements to capabilities provided by IMS functional components [20]. An overview of the PacketCable 2.0 security architecture can be found in [21].

Some of the major PacketCable 2.0 enhancements to the IMS include the following:

- Support for QoS for IMS-based applications on cable access networks, leveraging the PacketCable multimedia architecture;
- Support for signaling and media traversal of network address translation (NAT) and firewall (FW) devices, based on IETF mechanisms;
- Support for the ability to uniquely identify and communicate with an individual when multiple UEs are registered under the same public identity;
- Support for additional access signaling security and UE authentication mechanisms for PacketCable UEs;
- Support for provisioning, activation, configuration, and management of PacketCable UEs.

PacketCable 2.0 includes both the existing IMS logical components and reference points, and logical elements and reference points added to support PacketCable 2.0 requirements.

UEs connect to the network through the access domain. Interfaces and components present in the access domain are shown in Figure 3.8.

UE interactions with the network occur in the Access domain. Access methods are varied, and include DOCSIS and wireless technologies. Table 3.14 provides a high-level overview of the security architecture that results from the PacketCable 2.0 enhancements to IMS. Each access domain reference point, along with the security mechanism employed for that interface, is included.

Intradomain reference points and components are contained within a service provider's network, and consequently, a holistic security policy.

IMS defines the security of intradomain connections with the Zb interface, as described in Ref. [22]. Within IMS, integrity is required and confidentiality is optional when the Zb interface is implemented. IPsec ESP is used to provide security services for the Zb interface between intradomain components.

PacketCable 2.0 provisioning also requires an interface between the KDC and the provisioning server to map the eUE's Mac address to the assigned IP address and the fully qualified domain name. Message integrity, authentication, and privacy for this interface are provided within the Kerberos protocol.

Interdomain reference points connect the operator security domain with external partners and networks. These connections provide interworking between the operator's network and other service providers and networks, including the PSTN. Figure 3.9 shows the interdomain trust boundary.

IMS defines the security of interdomain connectivity with the Za interface, as described in Ref. [7]. Both integrity and confidentiality are required for the Za

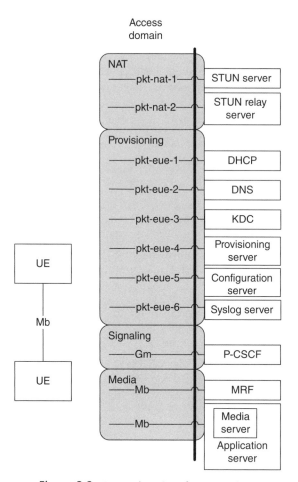

Figure 3.8. Access domain reference points.

interface, based on IPsec ESP. Interdomain traffic in IMS is required to pass through a security gateway (SEG). The SEG terminates reference point Za IPsec tunnels and enforces security policy on interdomain traffic flows. Figure 3.9 shows an architecture including the SEG functionality in the border element, but the SEG may be a separate element.

The PSTN gateway to PSTN reference point is secured using PSTN security mechanisms.

PacketCable 2.0 adds support for internetworking to PacketCable networks. The call management server (CMS) provides translation for PacketCable messaging. Security for the CMS reference point is detailed in Ref. [15].

3.8.2.2.2 User and UE Authentication. 3GPP IMS relies completely on credentials stored in a Universal Integrated Circuit Card (UICC) for access security. The UICC is a platform for security applications used for authentication and key agreement.

TABLE 3.14.   Access Domain Reference Points Description [20]

Reference Point	PacketCable Network Elements	Reference Point Security Description
pkt-nat-1	UE<->STUN relay server	STUN relay: STUN relay requests are authenticated and authorized within the STUN relay protocol itself.
pkt-nat-2	UE—external STUN server	STUN: Message integrity is provided by STUN mechanisms.
pkt-eue-1	eUE—DHCP	DHCP: Security for this interface is out-of-scope. Security threats are mitigated via follow-on Kerberos procedures in the secure provisioning flow.
pkt-eue-2	eUE—DNS	DNS: Security for this interface is out-of-scope. Security threats are mitigated via follow-on Kerberos procedures in the secure provisioning flow.
pkt-eue-3	eUE—KDC	Kerberos: Authentication, message integrity, and privacy are provided within the Kerberos protocol.
pkt-eue-4	eUE—provisioning server	Kerberos, SNMP: Authentication, message integrity, and privacy (optional in SNMPv3) are provided within the Kerberos and SNMP protocols in the secure provisioning flow only.
pkt-eue-5	eUE—configuration server	TFTP, HTTP (optional): Message integrity of the configuration file contents, and optional privacy, is provided via pkt-eue-4 signaling in the secure provisioning flow only.
pkt-eue-6	eUE—syslog server	Syslog: Security for this interface is out-of-scope.
Mb	UE—UE UE—media Server UE—MG UE—E-MTA UE—MRF	RTP: Media security is out of scope for this specification.

PacketCable has a requirement to support multiple types of UEs, such as software UEs, that will not contain or have access to UICCs.

Reference [23] describes the IMS approach to authentication and establishing transport security between the UE and the P-CSCF. The IMS uses a combination of IPsec for integrity and optional confidentiality, and IMS-AKA for authentication. To meet the IMS requirements of minimal round trips, the security elements of the negotiation "piggy-back" on the SIP register messaging flow. IETF RFC 3329 [24] is used to negotiate security between the UE and the P-CSCF, and IMS-AKA [25] is used between

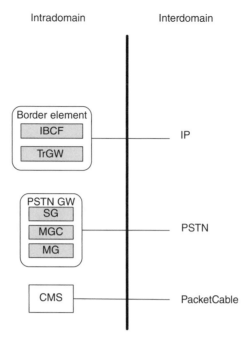

Figure 3.9. Internetwork domain reference points.

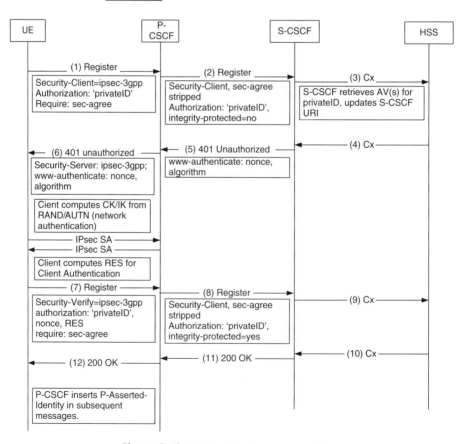

Figure 3.10. IMS registration message flow.

the UE and the S-CSCF to perform mutual authentication. IETF RFC 2617 [26] is extended to pass authentication data from the UE to the S-CSCF. The communications between the UE and the P-CSCF and the communications between the UE and the S-CSCF are related in that the keying material for the SAs between the UE and the P-CSCF are computed from the long-term shared secret stored in the Home Subscriber Server (HSS) and the UICC in the UE. Figure 3.10 shows the high-level message flows for authentication during registration. Some elements and messages are not displayed in order to simplify discussion.

For authentication during registration, the following basic steps occur:

1. The UE sends a register request to the P-CSCF. The message includes an RFC 3329 Security-Client header, which includes the security mechanisms the UE supports. IMS mandates "ipsec-3gpp." The message also includes an authorize header that includes the private identity of the subscriber.

2. The P-CSCF strips the security agreement headers, inserts "integrity-protected = no" in the authorized header, and forwards the register request to the appropriate I-CSCF, which forwards the request to the appropriate S-CSCF of the subscribers home network.

3. The S-CSCF contacts the HSS to update the S-CSCF URI for that user, and if necessary, request one or more authentication vectors.

4. The HSS returns one or more authentication vectors if requested. The authentication vectors provide the necessary data for the S-CSCF to create a www-authenticate header and challenge the user.

5. The S-CSCF creates and sends a SIP 401 (unauthorized) response, containing a WWW-Authenticate header that includes a challenge. This response is routed back to the P-CSCF.

6. The P-CSCF strips the integrity key (IK) and the confidentiality key (CK) from the 401 response to use for IPsec SAs between the P-CSCF and the UE, and sends the rest of the response to the UE.

7. Upon receiving the challenge message, the UE determines the validity of the received authentication challenge. The UE sets up SAs with the P-CSCF using the IK and CK that was derived from the data sent by the HSS, utilizing the long-term shared key in its UICC. The UE then calculates a response (RES) and sends a second register request with an Authorization header including the challenge response.

8. The P-CSCF strips the security agreement headers, inserts "integrity-protected = yes" in the authorize header, and forwards to the appropriate I-CSCF, which forwards to the appropriate S-CSCF.

9. The S-CSCF compares the authentication challenge response received from the UE with the expected response received from the HSS. If they match, the S-CSCF updates HSS data using the Cx interface.

10. The HSS provides the S-CSCF with subscriber data over the Cx interface, including service profiles, which contain initial filter criteria.

11. The S-CSCF forwards a 200 OK response to the UE. The 200 OK contains a P-Associated-URI header, which includes the list of public user identities that are associated to the public user identity under registration.

12. The P-CSCF forwards the 200 OK to the UE. Because the user has now been authenticated, and there is an existing SA between the P-CSCF and the UE, the P-CSCF inserts a P-Asserted-Identity header in all subsequent messages from that UE.

PacketCable 2.0 security has requirements to support UEs and authentication schemes not considered in the IMS architecture, as well as additional transport security mechanisms. PacketCable 2.0 enhances the IMS specifications in several areas in order to support these requirements.

The PacketCable 2.0 architecture supports the following authentication mechanisms:

- IMS AKA;
- SIP digest authentication;
- Certificate bootstrapping.

The architecture must also accommodate UEs with multiple authentication credentials. For example, a UE may have a certificate for accessing services while on a cable network, and a UICC for accessing services while on a cellular network.

A subscriber may have multiple credentials. A subscriber may have multiple UEs, with different capabilities related to those credentials. For example, a subscriber may have an MTA with a certificate for home use, and a UICC-based UE for traveling.

IMS AKA authentication with UICC credentials will continue to operate as described in 3GPP specifications.

IMS also supports SIP authentication, which is described in Ref. [23]. SIP authentication uses a challenge–response framework for authentication of SIP messages and access to services. In this approach, a user is challenged to prove their identity, either during registration or during other SIP dialogues initiations.

SIP authentication is handled in a similar manner to IMS AKA, and follows IETF RFC 3261 [27] and IETF RFC 2617 [26]. This approach minimizes impact to the existing IMS authentication flow by maintaining existing headers and round trips. Unlike IMS AKA, however, challenges are not precomputed. In order to maximize the security of SIP digest authentication, cnonces and qop "auth" directives are used, which requires challenges to be computed in real-time at the S-CSCF.

Figure 3.11 shows the message flow for SIP-based authentication during a registration.

For SIP digest authentication during registration, the following basic steps occur. IETF RFC 3329 [24] headers and other SIP header content is not shown for simplicity.

1. The UE sends a register request to the P-CSCF. The message includes an Authorization header, which includes the private identity of the subscriber.

2. The P-CSCF forwards the register request to the appropriate I-CSCF, which forwards the request to the appropriate S-CSCF of the subscriber's home network.

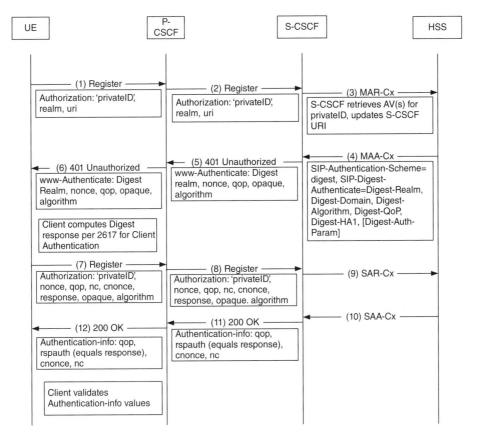

Figure 3.11. SIP digest authentication.

3. The S-CSCF contacts the HSS using a MAR command toward the HSS on the Cx interface. The MAR message includes the private identity of the subscriber, the S-CSCF information, and the number of authentication vectors requested. This information is used by the HSS to update the S-CSCF URI for the private identity and to deliver the correct authentication vector information to the S-CSCF.

4. The HSS returns an MAA message on the Cx interface. The MAA message includes the public identities and authentication vectors for that subscriber.

5. The S-CSCF creates a SIP 401 (Unauthorized) response, which includes a challenge in the WWW-Authenticate header field and other fields.

6. This response is routed back to the I-CSCF, then to the P-CSCF, and then to the UE.

7. Once the UE receives the challenge, the UE calculates the response based on items in the WWW-Authenticate header and additional items (e.g., cnonce) generated by the UE. The UE sends a second register request with the Authorization header.

8. The P-CSCF forwards the message to the appropriate I-CSCF, which forwards to the appropriate S-CSCF.

9. Upon receiving the second register from the UE, the S-CSCF calculates the challenge in the same manner as the UE, in order to compare the two results and thus authenticate the subscriber. Using parameters from the HSS such as HA1, and the parameters from the Authorization header such as cnonce, the S-CSCF computes the challenge response. The computation is performed in a manner consistent with the qop parameter with the value of "auth."

10. If the two challenge results are identical, the S-CSCF performs an SAR procedure on the Cx interface, informing the HSS the user is registered and requesting the user profile.

11. The HSS returns an SAA message to the S-CSCF containing the user profile, which includes, among other things, the collection of all the public user identities allocated for authentication of the private user identity, as well as the initial filter criteria.

12. The S-CSCF sends a 200 OK response to the register request. The response includes an Authentication-Info header, which allows the UE to authenticate the network, or S-CSCF. The 200 OK message is forwarded to the UE.

13. The 200 OK is routed to the appropriate P-CSCF, and then to the UE.

14. The UE validates the rspauth value, to authenticate the network, or S-CSCF.

Because the user has now been authenticated and there is an existing SA between the P-CSCF and the UE, the P-CSCF inserts a P-Asserted-Identity header in all subsequent messages from that UE. In the case that signaling security is disabled, the S-CSCF inserts P-Asserted-Identity after successful authentication.

Adding support for SIP digest impacts the IMS specifications in the following ways:

- New digest algorithms are allowed to be present in the WWW-Authenticate and Authorization headers.
- The HSS must compute and store new types of data elements.
- UEs must be able to support and compute new types of digest responses.
- The home network (or S-CSCF) authenticates to the UE by including an Authentication-Info header in the 2xx response following a successful authentication of the UE.

PacketCable 2.0 embedded UEs (eUEs) contain digital certificates; however, SIP [27] does not define an authentication solution for certificates. PacketCable 2.0 security defines procedures for the UE to bootstrap IMS credentials using an X.509 certificate. This is termed "certificate bootstrapping."

The eUE connects to a certificate bootstrapping server and performs mutually authenticated TLS procedures. Once completed, the eUE is provided with IMS credentials. The eUE can then authenticate through normal registration procedures using the bootstrapped credentials.

3.8.2.2.3 Signaling Security.
The IMS defines IPsec and TLS for the secure signaling between UEs and edge proxies. The UICC provides credentials for authentication

and IPsec. The security mechanism is negotiated using [24] SIP Security Agreement. TLS is an option for signaling security between the UE and the P-CSCF. The use of TLS by the UE is optional, and is based on the following advantages:TLS is the recommended security mechanism specified in Ref. [27].

- There is a general shift towards the use of TCP to better handle longer messages.
- TLS supports NAT traversal at the protocol layer.
- TLS is implemented at the application level instead of the kernel level, which provides some advantages such as easier support in multiple environments.

Adding support for TLS for signaling leads to the consideration of TLS credentials.

- Mutually authenticated TLS—UE and server both provide certificates when establishing signaling security. The server must validate the UE certificate, and the UE must validate the server certificate. Mutual authentication provides a high degree of security.
- Server-side authentication—Only the server provides a certificate when establishing signaling security. This approach avoids the extra computational overhead of a PKI operation on the UE. It provides a medium level of security, with lower CPU requirements on the UE and may be used to secure HTTP digest sessions.

Both of these models require the P-CSCF and the UE to support PKI features, such as certificate validation and certificate management. As not all UEs utilize certificates, only server-side authentication is supported in PacketCable.

Adding support for TLS also leads to the consideration of TLS port assignments and TLS connection management. PacketCable will use the standard SIP ports for UDP, TCP, and TLS as defaults. UEs negotiating TLS connect to the SIPS port of 5061. Otherwise, UEs use the standard SIP UDP/TCP port of 5060. Operators may configure other ports for requests.

Figure 3.12 shows signaling security negotiation during a successful register dialogue. Only signaling security headers are shown for simplicity.

To support TLS for signaling security between the UE and the P-CSCF, the IMS specifications must be enhanced to allow TLS as an optional SIP security mechanism to be negotiated. IETF RFC 3329 [24] includes TLS as a security mechanism that can be negotiated; thus, the only change is to IMS specifications.

While not recommended, unless signaling messages are protected by some other mechanism, signaling security may be disabled at the P-CSCF. By disabling signaling security, UEs and the network may be exposed to many threats especially when combined with a weaker form of authentication, such as SIP digest.

When TLS is not used with SIP digest, the P-CSCF creates an IP association, which maps the IP address and port used during registration to the public identities provided by the S-CSCF. This information is used to police subsequent requests (e.g., INVITEs) to ensure only authorized SIP identities are used.

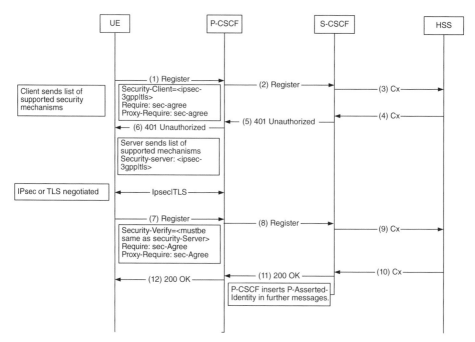

Figure 3.12. Transport security.

The PacketCable SIP Signaling Technical Report [28] and the PacketCable [29] delta specification contain detailed information on the procedures for disabling signaling security. The major difference in procedures for disabling signaling security is nonregister dialog requests should be challenged and IP associations are used to add an additional layer of security.

3.8.2.2.4 Identity Assertion, and Configuration and Management Security. PacketCable environments require a way for trusted network elements to convey the identity of subscribers to other elements or services, and to remove the identity when communicating with untrusted networks. Identity assertion is the mechanism by which elements and services can trust the identity of a user.

As described in Ref. [29], IMS assigns the task of identity assertion to P-CSCFs for all SIP messages. Once the IPsec SAs are established and the subscriber is authenticated, the P-CSCF asserts the identity of the subscriber. By monitoring SIP messaging toward the UE, the P-CSCF observes the 200 OK message from the subscribers S-CSCF. This information and the presence of SAs to the UE allow the P-CSCF to substantiate successful authentication of the UE.

PacketCable 2.0 enhances IMS with the following requirement:

- A P-CSCF with an established TLS session or IP association with a UE that observes a 200 OK response from the S-CSCF for that subscriber can assert the identity of the public identity used by that UE.

Configuration security for embedded UEs is provided only in the secure provisioning flow. It is accomplished by establishing SNMPv3 via the Kerberos protocol with PacketCable-specific PKINIT extensions. This provides authentication, message integrity, and optionally privacy.

In order to provide security protection for management information for devices that are not behind NATs, the user-based security model (USM) [30] and the view-based access control model (VACM) [31] features of SNMPv3 are supported. USM provides authentication, integrity, and privacy services for SNMP through the specification of two cryptographic functions: authentication and encryption. VACM provides further security protection to management information by controlling access to managed objects.

3.9 CONCLUSION

In this chapter, we have described the OAMPT for residential services based on the TMN FCAPS model. Once a service order is received from a customer, the service is provisioned and activated. After that, measurement data is collected periodically and customers are billed accordingly.

Using TMN FCAPS model for managing the cable network, individual interfaces have been defined and standardized, such as the DOCSIS operations support system interface standard between the network layer and network management layer for devices including the CM, CMTS, and CCAP. Various management servers in the MSO's back-office interface with these devices to perform the FCAPS functions. For example, a TFTP configuration file server is used to provision the CMs on the MSO's network. Each management function of "FCAPS" has unique requirements, protocols, and corresponding interfaces that are often specified in the CableLabs specifications.

Fault management for residential services includes the functions of alarm surveillance, fault localization, fault correction, and testing. The DOCSIS family of specifications defines a large set of event and alarm message definitions covering various functional areas, at the protocol level, feature level, management level, and beyond. MSOs have full configuration control over what types, severities, and quantity of events and alarms they want to see out of the network layer from cable devices. Fault localization can be performed using a number of DOCSIS-specified toolkits, including proactive network maintenance feature sets, spectrum analysis measurements, and a collection of other metrics that can be gathered from the network layer. Correcting faults, once isolated, is a wide area, as any number of parameters could be the root cause. Faults could be due to user error, MSO operator error, hardware failure, software failure, configuration errors, and so on. Once the fault has been correcting, additional testing should be performed to validate the fault has been mitigated. Testing methods also include proactive network maintenance, spectrum measurements, bandwidth measurements, and various other metrics provided by the DOCSIS specifications and vendor providers.

Performance management for residential services includes the functions of threshold crossing alerts relating to performance metrics, collecting performance metrics on a regular interval basis, and performing data analytics to assess the service quality and if service degradation may be occurring. The DOCSIS family of specifications defines a large set of performance metrics covering many different functional areas, at the protocol level, feature level, management level, and beyond. MSOs also have a dedicated protocol, IPDR/SP, for bulk data collection of performance metrics on a time interval basis (e.g., every 15 min).

Billing interfaces are complex. These systems have grown up to meet specific needs of the business and each vendor would profess to have an industry standard.

Having good network security is critical for providing a good subscriber experience and helping prevent theft of service. DOCSIS and PacketCable specifications define many security features that address the most significant threats and help protect residential high-speed Internet and Voice over Internet Protocol telephony services. When used properly operators can focus on creating new services instead of fighting hackers and responding to unhappy customers.

REFERENCES

1. ITU-T Recommendation M.3400 (02/2000), *TMN management function.*
2. Operations Support System Interface Specification, CM-SP-OSSIv3.0-I21-130404, April 4, 2013, Cable Television Laboratories, Inc.
3. MAC and Upper Layer Protocols Specification, CM-SP-MULPIv3.0-I22-130808, August 8, 2013, Cable Television Laboratories, Inc.
4. Operations Support System Interface Specification, CM-SP-OSSIv3.0-I21-130404, April 4, 2013, Cable Television Laboratories, Inc.
5. Operations Support System Interface Specification, CM-SP-CCAP-OSSI-I05-130808, August 8, 2013, Cable Television Laboratories, Inc.
6. MAC and Upper Layer Protocols Specification, CM-SP-MULPIv3.1-I01-131029, October 29, 2013, Cable Television Laboratories, Inc.
7. Physical Layer Specification, CM-SP-PHYv3.1-I01-131029, October 29, 2013, Cable Television Laboratories, Inc.
8. MAC and Upper Layer Protocols Specification, CM-SP-MULPIv3.0-I22-130808, August 8, 2013, Cable Television Laboratories, Inc.
9. Security Specification, CM-SP-SECv3.0-I15-130808, August 8, 2013, Cable Television Laboratories, Inc.
10. ITU-T Recommendation M.3400 (02/2000), *TMN management function.*
11. Data-Over-Cable Service Interface Specifications, Media Access Control and Upper Layer Protocols Interface Specification, CM-SP-MULPIv3.0, Cable Television Laboratories, Inc.
12. Data-Over-Cable Service Interface Specifications, Operations Support System Interface Specification, CM-SP-OSSIv3.0, Cable Television Laboratories, Inc.
13. Data-Over-Cable Service Interface Specifications, Security Specification, DOCSIS 3.0 Specification, CM-SP-SECv3.0, Cable Television laboratories, Inc.
14. IETF RFC 4131—Management Information Base for Data Over Cable Service Interface Specification (DOCSIS) Cable Modems and Cable Modem Termination Systems for Baseline Privacy Plus.

15. PacketCable 1.5 Security Specification, PKT-SP-SEC1.5, June 24, 2009, Cable Television Laboratories, Inc.

16. PacketCable Operator Security Guide, PKT-OSG, Cable Television Laboratories, Inc.

17. PacketCable 1.5 MTA Device Provisioning Specification, PKT-SP-PROV1.5, June 24, 2009, Cable Television Laboratories, Inc.

18. PacketCable 1.5 MTA MIB Specification, PKT-SP-MIB-MTA1.5, January 28, 2005, Cable Television Laboratories, Inc.

19. 3rd Generation Partnership Project; Technical Specification Group Services and Systems Aspects, Network architecture (Release 7), June 2007.

20. PacketCable Architecture Framework Technical Report, PKT-TR-ARCH-FRM-V05-080425, April 25, 2008, Cable Television Laboratories, Inc.

21. PacketCable 2.0 Security Technical Report, PKT-TR-SEC, Cable Television Laboratories, Inc.

22. 3rd Generation Partnership Project; Network Domain Security Specification, IP network layer security (Release 7), September 2007.

23. PacketCable Access Security for IP-Based Services Specification, 3GPP TS 33.203, PKT-SP-33.203-I04-0880425, April 25, 2008, Cable Television Laboratories, Inc.

24. IETF RFC 3329 (January 2003): "Security Mechanism Agreement for the Session Initiation Protocol (SIP)."

25. IETF RFC 3310 (September 2002): "Hypertext Transfer Protocol (HTTP) Digest Authentication Using Authentication and Key Agreement (AKA)."

26. IETF RFC 2617 (June 1999): "HTTP Authentication: Basic and Digest Access Authentication."

27. IETF RFC 3261 (June 2002): "SIP: Session Initiation Protocol."

28. PacketCable SIP Signaling Technical Report, PKT-TR-SIP-V04-071106, November 6, 2007, Cable Television Laboratories, Inc.

29. PacketCable SIP and SDP Stage 3 Specification, 3GPP TS 24.229, PKT-SP-24.229-I04-0800425, April 25, 2008, Cable Television Laboratories, Inc.

30. IETF STD 62 (RFC 3414) December 2002: "User-based Security Model (USM) for Version 3 of the Simple Network Management Protocol (SNMPv3)."

31. IETF STD 62 (RFC 3415) December 2002: "View-based Access Control Model (VACM) for the Simple Network Management Protocol (SNMP)."

4

BUSINESS NETWORK ARCHITECTURES AND SERVICES

Mehmet Toy, Curtis Knittle and David Hancock

4.1 INTRODUCTION

Businesses may be categorized as small, medium, and large. A business with less than 50 employees is considered a small business. Medium business is a business with less than 500 employees. A business over 500 employees is defines as a large business. As in residential services, business services can be voice, data, and video services. The nature of services can be different depending on the size of the business. For example, voice services for small- and medium-sized enterprises (SMEs) might involve interfacing private branch exchanges (PBXs) at customer premises and managing voice mailboxes, while voice services for a large corporation might involve providing a private network serving many locations. Furthermore, services for large corporations require better availability, better performance, and multiple classes of services.

MSOs have been mostly offering services to SMEs. MSOs have begun offering services to large businesses as well. Metro Ethernet services in the form of private line, virtual private line (VPL), and multipoint-to-multipoint services are among them.

MSOs and their vendors, and CableLabs jointly defined systems to optimize MSO networks and improve quality and rates of service offerings. DOCSIS 3.1, DOCSIS provisioning of EPON (DPoE), EPON protocol over coax (EPoC), Converged Cable Access Platform (CCAP), and Wi-Fi systems are among them.

Cable Networks, Services, and Management, First Edition. Edited by Mehmet Toy.
© 2015 The Institute of Electrical and Electronics Engineers, Inc. Published 2015 by John Wiley & Sons, Inc.

In the following sections, we will describe the business network architectures and services for MSOs.

4.2 METRO ETHERNET ARCHITECTURE AND SERVICES

Ethernet is becoming a dominant technology in designing today's Information Communications Networks (ICNs). With IEEE 802.1P and 1Q standards, bandwidth partitioning, user segregation, and traffic prioritization are possible. Carrier Ethernet services revenue is expected to be $62B in 2018.

In 2001, Metro Ethernet Forum (MEF) was formed to define Ethernet-based services. MEF defined Carrier Ethernet for business users as a ubiquitous, standardized, carrier-class service and network defined by five attributes: standardized services, quality-of-service (QoS), service management, reliability, and scalability.

Ethernet was the technology for LAN connectivity in enterprise networks only. However, with the introduction of Carrier-Class Ethernet, Ethernet is currently being used as a service in all carriers networks. It simplifies network operations and is cost-effective due to widespread usage of Ethernet interfaces. Service providers (SPs) offer various commercial services over Carrier Ethernet such as VPN, Internet access, and phone services

With Ethernet, the same protocol is being used for LAN and MAN, resulting in no protocol conversion between LAN and WAN. This results in lower equipment and operational cost [1].

Today, Carrier Ethernet scales to support large number of customers to adequately address metropolitan and regional distances, supports 99.999% network availability, with a range of protection mechanisms capable of sub-50 ms recovery. Service management systems are available to provision new services, manage service-level agreements (SLAs) and troubleshoot the network under fault conditions.

Carrier Ethernet is connection-oriented by disabling unpredictable functions such as media access control (MAC) learning, Spanning Tree Protocol, and "broadcast of unknown." It is scalable from 10 Mbps to 100 Gbps with finer granularity as low as 56 Kbps and 1 Mbps. With QoS and synchronization capabilities, applications with strict performance requirements such as mobile backhaul are being supported. Port- and connection-level fault monitoring, detection and isolation, performance monitoring and statistics, and protection failover are being supported.

Some SPs concern the scalability of Carrier Ethernet. There is an effort by Cloud Ethernet Forum (CEF) to increase scalability of Carrier Ethernet by using provider backbone bridge (PBB) and provide backbone transport (PBT).

MSOs offer various Carrier Ethernet[1] commercial services. They use service OAM capabilities of Metro Ethernet extensively. Automation of Metro Ethernet equipment and service provisioning is underway.

---

[1] Metro Ethernet and Carrier Ethernet are used interchangeably in the book.

4.2.1 Architecture

The basic network reference model of a Carrier Ethernet Network (CEN) is depicted in Figure 4.1 [2]. Two major functional components are involved: the subscriber/customer-edge equipment and CEN (or MEN) transport infrastructure.

User Network Interface (UNI) consists of a client side, referred to as the UNI-C, and a network side, referred to as the UNI-N (Figure 4.2). UNI is used to refer to these two functional elements and the data, management, and control plane functions associated with them. Reference point T, also referred to as the UNI reference point, demarcates the boundaries between the CEN and a customer network.

An Ethernet flow represents a particular unidirectional stream of Ethernet frames that share a common treatment for the purpose of transfer steering across the CEN. The Ethernet Virtual Connection (EVC) is the architecture construct that supports the association of UNI reference points for the delivering an Ethernet flow between subscriber sites across the MEN. There may be one or more subscriber flows mapped to a particular EVC.

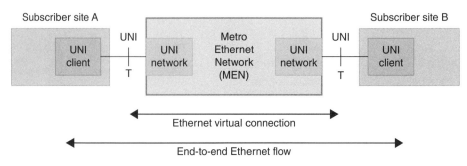

Figure 4.1. Basic network reference model [2].

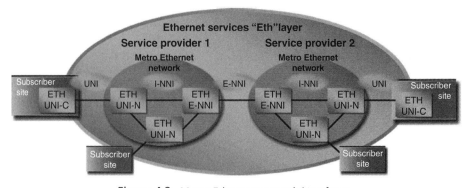

Figure 4.2. Metro Ethernet network interfaces.

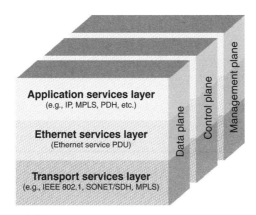

Figure 4.3. Protocol stack of Metro Ethernet networks [2].

Protocol stack for a MEN can be represented as in Figure 4.3. Data plane, control plane, and management plane cross all these layers. Packets of application layer are represented in the form of Ethernet Service Layer PDUs. In turn, Ethernet service PDUs are converted into transport layer frames or packets, depending on whether transport protocol is Ethernet, or SONET/SDH, or MPLS, and so on

Application layer supports applications carried on Ethernet services across the MEN. Ethernet service layer (ETH layer) is responsible for the instantiation of Ethernet MAC-oriented connectivity services and the delivery of Ethernet service frames. The transport layer (TRAN layer) supports connectivity among ETH layer functional elements in a service independent manner. Various layer network technologies and interconnect approaches may be used to support the transport requirements for the Ethernet services layer.

4.2.2 ETH Layer Characteristic Information (Ethernet Frame)

The characteristic information (ETH_CI), which is the Ethernet frame, exchanged over ETH layer [1, 3]. It consists of preamble, start frame delimiter (SFD), destination MAC address (DA), source MAC address (SA), (optional) 802.1QTag, Ethernet length/type (EtherType), user data, padding if required, frame check sequence (FCS), and extension field that is required only for 100 Mbps half-duplex operation (Figure 4.4).

A cyclic redundancy check (CRC) is used by the transmit and receive algorithms to generate a CRC value for the FCS field containing a four-octet (32-bit) CRC value that is computed as a function of the contents of the source address, destination address, length, data, and pad.

The optional four-octet 802.1QTag comprises a two-octet 802.1QTagType and the tag control information (TCI) that contains 3-bit of CoS/priority information, the single-bit canonical format indicator (CFI), and the 12-bit VLAN identifier (VLAN ID). CFI is

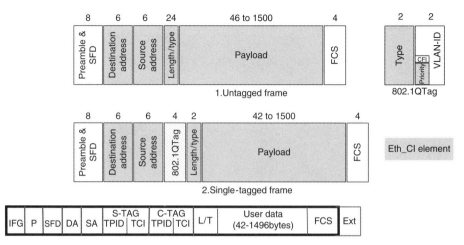

Figure 4.4. Ethernet frame.

always set to zero for Ethernet switches. Figure 4.4 and 4.5 illustrate the ETH_CI for common IEEE 802.3–2002/2005 compliant frame formats.

ETH layer PDUs that are the data frames exchange the ETH_CI across standardized ETH layer interfaces and associated reference points such as UNIs. The Ethernet service frame is the ETH layer PDU exchanged across the UNI.

- IFG: interframe gap, 12 bytes;
- TCI = VID + CFI/DE + PCP), 2 bytes;
- S-Tag TPID: service tag identifier, 2 bytes (0x88-A8);
- C-Tag TPID: customer tag identifier, 2 bytes (0x81-00);
- L/T (length/type): length of frame or data type, 2 bytes
  (L/T field of a tagged MAC frame always uses the type interpretation, and contains the 802.1Q protocol type: a constant equal to 0x81-00);
- FCS, 4 bytes;
- Ext: extension field.
- Preamble and start of frame delimiter (P/SFD): Alternate ones and zeros for the preamble, 11010101 for the SFD. This allows for receiver synchronization and marks the start of frame; 8 bytes (P-7 bytes, SFD-1 byte).

Class of service (CoS)/priority information associated with an ETH layer PDU can be conveyed explicitly in the 3-bit priority field.

The IEEE 802.1Q standard adds four additional bytes to the standard IEEE 802.3 Ethernet frame and is referred to as the VLAN tag (Figure 4.4). Two bytes are used for the tag protocol identifier (TPID) and the other 2 bytes for TCI. The TCI field is divided into PCP, CFI, and VID.

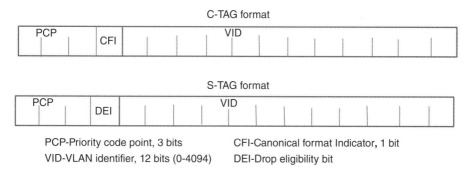

PCP-Priority code point, 3 bits   CFI-Canonical format Indicator, 1 bit
VID-VLAN identifier, 12 bits (0-4094)   DEI-Drop eligibility bit

Figure 4.5. C-Tag and S-Tag.

TABLE 4.1.   MAC Addresses for L2 Control Protocols

MAC Addresses5	Description
01-80-C2-00-00-00 through 01-80-C2-00-00-0F	Bridge block of protocols
01-80-C2-00-00-20 through 01-80-C2-00-00-2F	GARP block of protocols
01-80-C2-00-00-10	All bridges protocol

IEEE 802.1ad introduces a service tag, S-Tag to identify the Service. The subscriber's VLAN Tag (C-VLAN Tag) remains intact and is not altered by the SP's anywhere within the provider's network. TPID of 0x88a8 is defined for S-Tag.

The service frame is just a regular Ethernet frame beginning with the first bit of the destination address through the last bit of the FCS. A service frame that contains an IEEE 802.1Q tag can be up to 1522 bytes long and a service frame that does not contain an IEEE 802.1Q tag can be up to 1518 bytes.

A service frame can be a unicast, multicast, broadcast, and Layer 2 Control Protocol (L2CP) frame. Unicast service frame is a service frame that has a unicast destination MAC address. Multicast service frame is a service frame that has a multicast destination MAC address. Broadcast service frame is a service frame with the broadcast destination MAC address. L2CP service frame is a frame whose destination MAC address is one of the addresses listed in Table 4.1, is treated as L2CP service frame.

Some L2CPs share the same destination MAC address and are identified by additional fields such as the Ethertype and a protocol identifier. Therefore, disposition of service frames carrying L2CP may be different for different protocols that use the same destination MAC address.

Service frame disposition can be discarding, tunneling, delivering unconditionally, and delivering conditionally where tunneling only applies to L2CP service frames. For example, all ingress service frames with an invalid FCS must be discarded by the CEN. For a point-to-point EVC, service frames with correct FCS are delivered across the egress UNI unconditionally.

4.2.3 User-to-Network Interface

The standard user interface to a CEN is called User-to-Network Interface (UNI) [2], which is a dedicated physical demarcation point between the responsibility of the SP and the responsibility of a subscriber. As mentioned earlier, UNI functions are distributed between CPE and CEN, as UNI-C and UN-N, respectively. UNI-C executes the processes of the customer side while UNI-N executes the processes of the network side. CPE may not support UNI-C functions and use a Network Interface Device (NID) to support the demarcation point between the CPE and the MEN (Figure 4.6).

The UNI consists of a data plane, a control plane, and a management plane. The data plane defines Ethernet frames, tagging, and traffic management. The management plane defines provisioning, device discovery, protection, and OAM (operation, administration, and maintenance). The control plane is expected to define signaling and control. There is no attempt in MEF to define signaling, but the controlling some of EVC and UNI attributes are being addressed in MEF.

The physical medium for UNI can be copper, coax, or fiber. The operating speeds can be 1–10 Mbps, 100 Mbps, 1 Gbps, or 10 Gbps where 10 Gbps rate is only supported over fiber. MSOs offer Metro Ethernet services over cable modem (CM) with coax UNI, and over demarcation device with copper and fiber UNI.

Service frames are Ethernet frames that are exchanged between the MEN and the customer edge across the UNI. A service frame sent from the Metro Ethernet network to the customer edge at a UNI is called an egress service frame. A service frame sent from the customer edge to the Metro Ethernet network at a UNI is called an ingress service frame.

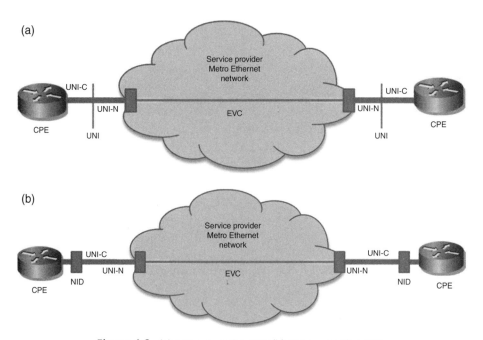

Figure 4.6. (a) CPE supporting UNI (b) NID supporting UNI.

There are three types of UNIs: UNI Type 1, Type 2, and Type 3. UNI Type 1 is divided into two categories:

- **Type 1.1:** Nonmultiplexed UNI for services like Ethernet private line (EPL);
- **Type 1.2:** Multiplexed UNI for Services like Ethernet VPL (EVPL).

UNI Type 2 supports an automated implementation model allowing UNI-C to retrieve EVC status and configuration information from UNI-N, enhanced UNI attributes, and additional fault management and protection functionalities. UNI Type 2 is divided as UNI Type 2.1 and Type 2.2. The Type 2.1 includes service OAM, enhanced UNI attributes such as bandwidth profile per egress UNI and L2CP handling mandatory features, and Link OAM, port protection via Link Aggregation protocol and E-LMI optional features. In Type 2.2, the optional features of Type 2.1 will become the mandatory features of UNI.

UNI Type 3 has not yet been defined.

4.2.4 Ethernet Virtual Connection

An EVC [1, 4] is a logical representation of an Ethernet service between two or more UNIs and establishes a communication relationship between UNIs. An UNI may contain one or more EVCs (Figure 4.7).

While connecting two or more UNIs, EVC prevents data transfer between subscriber sites that are not part of the same UNI. The service frames cannot leak into or out of an EVC.

Point-to-point EVC (Figure 4.7) supports communication between only two UNIs. All ingress service frames at one UNI, with the possible exception of L2CP messages are typically delivered to the other UNI.

Multipoint-to-multipoint (P2MP) EVC (Figure 4.13) supports any-to-any communication between two or more UNIs. As the number of users grow in the EVC, additional UNIs can be added.

In a multipoint-to-multipoint EVC, a single broadcast or multicast ingress service frame, that is determined from the destination MAC address, at a given UNI is replicated in the MEN, and one copy would be delivered to each of the other UNIs in the EVC. Broadcast service frame is defined by all "1"s destination MAC address.

Unicast service frames with unicast destination MAC addresses can be replicated and delivered to all other UNIs in the EVC. This makes the EVC behave like a shared-media Ethernet. Another approach is having the MEN to learn which MAC addresses are "behind" which UNIs by observing the source MAC addresses in service frames and delivering a service frame to only the appropriate UNI when it learns the destination MAC address. When it has not yet learned the destination MAC address, it replicates the service frame and delivers it to all other UNIs in the EVC. In this case, the MEN behaves like a MAC learning bridge.

P2MP EVC that is also called as rooted-multipoint EVC that supports communication between two or more UNIs, but does not support any-to-any communication. UNIs

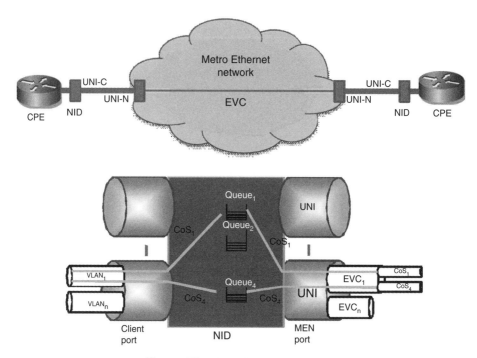

Figure 4.7. UNI and EVC relationship.

are designated as root or leaf. Transmissions from the root are delivered to the leaves, and transmissions from the leaves are delivered to the root(s). No communication can occur between the leaves or between the roots. This EVC can be used to create a hub-and-spoke communication arrangement. It is an essential component of the E-Tree service.

In a given UNI, subscriber flows may be mapped into one or more EVCs. When multiple subscriber flows mapped into a single EVC, there could be multiple bandwidth profiles associated with the EVC where there is a bandwidth profile per CoS instance.

Similarly, multiple CE-VLAN IDs can be mapped to one EVC where each CE-VLAN ID may belong to a different service (i.e., service multiplexing). This configuration is called a bundling map. When there is a bundling map, the EVC must have the CE-VLAN ID preservation that allows the subscriber to use the same VLAN values at all sites.

4.2.5 UNI and EVC Attributes

The key UNI/EVC attributes are as follows:

- **Service Frame Disposition:** An ingress service frame can be discarded, delivered conditionally, or delivered unconditionally. For example, frames with FCS error or containing a particular L2CP can be discarded. In an E-Line service, all frames with no FCS error can be delivered unconditionally.

- **Service Frame Transparency:** Transparency implies minimal interaction with client's data frames, including associated management and control traffic between the subscriber's routers and switches. An EPL service provides a high degree of transparency for service frames between the UNIs it interconnects such that the service frame's header and payload are identical at both the source and destination UNI when a service frame is delivered. An EVPL need not provide as much transparency of service frames as with an EPL.
- **CE-VLAN Tag (C-Tag) Preservation:** Customer-edge VLAN (CE-VLAN) Tag consists of CE-VLAN ID and priority code points (PCPs) that can be preserved for operational simplicity. The ingress and egress service frames will have the same CoS.
- **L2CP Processing Service:** An L2CP frame is either discarded, tunneled or peered for a given EVC at a given UNI. When an L2CP is tunneled, the service frame at each egress UNI is identical to the corresponding ingress service frame. When an L2CP is peered, it is processed in the network.
- **CoS Identifier:** The CoS identifier can be based on EVC, PCP, or DSCP. All frames with the same CoS identifier can be mapped to the same EVC. Therefore, EVC ID will identify CoS as well. EVC ID-CoS mapping needs to be stored in operation systems. PCP field of C-Tag or S-Tag can be used to identify CoS. If the frame is untagged, it **will** have the same CoS identifier as a frame with PCP field$=0$. Similarly, DSCP which is contained in TOS byte of IPv4 [RFC2474 and RFC2475] and Traffic Class octet in IPv6 can be used to identify CoS.
- **EVC Performance:** The EVC performance attributes specify the frame delivery performance for green color frames of a particular CoS, arriving at the ingress UNI during a time interval T. These attributes are frame delay (FD), interframe delay variation (IFDV), mean FD (MFD), FD range (FDR), frame loss ratio (FLR), and availability. The users choose to use either FD and IFDV or MFD and FDR.
- **EVC ID:** The EVC is identified by an EVC ID. MEF defines it as an arbitrary string administered by the SP to identify the EVC within the MEN. The EVC ID is not carried in any field in the service frame, but stored in operation systems. Alliance for Telecommunications Industry Solutions (ATIS) further defines EVC ID as 28 alpha/numeric characters [1],

<Prefix>. <Service Code/Modifier>.<Serial Number>.<Company ID>. <Segment Number>
where

- ○ **Prefix** is a non-standard code populated according to the special services circuit coding methodology of each carrier. It is 1–2 alpha/numeric characters that may be used to identify a regional network of the carrier's nationwide network.
- ○ **Service Code** is a standardized code that represents a tariff offering that requires special services circuit provisioning. It is two alpha/numeric characters (Table 4.2).
- ○ **Service Code Modifier** is a standardized code that designates the jurisdiction, networking application, and additional technical information of the service identified in the service code. It is two alpha/numeric characters (Table 4.2).

TABLE 4.2.　EVC Types and Service Codes/Modifiers

EVC Types	Service Code/Modifier
Point-to-Point	VLXP
Multipoint-to-Multipoint	VLXM

TABLE 4.3.　UNI Port Speeds and Associated Service Codes/Modifiers

UNI	Service Codes/Modifiers	
Port Speed	Inter	Intra
10 M	KDGS	KDFS
100 M	KEGS	KEFS
1 G	KFGS	KFFS
10 G	KGGS	KGFS

∘ **Serial Number** is a serial number type code that uniquely identifies a special services circuit having the same prefix, service code, and service code modifier within a network. It is 1–6 numeric characters.
∘ **Suffix** is a serial number type code that relates a group of special services circuits having the same service code for the same customer, and with similar termination equipment at each end. It is 1–3 numeric characters.
∘ **Assigning Company ID** is a standardized code that uniquely identifies the carrier assigning the circuit identification. It is 2–4 alpha characters.
∘ **Segment Number** is a serial number type code that uniquely identifies each termination point of a special services circuit, when the circuit has more than two termination points as in multipoint circuits. It is 1–3 alpha/numeric characters.
  For example, 12.VLXP.123456..OB. represents carrier OB's point-to-point connection in region 12 where suffix before the company code and segment number after the company code are not utilized.
• **UNI ID** is the UNI identified by an UNI ID. MEF defines UNI ID as arbitrary text string. The UNI ID must be unique among all UNIs for the MEN.

The ATIS standards define it as
<Prefix>.　<Service　Code/Modifier>.<Serial　Number>.<Company　ID>. <Segment Number>

where service code/modifier as given in Table 4.3.
  For example, **15.KFGS.123457..BO.** represents a 1 Gbps UNI in intrastate region 15 of the carrier BO.

4.2.6 External Network-to-Network Interface

An EVC between two UNIs may have to travel multiple operator MENs. One operator/ SP may not be able to support all UNIs of a subscriber. As a result, MENs need to interface each other. This interface between two MENs is called the External Network-to-Network Interface (ENNI) or ENNI, depicted in Figure 4.8.

An ENNI can be implemented with one or more physical links for protection. When there are multiple links between ENNI gateways, the Link Aggregation Group (LAG)/Link Aggregation Control Protocol (LACP) mechanism is commonly set for the protection.

Similar to UNI, ENNI consists of a data plane, a control plane and a management plane. The data plane defines Ethernet frames, tagging, and traffic management. The management plane defines provisioning, device discovery, protection, and OAM. The control plane is expected to define signaling and control, which is not addressed in the standards yet.

The physical medium for ENNI can be copper, coax, or fiber. However, the fiber interface between ENNI gateways are more common. The operating speeds can be 1 or 10 Gbps.

Similar to UNI, ENNI has a bandwidth profile. The parameters of the bandwidth profile are *CIR* in bps, *CBS* in bytes that is greater than or equal to the largest MTU size allowed for the ENNI frames, *EIR* in bps, *EBS* in bytes that is > the largest MTU size allowed for the ENNI frames, Coupling Flag (CF), and Color Mode (CMd). The CF has value 0 or 1 and CMd has value "color-blind" or "color-aware."

4.2.7 Operator Virtual Connection

An EVC travelling multiple operator MENs is realized by concatenating EVC components in each operator network, which is called operator virtual connections (OVCs). For example, in Figure 4.8, EVC is constructed by concatenating OVC 2 in MEN 1 with OVC 1 in MEN 2

Each OVC end point (EP) is associated with either a UNI or an ENNI and at least one OVC EP associated by an OVC at an ENNI.

Hairpin switching occurs when an OVC associates two or more OVC EPs at a given ENNI. An ingress S-Tagged ENNI frame at a given ENNI results in an egress S-Tagged ENNI frame with a different S-VLAN ID value at the given ENNI (Figure 4.9).

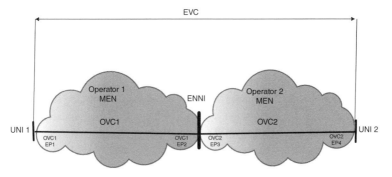

Figure 4.8. ENNI, EVC, and OVC.

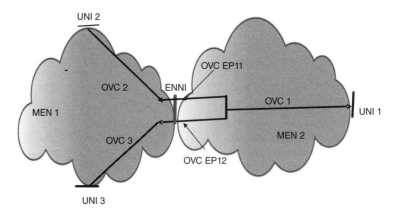

Figure 4.9. Example of supporting multiple OEPs for one OVC.

An OVC EP represents the OVC termination point at an External Interface such as ENNI or UNI. The EP Map specifies the relationship between S-Tagged ENNI Frame and OVC EP within an Operator MEN. An ingress S-Tagged ENNI Frame that is not mapped to an existing EP will be discarded.

Multiple S-VLAN IDs can be bundled and mapped to an OVC EP. In that case, the OVC EP should be configured to preserve S-VLAN IDs. S-VLAN ID preservation cannot be supported in hairpin switching configuration.

When there is Bundling, frames originated by more than one Subscriber may be carried by the OVC resulting in duplicate MAC addresses. To avoid possible problems due to this duplication, MAC Address learning may be disabled on the OVC.

There are service attributes for each instance of an OVC EP at a given ENNI:

- **OVC EP Identifier** which is a string that is unique across the Operator MEN
- **Class of Service Identifier** that identifies class of service is determined for an ENNI frame at each ENNI. The OVC associating the EP and the S-Tag priority code point.
- **Ingress Bandwidth Profile per OVC EP** consists of ingress policing parameters<CIR, CBS, EIR, EBS, CF, CMd>. All ingress ENNI Frames are mapped to the OVC EP. The CMd for the bandwidth profile algorithm is color-aware.
- **Ingress Bandwidth Profile per ENNI Class of Service Identifier** consists of ingress policing parameters<CIR, CBS, EIR, EBS, CF, CMd>. Bandwidth profile algorithm **is** color-aware.
- **Egress Bandwidth Profile per EP** consists of Egress policing parameters<CIR, CBS, EIR, EBS, CF, CMd>. All egress ENNI Frames are mapped to the OVC EP. Bandwidth profile algorithm **is** color-aware.
- **Egress Bandwidth Profile per ENNI Class of Service Identifier** consists of egress policing parameters<CIR, CBS, EIR, EBS, CF, CMd>. Bandwidth Profile Algorithm **is** color-aware.

4.2.8 VUNI/RUNI

Ordering and maintaining EVCs crossing an ENNI can be cumbersome due to coordination between entities of two or more carriers. In order to somewhat simplify this process, the operator responsible from the end-to-end EVC may order a tunnel between its remote user that is connected to another operator's network and its ENNI gateway. The tunnel can accommodate multiple OVCs, instead of ordering the OVCs one-by-one. This access of the remote user to the SP's network is called UNI tunnel access (UTA). The remote user end of the tunnel is called remote UNI (RUNI), while the SP end of this tunnel is called virtual UNI (VUNI) (Figure 4.10) [5].

A single bandwidth profile and CoS may be applied at this remote UNI OVC EP. At the UTA OVC EP at the network operator's side of the ENNI, an S-VLAN ID is used to map ENNI Frames to the OVC EP supporting the UTA, and applies a UTA specific single bandwidth profile and CoS.

Figure 4.11 shows an example where the SP's subscriber has a UNI requiring connectivity by through an access provider (AP), in addition to UNI W and UNI X that the

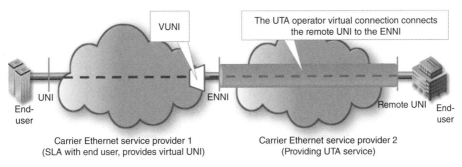

Figure 4.10. UNI Tunnel Access (UTA) enabling EVC service frames associated with a remote user's UNI to be tunneled through an Off-Net providers' network [5].

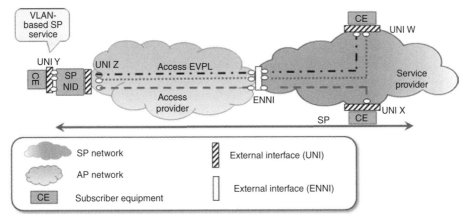

Figure 4.11. Network topology with SP placing its own NID.

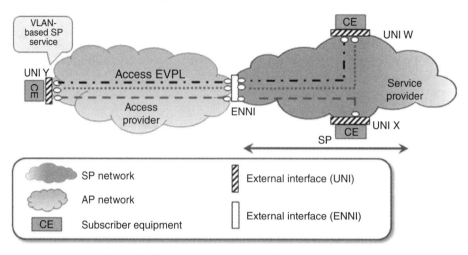

Figure 4.12. Network topology with SP relying on AP.

SP supports directly. The SP may deploy a device at the subscriber's location to the left of UNI Z, as shown in the Figure 4.11, in addition to a local device that the AP has deployed to the right of UNI Z. The SP's device provides a UNI to the subscriber.

Figure 4.10 [5]: UNI tunnel access (UTA) enabling EVC service frames associated with a remote user's UNI to be tunneled through an off-net providers' network.

This approach provides the SP with a direct interface to the subscriber using the SP's device, and the ability to test, monitor the performance, and instantiate key service attributes without coordinating with the AP in a manner similar to what it offers at in-network UNIs.

However, for the SP to place a device at the subscriber's location may be difficult for reasons ranging from lack of installation personnel to difficulties in getting approval to place equipment in another country. In addition, each device introduces another failure point and another possibility for mis-configuration or errored performance that complicates any troubleshooting effort.

An alternative approach is for the SP to not place a device at the subscriber's UNI, but instead to rely on the AP to provide all required functions (Figure 4.12). The AP is responsible for deployment of the sole-edge device and resolving issues associated with it. However, with this approach, coordination between SP and AP is required for many functions such as collecting data for troubleshooting. The response would be much slower for the SP than querying its own device.

4.2.9 Class of Service

MEF [6] defined three class of services to differentiate real-time and non-real time applications over Metro Ethernet networks and provide differentiated performance according to the needs of each application: H, M, and L. Class H is intended for real-time applications with tight delay/jitter constraints such as VoIP. Class M is intended for time-critical. Class L is intended for non-time-critical data such as e-mail. Each has its own performance parameters as shown in Tables 4.4 and 4.5 [1, 7].

TABLE 4.4. Parameters of H, M, and L Classes

Performance Attribute	H	M	L
Frame delay (FD)	>99.9th percentile of FD values monitored over a month	>99th percentile of FD values monitored over a month	>95th percentile of FD values monitored over a Month
Mean frame delay (MFD)	MFD values calculated over a month	MFD values calculated over a month	MFD values calculated over a month
Interframe delay variation (IFDV)	>99.9th percentile of FD values monitored over a month, with 1 s between frame pairs	>99th percentile of FD values monitored over a month, with 1 s between frame pairs or not supported	Not supported
Frame delay range (FDR)	>99th percentile corresponding to minimum delay	>99th percentile corresponding to minimum delay or not supported	Not supported
Frame loss ratio (FLR)	FLR values calculated over a month	FLR values calculated over a month	FLR values calculated over a month
Availability	Undefined	Undefined	Undefined
High loss interval	Undefined	Undefined	Undefined
Consecutive high loss interval	Undefined	Undefined	Undefined

TABLE 4.5. An Example of PCP—CoS Mapping

EVC ID	Tagged or Untagged	PCP Values	CoS
1	Tagged	2,5,6,7	Gold
		0,3,4	Silver
		1	Discard
	Untagged	Traffic is treated as PCP=0	Silver
2	Tagged	5	Best effort
		0,1,2,3,4,6,7	Gold
	Untagged	Traffic is treated as PCP=0	Gold

CoS flows are policed according to bandwidth profile parameters, committed information rate (CIR), excess information rate (EIR), committed burst size (CBS), excess burst size (EBS), coupling flag (CF), and (CMd). In order to rate limit traffic, policers employ the leaky bucket method which supports single-rate three-color marking (SRTCM) [8] or two-rate three-color marking (TRTCM) [9].

CIR defines the average rate (bps) of ingress traffic up to which the device guarantees the delivery of the frames while meeting the performance objectives related to traffic loss, frame delay, and frame delay variations. EIR defines the excess rate (bps) of ingress traffic up to which the device may deliver the frames without any performance objectives. EIR traffic may be oversubscribed on the WAN interface and may be dropped if the network cannot support it. Traffic offered above the EIR is dropped at the UNI. CBS defines the maximum number of bytes available for a committed burst of ingress frames sent at the UNI line rate for which the device guarantees delivery. EBS defines the maximum number of bytes available for a burst of excess ingress frames sent at the UNI line rate for which the device may deliver. The CF is used to share bandwidth between CIR and EIR settings on a flow.

Policers can operate in color-aware mode or color-blind mode. For example, supporting color-aware mode is mandatory for ENNI, while it is not the case for UNI. Traffic is less than or equal to CIR is marked as green and transported within performance constraints. On the other hand, traffic that exceeds the CIR setting, but is less than or equal to the EIR, is marked in color-aware mode as yellow and transported if bandwidth is available. There are no performance constraints for Yellow frames. The policing algorithm determines whether a customer frame is accepted from the UNI and if it is to be marked green or yellow. The algorithm is executed for each received frame.

Through combinations of CIR, EIR, CBS, and EBS settings of bandwidth profile, the system can be tuned to match bandwidth, traffic loss, latency, and jitter requirements for each traffic class. CBS and EBS values depend on various parameters, including CIR, EIR, loss and the delay to be introduced. In [8], CBS is required to be greater than 12176 bytes.

Bandwidth profile for CoS Label L allows CIR or EIR = 0. When CIR = 0, there will be no performance objectives, while the case of CIR > 0 will require conformance with performance attribute objectives.

Service frame delivery performance is specified for all service frames transported within an EVC with a particular COS. A CoS instance for a service frame is identified either by the EVC or by the combination of the EVC and the priority code point field (PC) in tagged service frames or differentiated service code point (DSCP) values.

If all frames with the same CoS Identifier such as gold, silver, or best effort are mapped to the same EVC, then EVC ID will identify CoS as well. If the frame is untagged, it will have the same CoS identifier as a frame with PCP = 0, as shown in Table 4.4.

Similarly, DSCP in TOS byte of IPv4 and Traffic Class octet in IPv6 can be used to identify CoS as shown in Table 4.6.

At ENNI, DEI as well as PCP can be used to represent frame colors. Table 4.7 provides an example mapping of PCP and DSCP vales to H, M, and L classes with colors.

TABLE 4.6.   DSCP—CoS Mapping

EVC ID	Tagged or Untagged	DSCP Values	CoS
10	Tagged	46	Gold
		26	Silver
		10	Discard
	Untagged	CoS per agreement	Silver
20	Tagged	10	Best effort
		46	Gold
	Untagged	CoS per agreement	Gold

TABLE 4.7. MEF Three CoS Bandwidth Constraints, and PCP and DSCP Mapping [7]

CoS Label	Ingress EI Bandwidth Profile Con-straints	CoS and Color Identifiers						CoS-only Identifiers			Example Applications
		C-Tag PCP		PHB (DSCP)		S-Tag PCP (CoS w/DEI)		C-Tag PCP	PHB (DSCP)	S-Tag PCP	
		Color Green	Color Yellow	Color Green	Color Yellow	Color Green	Color Yellow				
H	CIR>0 EIR≥0 CF=0	5	Undefined	EF (46)	Undefined	5	Undefined	5	EF (46)	5	VoIP and mobile backhaul control
M	CIR>0 EIR≥0	3	2	AF31 (26)	AF32 (28) or AF33 (30)	3	2	2–3	AF31-33 (26, 28, 30)	2–3	Near-real-time or Critical data apps
L	CIR≥0 EIR≥0	1	0	AF11 (10)	AF12 (12), AF13 (14) or Default (0)	1	0	0–1	AF11–13 (10, 12, 14) or Default (0)	0–1	Non-critical data apps

For a given EVC, SP and subscriber agree for specific performance parameters that are called SLA consisting of delay, jitter, loss, and availability. MEF [7] defines four performance tiers (PTs) that associates SLAs with distance:

- PT1 (metro PT) for typical Metro distances (<250 km, 2 ms propagation delay);
- PT2 (regional PT) for typical regional distances (<1200 km, 8 ms propagation delay);
- PT3 (continental PT) for typical national/continental distances (<7000 km, 44 ms propagation delay);
- PT4 (global PT) for typical global/intercontinental distances (<27500 km, 172 ms propagation delay).

Tables 4.7, 4.8, 4.9, 4.10 and 4.11 list values of performance parameters for these PTs. An user or carrier may choose to support FD or MFD for delay, and IFDV or FDR for jitter in these PTs.

TABLE 4.8. SLAs for Point-to-Point EVCs of Three Service Categories (i.e., CoS), Spanning No More Than 250 Network km with 2 ms Propagation Delay of Metro Distances

SLA Parameters Measured over a Month	H	M	L
FD	<8 ms for 99.9th percentile	<20 ms for 99th percentile	<37 ms for 95th percentile
MFD			
IFDV	<2 ms for 99.9th percentile with 1 s pair interval	<8 ms for 99th percentile with 1 s pair interval	N/S
FDR			
FLR	$<10^{-5}$	$<10^{-5}$	$<10^{-3}$
Availability	Undefined	Undefined	Undefined

TABLE 4.9. SLAs for Point-to-Point EVCs of Three Service Categories (i.e., CoS), Spanning No More Than 1200 km with 8 ms Propagation Delay of Regional Distances

SLA Parameters Measured over a Month	H	M	L
FD (ms)	25 for 99.9th percentile	75 for 99th percentile	125 for 95th percentile
MFD			
IFDV (ms)	8 for 99.9th percentile with 1 s pair interval	40 for 99th percentile 1 sec pair interval	N/S
FDR			
FLR	10^{-5}	10^{-5}	10^{-3}
Availability	Undefined	Undefined	Undefined

TABLE 4.10. SLAs for Point-to-Point EVCs of Three Service Categories (i.e., CoS), Spanning No More Than 7000 km with 44 ms Propagation Delay of National/Continental Distances

SLA Parameters Measured over a Month	H	M	L
FD (ms) **MFD**	77 for 99.9th percentile	115 for 99th percentile	230 for 95th percentile
IFDV (ms)	10 for 99.9th percentile with 1 s pair interval	40 for 99th percentile with 1 s pair interval	Not Specified
FDR			
FLR	10^{-4}	10^{-4}	10^{-3}
Availability	Undefined	Undefined	Undefined

TABLE 4.11 SLAs for Point-to-Point EVCs of Three Service Categories (i.e., CoS), Spanning No More Than 27,500 km with 44 ms Propagation Delay of Global/Intercontinental Distances

SLA Parameters Measured over a Month	H	M	L
FD (ms) **MFD**	230 for 99.9th percentile	250 for 99th percentile	390 for 95th percentile
IFDV (ms)	32 for 99.9th percentile with 1 s pair interval	40 for 99th percentile with 1 s pair interval	Not specified
FDR			
FLR	10^{-4}	10^{-4}	10^{-3}
Availability	Undefined	Undefined	Undefined

4.2.10 Services

Carrier Ethernet services are defined in terms of what is seen by customer-edge (CE) and independent from technology inside Metro Ethernet network. UNI is the physical demarcation point between a subscriber and SP, while ENNI is the demarcation point between two SPs. If it is between SP internal networks, it is called I-NNI. The CE and MEN exchange service frames across the UNI where the frame transmitted toward the SP is called an ingress service frame and the frame transmitted toward the subscriber is called an egress service frame.

EVC, which is the logical representation of an Ethernet service and defined between two or more UNIs, can be point-to-point, P2MP (or rooted multipoint), or multipoint-to-multipoint as depicted in Figure 4.13.

Carrier Ethernet services built over these EVCs are listed in Table 4.12.

These services can support applications with strict performance requirements such as voice and best-effort performance requirements such as Internet access. Their Ethernet delivery is independent from underlying technologies. At the access, Carrier Ethernet service may ride over TDM, SONET, Ethernet, and passive optical network

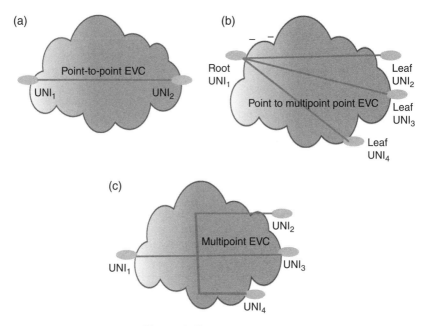

Figure 4.13. EVC types.

TABLE 4.12   Carrier Ethernet Services

Service Type	Port-Based (All-to-One Bundling)	VLAN-Based (Service Multiplexed)
E-Line over point-to-point EVC	Ethernet private line (EPL)	Ethernet virtual private line (EVPL)
E-LAN over multipoint-to - multipoint EVC	Ethernet private LAN (EP-LAN)	EVP LAN (EVP-LAN)
E-Tree over rooted multipoint EVC	Ethernet private tree (EP-Tree)	EVP Tree (EVP-Tree)

(PON) technologies. At the backbone, Carrier Ethernet service may ride over IP/ MPLS network.

E-Line service can be divided into EPLs and VPLs. The EPL service is between two UNI ports where there is no multiplexing at UNI and provides transparency for service frames (Figure 4.14). On the other hand, Ethernet VPL may involve more than two UNI ports where there is multiplexing at UNI. E-Line service can support a specific performance assurance as well as no-performance assurance.

EPL is used to replace a TDM, FR and ATM private lines. There is Single EVC per UNI. For cases where EVC speed is less than the UNI speed, the CE is expected to shape traffic.

An EVPL (Figure 4.15) can be used to create services similar to the EPL. As a result of EVC multiplexing, an EVPL does not provide as much transparency of service frames as with an EPL since some of service frames may be sent to one EVC, while other service frames may be sent to other EVCs.

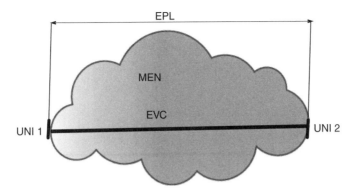

Figure 4.14. EPL example.

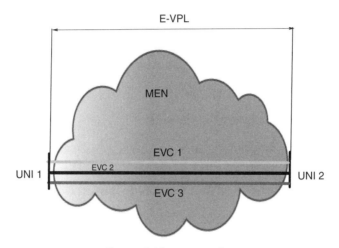

Figure 4.15. EVPL service.

E-LAN service (Figure 4.16) is used to create multipoint L2 VPNs and transparent LAN service with a specific performance assurance as well as no-performance assurance. It is foundation for Internet Protocol television (IPTV) and multicast networks.

In the EP-LAN service (Figure 4.16), CE-VLAN Tag can be preserved and L2CPs can be tunneled. This provides additional flexibility to users by reducing the coordination with SP.

EVCs can be multiplexed on the same UNI to create EVP-LAN services (Figure 4.17). Furthermore, E-LAN and E-Line services can be multiplexed in the same UNI where some of the EVCs can be part of the E-LAN, while the others can be part of E-Line services. Furthermore, UNIs can be added and removed from E-LAN service without disturbing the users on the E-LAN.

E-Tree service depicted in Figures 4.18 and 4.19 is also a subset of E-LAN service where traffic from any "leaf" UNI can be sent/received to/from "Root" UNI(s) but never being forwarded to other "Leaf" UNIs. The service issued for applications requiring

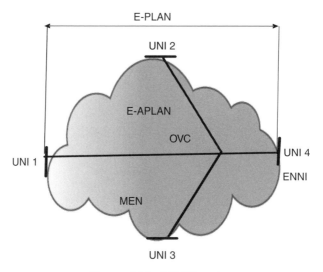

Figure 4.16. E-PLAN service.

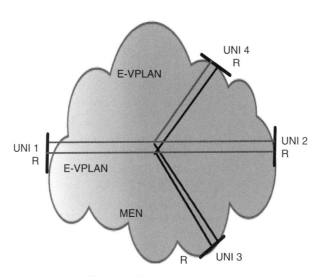

Figure 4.17. E-VPLAN service.

P2MP topology such as video on demand, Internet access and mobile backhaul. The EP-Tree services has the same benefits that EP-LAN has.

4.2.11 Access EPL and EVPL

When the SP does not have a presence at a particular customer location, the SP needs to buy an access circuit from a local operator that can serve customers at this location. The services provided by the access circuit are called E-Access services. The E-Access

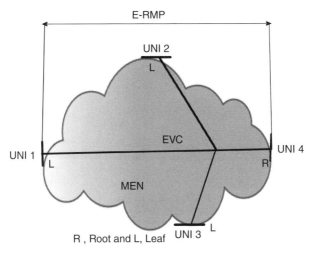

Figure 4.18. EP-Tree service.

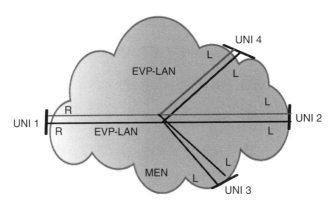

Figure 4.19. EVP-Tree Service.

services defined by MEF are Access EPL (Access EPL) service and Access Ethernet VPL (Access EVPL) service [10].

An Access EPL service uses a point-to-point OVC associating one OVC EP at a UNI and one OVC EP at an ENNI. This service can provide a high degree of transparency for frames between the EIs it interconnects such that the frame's header and payload upon ingress at the UNI is delivered unchanged to the ENNI, with the addition of an S-VLAN tag. The frame's header and payload upon ingress at the ENNI is delivered unchanged to the UNI except for the removal of the S-VLAN tag. The FCS for the frame is recalculated when an S-VLAN tag is inserted or removed. Figure 4.20 below shows the Access EPL service.

A SP can use the Access EPL service from an AP to deliver the port-based Ethernet services, EPL, and Ethernet Private LAN (EP-LAN). Although MEF33 [10] does not

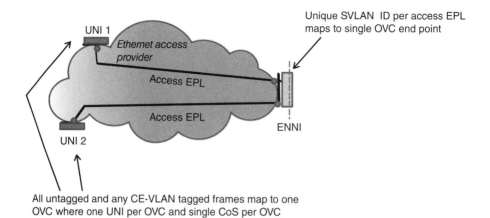

Figure 4.20. Access EPL service [10].

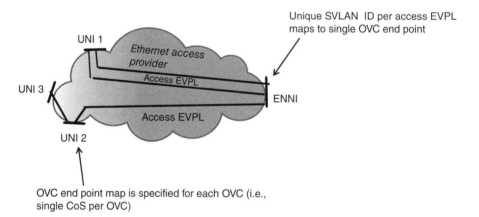

Figure 4.21. Access EVPL service [10].

address delivery of Ethernet Private Tree (EP-Tree) services, delivering EP-Tree services using Access EPL and EVPL are possible.

The SP and AP need to coordinate the value of the S-VLAN ID at the ENNI and other service attributes, but there is no need for coordination between the subscriber and SP on a detailed CE-VLAN ID/EVC map for each UNI because all service frames at the UNI are mapped to a single OVC EP.

An Access EVPL service uses a point-to-point OVC that associates a UNI OVC EP and an ENNI OVC EP. With Access EVPL, a UNI can support multiple service instances and mix of Access and EVC services, which is not possible with Access EPL. However, an Access EVPL (Figure 4.21) does not provide as much transparency of service frames as with an Access EPL. The OVC EP map determines which CE-VLANs are mapped to OVCs or dropped.

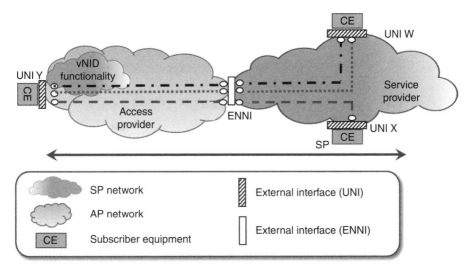

Figure 4.22. Network topology with AP offering vNID functionality.

4.2.12 E-Access Services with Virtual Network Interface Device

As in all MEF services, an Network Interface Device (NID) at customer location is needed to deliver E-Access services. The NID can be provided by the SP or by AP. Placing a NID at a customer premises and maintaining it are costly for a SP that has no presence at this location. Therefore, the concept of having a NID provided by AP that can allow the SP to exercise certain degree of control which is called virtual NID (vNID) is being developed by MEF.

A network topology with vNID is depicted in Figure 4.22.

The AP has equipment at customer premises to support UNI functions. In the AP network, a set of managed objects supporting the service are identified for the SP to interact with (e.g., "read" and "write") and provide the SP with the service management functionalities that an SP's NID would have provided.

Figure 4.11 shows the topology without a vNID where the SP deploys a device at the subscriber's location to the left of UNI Y. This device could be in addition to a local device that the AP has deployed to the right of UNI Y. Issues with this approach are described earlier.

4.2.13 OVC Services

In order for an SP to build an end-to-end EVC Service, OVC Services are being worked by MEF [11]. In Figure 4.23, a subscriber buys a multipoint Ethernet service, E-LAN service, to interconnect its four sites, which are located in four different cities. The SP is responsible for connecting the sites and providing SLAs. The subscriber does not need to know about the Operator networks and the OVCs and the SP does not need to own any of the networks.

The EVC provides connectivity to the four UNIs at the four sites. Let's assume that the SP subcontracts the building of the service to four operators and ENNIs are already deployed to interconnect the operators' networks (Figure 4.24).

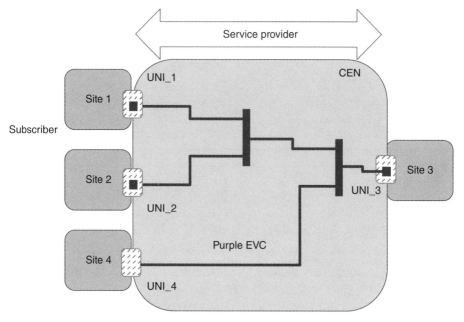

Figure 4.23. E-LAN service from a SP.

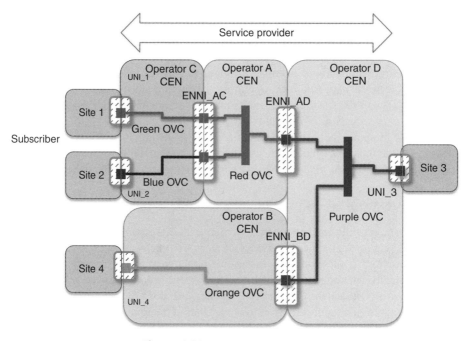

Figure 4.24. E-LAN service from a SP.

TABLE 4.13. Basic OVC Services

OVC Type	Basic OVC Service
Point-to-point	Service 1
Multipoint-to-multipoint	Service 2
Rooted multipoint	Service 3

For example, in Figure 4.24, the SP designs and builds the end-to-end E-LAN service from components provided by the four operators. Operator C is responsible for providing access service to Sites 1 and 2, and a point-to-point OVC from each UNI to ENNI_AC. Operator B is responsible for providing access service to Site 4, and a point-to-point OVC from the UNI to ENNI_BD. Operator D is responsible for providing access service to site 3, and a multipoint-to-multipoint OVC to interconnect UNI_3 with ENNI_AD and ENNI_BD. Operator A is responsible for the transit service, a multipoint-to-multipoint OVC interconnecting ENNI_AC with ENNI_AD.

Three basic OVC services are being studied as show in Table 4.13.

4.3 DPoE ARCHITECTURE AND SERVICES

A PON is characterized by having a shared optical fiber network that supports a single downstream transmitter and multiple simultaneous upstream transmitters in what is commonly referred to as a P2MP topology. The downstream transmitter, an optical line terminal (OLT), is typically located in the headend or hub of a SP's network. Optical network units (ONUs) located on the customer premise terminate the PON. A passive optical splitter/combiner enables the fiber to be divided to reach multiple customer locations. This network topology is shown in Figure 4.25.

The nature of the P2MP topology and a single downstream transmitter is such that all ONUs receive the same downstream transmissions. Through a combination of frame addressing and MAC layer assignment of identifiers to the ONU, each ONU only consumes frames that it is supposed to consume. All other frames are discarded by the ONU (see Figure 4.25a).

Multiple ONUs transmitting upstream on the same fiber requires a coordination between the ONUs such that their transmissions do not overlap when they reach the OLT. ONU scheduling, or dynamic bandwidth allocation, is a principle function of the OLT in a PON network. The OLT schedules each ONU for upstream transmission and guarantees the frames do not overlap (see Figure 4.25b).

Cable operators favor PON technologies for business services for several key reasons including the following:

- Reduced CAPEX and OPEX-Optical devices are expected to be at lower cost compared their electronic counterparts to perform similar functions. Furthermore, network consisting of passive components requires much less maintenance.
- Better multiplexing and increased bandwidth—Up to 128 users can be multiplexed on a single fiber at 1 and 10G rates.

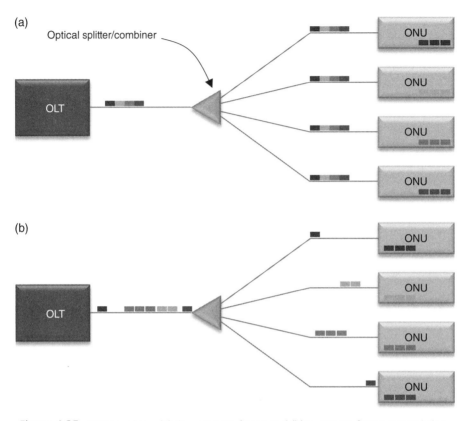

Figure 4.25. PON topology, (a) downstream frame and (b) upstream frame transmission.

- Bandwidth symmetry—The two PON standards can support 1 Gbps, 2.5 Gbps, and 10 Gbps, far more than can be provided by DOCSIS networks.
- QoS—Multiple priorities are supported.

There are two PON technologies used by cable operators: Gigabit PON (GPON) and Ethernet PON (EPON). GPON is defined in Ref. [12], while EPON is defined in Ref. [13]. Here we will not describe GPON.

4.3.1 DOCSIS Provisioning of EPON

Historically as cable operators deployed EPON to meet their service requirements, the EPON provisioning system was typically a manual process. Additionally, each EPON technology supplier had different methods and different interfaces for their respective product offering. As EPON deployments increased, and to avoid getting locked in to a particular EPON supplier, cable operators preferred to use a standardized provisioning interface that would scale well with increasing subscribers.

Cable operators already had a provisioning methodology that worked very well for them: it configured millions of devices every day, troubleshooting techniques were already proven, and operations staff knew how to use it. The provisioning method was generally referred to as DOCSIS provisioning.

The main objective of the DPoE specifications [14] is to allow for EPON access networks to be provisioned using the same provisioning servers and service concepts as those used for DOCSIS networks. In essence, the DPoE specifications cause an EPON network to mimic a DOCSIS network. Although the underlying EPON devices support a completely different access network technology, the network looks like a DOCSIS network to the back-office provisioning and network management systems. This coexistence of DOCSIS and EPON for access network solution is shown in Figure 4.26.

In a DPoE network the DPoE system is the heart of the network, much like the CMTS is the heart of the DOCSIS network. In addition to an EPON OLT, the DPoE system also contains routing functionality to support routing of IP packets, and to accommodate MPLS provider edge router functionality.

One challenge in meeting the main objective of the DPoE specifications is to make the ONU, which is inherently a L2 device, look like a CM, which is a layer-3 (L3) device. In other words, a device that is purposefully not IP addressable must act like a device which is IP addressable, because the back office servers expect to provision IP-addressable devices. The solution to this challenge was to create a vCM on the DPoE System that functions as an IP proxy for the D-ONU. When an ONU is powered on and registers with the DPoE System using the EPON registration process, a vCM is created to represent the ONU to the provisioning servers. The combination of ONU and vCM work together to provide the same functions and interfaces as a DOCSIS CM. The remaining components of the DPoE System mimic a CMTS, as shown in Figure 4.27.

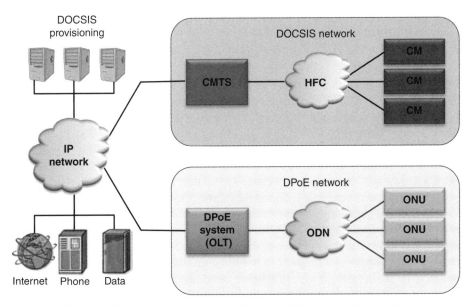

Figure 4.26. DOCSIS provisioning of both DOCSIS and DPoE networks.

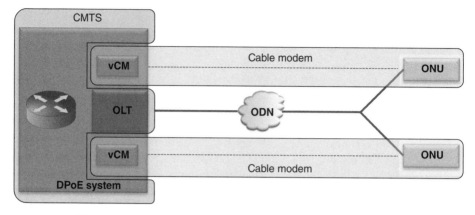

Figure 4.27. DOCSIS cable modem represented by vCM and ONU.

In addition to acting as an IP proxy on behalf of the ONU, another critical function of the vCM is to translate the service information in the configuration file to messages the ONU understands. These messages, called operations, administration, and maintenance (OAM) messages, are defined in IEEE 802.3 and extended in the DPoE specifications using the standardized method for extending IEEE 802.3 OAM. The set of messages are typically referred to as extended OAM (eOAM).

4.3.1.1 Device Provisioning Concepts

The DOCSIS provisioning process follows a consistent set of stages to initialize devices from power-up to service-ready. The four stages include (i) topology resolution and physical layer initialization, (ii) authentication and encryption initialization, (iii) IP initialization, and (iv) registration. These four stages, and the relative similarities between DOCSIS and DPoE specifications, are shown in Figure 4.28.

Topology resolution and physical layer initialization is substantially different in DPoE networks compared to DOCSIS networks. IEEE 802.3 defines a single downstream optical wavelength, making topology resolution relatively simple. Locating the downstream wavelength does not require frequency hunting—at power-up the ONU automatically tunes its optical receiver to the appropriate downstream wavelength. The first stage in DPoE specifications essentially amounts to EPON registration, which is outlined in detail in the IEEE 802.3 standard. Using the Multi-Point Control Protocol (MPCP), messages are exchanged between the DPoE System and ONU to establish a point-to-point (emulated) connection in the P2MP PON.

The second stage involves the setup of encryption based on the AES-128 cipher, followed by validation that the ONU can be trusted. Downstream encryption is mandatory, while encryption of the upstream channel is optional. Encryption involves the exchange of AES encryption keys that are used to encrypt the data. Authentication requires the ONU to send its X.509 digital certificate (including its RSA public key), along with the certificate of the issuing certification authority, to the DPoE system for validation.

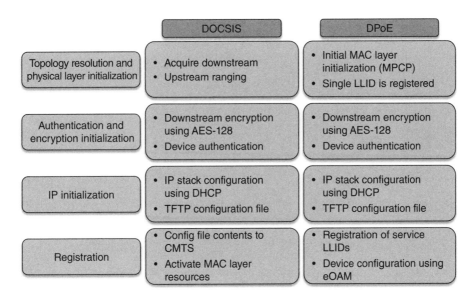

Figure 4.28. Four-step device provisioning process for DOCSIS and DPoE.

IP initialization occurs in the third stage. This involves the vCM acquiring an IP address and other relevant configuration information using the Dynamic Host Configuration Protocol (DHCP). With IP addressability, the vCM can proxy management messages on behalf of the ONU, essentially allowing the ONU to be managed via IP methods.

Two key parameters received by the vCM during the DHCP processing are the name of a configuration file and the IP address of a server which supports the Trivial File Transfer Protocol (TFTP). With the configuration filename and TFTP server, the vCM initiates a TFTP file download session to obtain the CM configuration file. The CM configuration file is a cornerstone of the DOCSIS provisioning process, and the DPoE specifications leverage the concept. The configuration file is a binary-encoded file that contains configuration parameters in type–length–value (TLV) format that are used by the vCM to configure the ONU. Examples of configuration parameters include classifiers for frame classification, or maximum sustained bit rate for a service flow (SF), both concepts will be described later.

The final registration stage involves the parsing of the configuration file by the vCM. Most likely, the configuration file will contain service parameters that the vCM must configure on the ONU. The configuration of the ONU requires the use of eOAM messages to properly set relevant service parameters.

Depending on the services to be supported by the ONU, the ONU may be required to perform EPON registration for additional MAC entities. Each MAC entity is represented by a MAC-level parameter known as a logical link identifier (LLID). Thus, if the ONU will support multiple services, the ONU will minimally register enough LLIDs to equal the number of services.

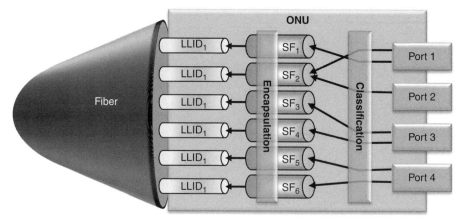

Figure 4.29. Service provisioning concepts as applied to an ONU.

4.3.1.2 Service Provisioning Concepts

DOCSIS specifications define several well-understood concepts that are consistently used to provision services on CMs. DPoE specifications leverage these concepts, and then define additional concepts that go beyond DOCSIS specifications. Figure 4.29 shows a high-level functional block diagram of an ONU to help explain these provisioning concepts and additional DPoE concepts. All of these concepts are provisioned via the CM configuration file.

In DOCSIS and DPoE environments, a SF is a logical provisioning object used to represent a unidirectional flow of Ethernet frames. Typically, an upstream SF is associated with the ONU and a downstream SF is associated with the DPoE System. Conceptually, an SF resembles a "bandwidth pipe" through which packets matching a specific classification criteria flow. SFs are useful provisioning objects because the QoS that should be provided to all packets that are part of the SF can be defined by providing QoS parameters for the SF in the configuration file.

Determining which SF an ingress frame belongs to relies on packet classification. Packet classification is the process of identifying packets based on configured matching criteria. The matching criteria are applied to packets ingressing a port on the ONU, and the set of packets that match the criteria necessarily belong to the same SF. Therefore, packets that belong to the same service are related and should be treated with the same QoS.

Packet classifiers are a well-known concept within DOCSIS provisioning. Typical matching criteria include L3 destination or source IP address, L4 destination or source port numbers, and many other L2-, L3-, or L4-header parameters. The DPoE specifications include the entire DOCSIS packet matching parameters, and then extend the set of packet matching parameters to include fields defined by the IEEE 802.1Q standard.

In the DOCSIS network, encapsulation (tagging) is the responsibility of the CMTS. If configured for tagging, the CMTS adds a tag to a frame prior to egressing the northbound interface. In DPoE networks, the encapsulation function is pushed to the edge and becomes a key capability of the ONU. For each SF provisioned in the configuration file, there may be one or more IEEE 802.1Q virtual local area network (VLAN) tags that

(CNUs), receive all data transmitted downstream, but only allow passing those frames that contain the properly assigned LLID. See Ref. [14] for more information on the use of LLID. For upstream transmission, similar to DOCSIS and EPON, the connected devices must be scheduled for upstream transmission to avoid collisions on the network. For EPoC, the scheduling methods are identical to those provided by MPCP in the EPON standard [14].

The CableLabs EPoC system specifications address the most desired network architecture in which the access network converts from a fiber network to a coaxial network at some point in the access network. This architecture, shown in Figure 4.31, uses a DPoE System (with OLT) in the headend, and one or more types of a FCU deployed in a node in the access network. Two types of FCUs are specified in the EPoC specifications: Repeater FCU (R-FCU) and Bridge FCU (B-FCU). While the physical topology of the network is identical for both an R-FCU and B-FCU, there are fundamental differences between the resulting networks.

The R-FCU is expected to perform physical layer conversion, which involves converting optical signaling from the fiber network to radio frequency (RF) signaling on the coaxial network, and vice versa. This resembles an optical-electrical or electrical-optical conversion found in other types of devices. The RF signaling uses orthogonal frequency division multiplexing (OFDM) to gain the most modulation efficiency over the coaxial network. The unique characteristic of the R-FCU is the resulting single MAC domain consisting of ONUs and CNUs, with all connected devices sharing the

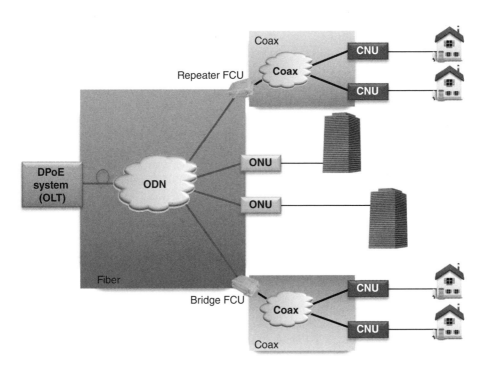

Figure 4.31. DPoE network architecture with EPoC fiber coax units.

same upstream bandwidth and scheduled by the same centrally located DPoE system (OLT). With a single scheduler controlling upstream access for ONUs and CNUs, the unified scheduling allows for common and consistent treatment across all services, regardless of the physical media used.

The B-FCU provides layer two bridging between the optical network and coaxial network by terminating the respective fiber and coaxial MAC domains. With an independent scheduler as an integral component of a B-FCU, the connected CNUs are scheduled independently from devices connected to the fiber network. The B-FCU is thus able to aggregate traffic on the coaxial network before bursting it upstream over the fiber network. The ramifications of separate MAC domains and the independent scheduling are an increase in frame delay and frame delay variation, but the result is more efficient use of fiber capacity.

The services supported over an EPoC network are identical to those supported over an EPON network. Specifically, IP (HSD) service will remain a fundamental service offering for both residential and commercial customers. Metro Ethernet services will also be supported depending on the service-level requirements of the customer. If service-level constraints are more stringent than the EPoC network can support, a fiber drop might be needed to accommodate the service requirements.

4.4.2 Migration Strategies

A cable operator's decision to deploy EPoC technology hinges on their strategy around FTTH, both in terms of timeline for FTTH deployment, as well as which PON technology will be chosen for FTTH services. The fundamental benefit of EPoC technology [19], at least as it relates to cable operator FTTH strategy, is the ability to invest in EPON headend equipment like DPoE Systems today to provide services to both business customers and residential customers, and delay the high cost of installing fiber through established neighborhoods and homes until the business case or competitive pressures make it inevitable.

The migration benefits provided by EPoC technology are easily understood with the help of Figure 4.32. Business customers, with their more stringent service requirements and higher returns on investment, typically provide strong motivation for cable operators to install fiber to the business. Installing DPoE equipment in the headend and on the customer premise allows an operator to provide high bandwidth, symmetric services to businesses. At the same time, installation of an R-FCU, for example, and replacing a CM in the home with a CNU, allows the cable operator to provide high-speed data services to both businesses and residential customers using the same headend equipment. This scenario is shown in Figure 4.32b.

At some point in the future, perhaps when competitive pressures dictate 1 Gbps residential service offerings, cable operators can install fiber media to the home and use the same headend equipment that was previously used to provide service over coaxial cable. Replacing the CNU with an ONU, and using the identical provisioning as was used for the CNU, the new FTTH customer equipment is provisioned with the same service offering. Figure 4.32b demonstrates this transition from coaxial customer to fiber customer.

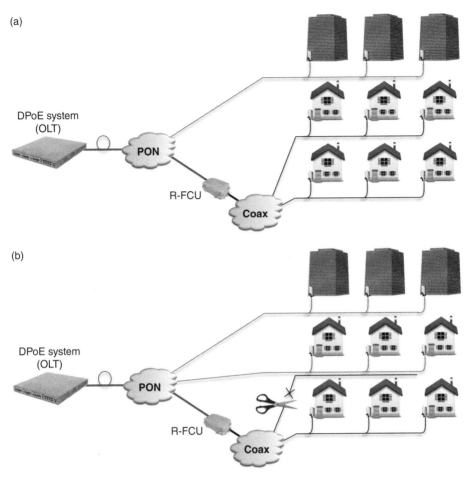

Figure 4.32. Migration to FTTH with EPoC technology.

4.5 BUSINESS VOICE

Building on their success in the residential voice market, MSOs turned their attention to the commercial sector, focusing on the small- and medium-sized business (SMB) segment of the market. This SMB segment was appealing for number of reasons; the installed cable plant covered a large portion of businesses within the MSOs footprint, and the PacketCable-based residential voice service-delivery platforms supported many of the functions required to provide business voice services.

4.5.1 The Business Communications Service Portfolio

A more accurate name for this service category would be "Business Real-Time Communications Services," since it includes both legacy business voice-calling services such as DID/DOD, hunt groups, paging, and so on plus advanced communications services, such as the following:

- **Media:** Hi-definition voice, video;
- **Collaboration Services:** video conferencing, desktop-sharing;
- **Mobility:** fixed/mobile convergence, remote office;
- **Unified Communications:** the integration of real-time media with other applications such as email, calendar, presence, and messaging.

There are two different business communications service models: Hosted IP-Centrex service, and SIP Trunking service. In the case of Hosted IP-Centrex, the SP owns and manages the service and the service delivery equipment; the business customer has no equipment to buy and simply pays a monthly subscription fee for the service. With SIP Trunking service, the business customer owns and operates the service delivery equipment (the SIP PBX and the user endpoint devices), while the SP simply provides the SIP Trunk that connects the SIP-PBX to the global telecom network.

4.5.2 Hosted IP-Centrex Service

Figure 4.33 shows the high-level architecture for Hosted IP-Centrex service. All call features are provided by (hosted by) the MSO network, and delivered over a coax or fiber access network to MSO-provided SIP endpoints located at the customer premises. The MSO owns the public user identities assigned to the business users. Each public user identity is assigned a unique database entry in the HSS that specifies the originating and terminating services assigned to that user. Each SIP endpoint in the enterprise network registers separately with the MSO network (i.e., each SIP endpoint appears as a UE to the IMS-based PacketCable 2.0 core). As originating and terminating calls are received for a user, the MSO network invokes the Business Voice Application Server to apply the services assigned to that user.

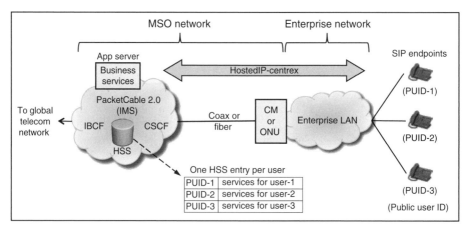

Figure 4.33. Hosted IP-Centrex architecture.

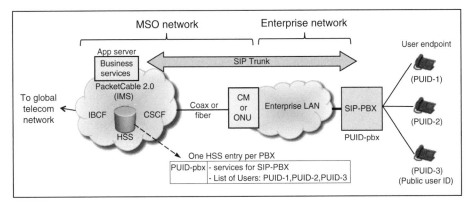

Figure 4.34. SIP Trunking architecture.

4.5.3 SIP Trunking Service

Figure 4.34 shows the high-level architecture for SIP Trunking service. In this case, all user communications services are provided by a customer-owned SIP-PBX to customer-owned endpoint devices (typically, SIP phones, but could be any type of endpoint—analog phone, H.323 phone, soft-phone, etc).

From the perspective of the global telecom network, the MSO network owns the public user identities assigned to the SIP-PBX users; that is, the global network routes calls addressed to these identities to the MSO network. However, the MSO delegates ownership of these identities to the SIP-PBX, assuming that the SIP-PBX will apply the originating and terminating services assigned to each user. The MSO network plays the role of a transit network when routing calls to/from the SIP-PBX; that is, it routes the calls without applying any per-user services. The MSO network does not act like a pure transit network however, since it may apply some minimal services across all calls to/from the enterprise; for example, originating or terminating call blocking, or time-of-day routing (say, when after-hour calls to a branch-office SIP-PBX are forwarded to the main office). To support this, the MSO network assigns a single public user identity to the SIP-PBX itself, and maintains a single HSS entry for this pubic user identity that specifies the subscription services to be applied to all calls to/from the SIP-PBX. The MSO network does not maintain HSS entries for the public user identities assigned to the individual SIP-PBX users.

4.5.4 IMS Enhancements to Support Business Voice Services

3GPP coined the term "Next Generation Corporate Network" (NGCN) to refer an enterprise network that contains a customer-owned SIP-PBX serving the enterprise users. SIP Trunking procedures to support NGCN's were added to IMS as part of the IMS-Corp work item; mostly in Release 8, with smaller enhancements being added in IMS Releases 9 through 12.

One of the challenges in modeling a SIP-PBX in a SIP-based network like IMS is that the SIP-PBX has characteristics of both a peer network and a user endpoint. The

SIP-PBX looks like a peer network in that it is a self-contained entity that is responsible for providing services to its multiple enterprise users. On the other hand, the SIP-PBX is similar to an endpoint, since the SP network can provide some (albeit limited) services to the SIP-PBX. Also, within the SMB market segment there can be thousands or even tens of thousands of SIP-PBXs connected to a single MSO network. Therefore, the process of establishing a connection between the PBX and the network must scale to large numbers as it does for the many individual user endpoints served by the network.

To support this "peer vs. endpoint" duality, IMS defined two NGCN models: the peering model and the subscription model. In the peering model, the SIP-PBX appears as a peer network to the IMS network, connecting to the network at the IBCF. For the subscription model, the SIP-PBX appears as a UE to the IMS network. In this case, the SIP-PBX registers with the IMS core via the P-CSCF, and receives originating and terminating services via the same IMS mechanisms that provide services to individual hosted users.

Several enhancements to the IMS core procedures were required to support SIP Trunking, including the following:

- Subscription model
  - ◦ Enhancements to the SIP registration and out-of-dialog request routing procedures as described in Section 4.3.10;
  - ◦ Changes to enable the served user identity to be different than the calling user identity (e.g., an originating call obtains the services assigned to the SIP-PBX identity, but displays the identity of the originating enterprise user);
  - ◦ Changes to enable the SIP-PBX to have more control over what calling user identity is delivered to the called user.
- Peering model
  - ◦ Enhancements to enable IMS services to be applied to the calls to/from the "peering" SIP-PBX connected via the IBCF (i.e., the IMS is acting as a transit network that can apply originating and terminating services to the peer network).

4.5.5 3GPP Defines SIP-PBX Registration/Routing Procedures for IMS

As part of the IMS-Corp work item, 3GPP updated the IMS SIP registration procedures specified in Ref. [20] to support SIP-PBXs, as shown in Figure 4.35. In this example, the SIP-PBX is serving three enterprise users who are assigned the following E.164 numbers: +13035551001, +13035551002, and +13035551003.

The SIP-PBX and the SP Network in Figure 4.35 are each pre-provisioned with the three public user identities assigned to the enterprise. One of these public user identities (sip:+1-3035551000@sp.com in our example) is designated as the primary identity that represents the SIP-PBX itself. The primary identity is marked in HSS with a special "loose routing" attribute to indicate to the SP network that it is a SIP-PBX, and therefore requires special processing. The HSS entry for the primary identity also contains the identities of the other enterprise users. The SIP-PBX registration procedure (messages

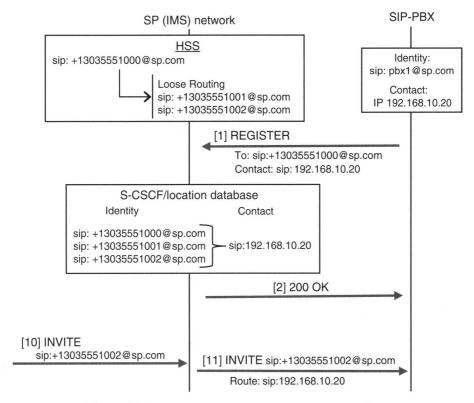

Figure 4.35. 3GPP IMS "loose route" registration procedure.

[20] and [21] of Figure 4.35) follows the IMS implicit registration procedures where a single registration transaction registers multiple public user identities. In our example, the SP network registrar function maps the three public user identities assigned to the enterprise to the single contact address of the SIP-PBX.

The normal practice for routing inbound out-of-dialog requests in the home network (based on RFC 3261 procedures defined in Ref. [21]) is to retarget the received Request-URI with the registered contact address of the target user. This is a problem for the SIP-PBX case, since multiple identities are sharing a single contact, and therefore retargeting wipes out information that would enable the SIP-PBX to identify the target user.

3GPP IMS solves this problem as shown in Figure 4.9. When the SP network receives the out-of-dialog request [21] for a public user identity that is marked for "loose routing," it does not overwrite the received Request-URI with the registered contact address. Instead, it applies SIP loose-routing procedures, where the ultimate destination identity is left intact in the Request-URI, and the SIP-PBX contact address is placed in a Route header. This preserves the identity of the target user in the Request-URI, so that when it receives [22] INVITE, the SIP-PBX can identify the target user.

The SIPconnect1.1 effort described in Section 4.5.6 resulted in the standardization of yet another SIP registration/routing procedure, defined in RFC 6140 [22]. CableLabs worked with other members of 3GPP to add support for RFC 6140 to IMS Release 12.

4.5.6 SIP Forum SIPconnect1.1

In a vacuum of standards, SIP-PBXs evolved many different profiles of SIP at the SIP Trunk interface to the SP network. This made it difficult for MSOs to scale their SIP-Trunk deployments to the large numbers that exist in the SMB market segment. Since the SIP-PBX was customer-owned, the MSO could not limit the set of SIP-PBXs connected to their network to a small well-behaved "certified" set of devices. Each new SIP-PBX make and model required extensive testing and updating before it could be turned up. One way to move SIP Trunks more toward a plug-and-play proposition was to document a standard SIP Trunk interface that all SIP-PBX vendors could adopt.

However, since SIP-PBX vendor participation and adoption of IMS is relatively low, this didn't move the ball very far down the field in terms of aligning SIP-PBXs to a common SIP Trunk interface. Therefore, the MSOs directed CableLabs to work with the SIP Forum in the development of a SIP Trunk interface profile, which eventually became the SIP Forum SIPconnect1.1 Recommendation [23]. The SIP Forum had good representation across the SIP-PBX vendor community, and so it was hoped that a standard profile from that organization had a good chance of being adopted.

SIPconnect1.1 defined two SIP Trunk modes; the registration mode and the static mode. These modes have parallels to the NGCN models described earlier, where the registration mode is similar to the NGCN subscription model, and the static mode is similar to the NGCN peering model. In registration mode, the MSO network learns the location of the SIP-PBX dynamically at run time via a variation of the SIP registration procedure. In the static mode, the SIP-PBX routing information is statically provisioned in the MSO network (similar to the process of establishing routes to peer networks).

Ideally, the SIPconnect1.1 procedures would align with the IMS procedures, but during development of SIPconnect1.1, concerns were raised with the loose-routing procedures defined for the SIP-PBX in IMS. The primary concern was that the IMS procedures created ambiguity on domain ownership; that is, as shown in Figure 4.35, SIP URIs with a domain of "sp.com" are owned either by the SP Network or the SIP-PBX based on the contents of the user-part of the URI. This was deemed to break basic domain routing principles, and so the SIP Forum asked the IETF to define a standard extension to the SIP registration procedures to support SIP-PBX registrations. The result of this IETF work was the "GIN" registration procedure defined in RFC 6140. CableLabs worked with other 3GPP member companies to add support for RFC 6140 to IMS Release 10.

The "GIN" registration procedure is shown in Figure 4.36.

Prior to registration, the SP network and the SIP-PBX are separately and independently provisioned with the same list of public user identities. The SP network HSS contains a single subscription entry for the public user identity "sip:pbx1@sp.com" that identifies the SIP-PBX itself. This subscription entry contains the list of public user identities of the individual SIP-PBX users.

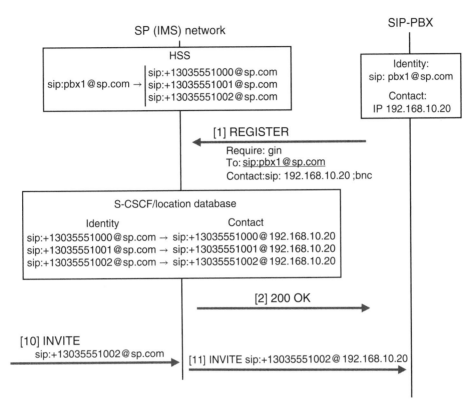

Figure 4.36. RFC 6140 "GIN" registration procedure.

The registration and request routing procedures in Figure 4.36 are as follows:

1. The SIP-PBX sends a REGISTER request containing its identity and contact address to the SP network. The REGISTER Require header field contains a new "gin" option tag indicating that this SIP-PBX requires support of the registration extension defined in RFC 6140. The Contact header field contains a new "bnc" URI parameter that indicates this is a bulk-number-contact address; that is, the special "gin" procedures defined in RFC 6140 should be applied to this contact address. On receiving this request, the registrar creates a unique identity-to-contact binding entry for each of the public user identities associated with the SIP-PBX identified in the REGISTER request. For each associated identity, the registrar builds a unique contact address that is created by adding the E.164 number of the associated identity to the user-part of the contact URI contained in the REGISTER request.

2. The SP network sends a 200 OK response to the SIP endpoint indicating that the REGISTER request completed successfully.

3. Later, the SP Network receives an INVITE request from the global network with a Request-URI containing one of the associated Public User Identities.

4. The SP Network finds the matching public user identity in the Location Database, and retargets the request by overwriting the received INVITE Request-URI with the registered contact of the target identity (i.e., by using the base RFC 3261 routing procedures). The Request-URI now contains sufficient information to identify both the target SIP-PBX and the target user; the host-part identifies the SIP-PBX and the user-part identifies the user served by that SIP-PBX.

4.5.7 PacketCable Enterprise SIP Gateway

The business voice marketplace introduces new challenges for MSOs that did not exist for residential deployments. Residential voice services are delivered to well-behaved SIP endpoints embedded in the CM, while business voice is delivered to stand-alone SIP endpoints located on a enterprise LAN behind a corporate NAT/Firewall. The simultaneous call load per CM increases substantially for business over residential. Corporate customers typically demand more stringent SLAs than residential customers.

To help mitigate these issues, the MSOs directed CableLabs to develop specifications for a device that sits at the demarcation point between the enterprise and SP networks. The result of this effort was the PacketCable Enterprise SIP Gateway Specification PKT-SP-ESG-I01-101103 [24].

Figure 4.37 shows the ESG deployed for hosted IP-Centrex service example. In this (typical) case the SP deploys two CMs; CM1 carries voice and provisioning traffic for the hosted SIP phones, while CM2 carries the remaining high-speed data traffic. The ESG and the Enterprise NAT/Firewall both sit on the demarcation point between the SP and enterprise network, providing NAT-traversal and access-control functions for all traffic between the SP and enterprise networks. The enterprise owns the enterprise NAT/firewall, and therefore controls the firewall rule-set for high-speed data to/from the SP network. The SP owns the ESG, and therefore controls the network address translator (NAT) traversal and access control rules for the traffic that supports the hosted voice service; namely the VoIP protocols (SIP, SDP, RTP, and RTCP) via the SIP ALG, and for the provisioning protocols (e.g., TR-069 CWMP) via the Provisioning Gateway.

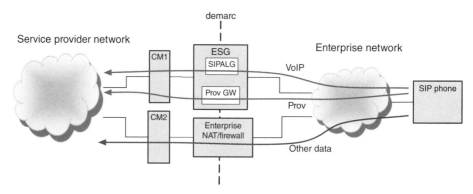

Figure 4.37. Typical hosted IP-Centrex deployment.

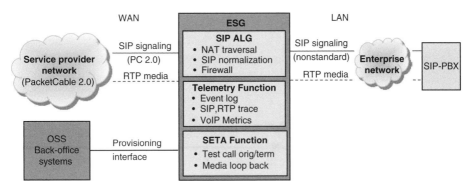

Figure 4.38. ESG functions.

Figure 4.38 shows a more detailed view of the internal functions supported by the ESG (in this example for SIP Trunking service to a SIP-PBX). The SIP ALG performs the functions of a SIP-aware NAT, translating IP addresses carried in the IP headers and in the SIP message payload between the enterprise private LAN addresses and the SP public WAN addresses. The SIP ALG performs SIP signal interworking to normalize the nonstandard SIP profile supported by the SIP-PBX on the LAN interface to SIP signaling compliant with PacketCable 2.0 on the WAN interface toward the SP network. Finally, the SIP-ALG supports a SIP-aware firewall that enforces operator-configured policies for SIP signaling, and allows RTP media to traverse between the enterprise and SP network only for valid sessions established via SIP.

The Telemetry Function collects and reports data associated with business voice traffic, such as VoIP Metrics, error event logs, and SIP and RTP traces. The SIP Endpoint Test Agent (SETA) can initiate and accept test calls under management control in order to verify the health of the ESG and its connectivity to the SP network. The operator can use the management interface to initiate test calls on demand, or at a programmed periodic interval. The SETA can also accept test calls, including RTP loopback calls (RTP packet reflector only). The Telemetry and SETA functions together enable the SP to detect problems early and resolve them quickly.

When deployed for hosted IP-Centrex service, the ESG can serve as a provisioning gateway at the demarcation point for provisioning traffic exchanged between the SP's provisioning servers and the enterprise SIP phones. Support of this function in the ESG provides operator-controlled access to the provisioning interface on the SIP endpoints and resolves the problem where the enterprise NAT/firewall is configured to block this provisioning traffic.

Depending on the deployment scenario, the provisioning gateway acts either as a network address translator (NAT), or a provisioning-aware application-level gateway (ALG). Two types of NATs are described: the traditional NAT and the twice NAT. Likewise, two types of provisioning ALGs are described; the HTTP ALG and the CWMP ALG.

Figure 4.39 shows a "class hierarchy" view of the different types of provisioning gateways supported by the ESG.

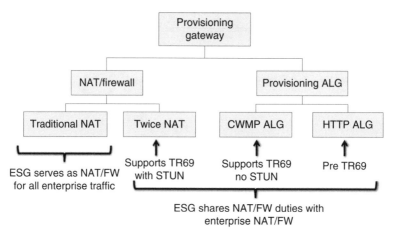

Figure 4.39. Provisioning gateway hierarchy diagram.

An MSO would choose to deploy the type of provisioning gateway that best suits their specific Hosted IP-Centrex service deployment scenario.

4.5.8 Support of ESG in Broadband Forum TR-069

Per direction from the MSOs, CableLabs worked with the Broadband Forum to align the ESG provisioning procedures with TR-069 "CPE WAN Management Protocol (CWMP)" [25]. This included contributing the ESG data model into the TR-104 Version-2 "Provisioning Parameters for VoIP CPE" [26]. TR-104 defines multiple data model objects that can be combined in different combinations to support a wide variety of CPE devices.

4.6 CONCLUSION

In this chapter, we have described MSO network architectures and services for commercial services.

Ethernet is becoming a dominant technology for all Information Communication Networks (ICNs) including enterprise networks. Both MSOs and telecom carriers offer E-Line and E-LAN services as well as E-Access services. Mobile backhaul is the dominant application for Metro Ethernet services. The growth of their customer base is phenomenal. We expect voice and video applications over Metro Ethernet services to grow rapidly as Metro Ethernet equipment prices go down.

Service OAM capabilities of Metro Ethernet are being used extensively. Automation of Metro Ethernet equipment and service provisioning is expected to be available soon.

In order to better multiplex low rates and leverage existing provisioning concepts and processes to provision and manage both residential and business services, DPoE was introduced by MSOs.

With EPoC technology, MSOs can commit to an FTTH technology that also provides services to business and residential customers at the same time. Investment in the headend equipment can be made based on the return on investment for business services, but the equipment can be leveraged for residential services, or businesses connected to the coaxial cable network.

For SMB market, MSOs offer SIP Trunking service to customer-owned SIP-PBXs, and Hosted IP-Centrex service that delivers traditional business voice features plus advanced real-time communications services such as video, fixed/mobile convergence, and unified communications.

The SMB market requires delivering service to stand-alone SIP endpoint devices connected to a complex customer LAN behind a customer-owned NAT/Firewall. MSOs have met these challenges by working with industry organizations to advance interface standards for the SIP Trunking interface, and by developing technology such as the Enterprise SIP Gateway to enable reliable delivery of real-time communications services in the more complex business networking environment to a more diverse set of endpoint devices.

REFERENCES

1. M. Toy, "Networks and Services: Carrier Ethernet, PBT, MPLS-TP and VPLS," J. Wiley, October 2012.
2. MEF 4: Metro Ethernet Network Architecture Framework—Part 1: Generic Framework, May 2004.
3. IEEE 802.1ad Local and Metropolitan Area Networks Virtual Bridged Local Area Networks, 2005.
4. MEF 11: User Network Interface (UNI) Requirements and Framework, November 2004.
5. MEF 28: External Network Network Interface (ENNI) Support for UNI Tunnel Access and Virtual UNI, October 2010.
6. MEF 43: Virtual NID (vNID) Functionality for E-Access Services, April 2014.
7. MEF 23.1: Carrier Ethernet Class of Service—Phase 2, January 2012.
8. RFC 2697, A Single Rate Three Color Marker, September, 1999.
9. RFC 2698, A Two Rate Three Color Marker, September, 1999.
10. MEF 33: Ethernet Access Services Definition, January 2012.
11. MEF Draft-OVC Services Definitions, May 2013.
12. ITU-T Recommendation G.984 series and their amendments: Gigabit-capable Passive Optical Networks.
13. IEEE 802.3™, IEEE Standard for Ethernet, December 28, 2012.
14. DOCSIS Provisioning of EPON specifications, CableLabs. www.cablelabs.com.
15. DOCSIS Provisioning of EPON, OAM Extensions Specification, DPoE-SP-OAMv1.0, Cable Television Laboratories, Inc.
16. IEEE Std. 1588™-2008, IEEE Standard for a Precision Clock Synchronization Protocol for Networked Measurement and Control Systems.
17. EPON Protocol Over Coax System Specifications, CableLabs, www.cablelabs.com.

18. MEF 6: 1-Ethernet Services Definitions—Phase 2, April 2008.

19. IEEE P802.3bn, Draft Standard for Ethernet Amendment X: Physical Layer Specifications and Management Parameters for Ethernet Passive Optical Networks Protocol over Coax.

20. 3GPP TS 24.229: "IP multimedia call control protocol based on Session Initiation Protocol (SIP) and Session Description Protocol (SDP); Stage 3."

21. IETF RFC 3261, SIP: Session Initiation Protocol, June 2002.

22. SIP Forum, SIPconnect1.1 Technical Recommendation, 2011.

23. IETF RFC 6140, Registration for Multiple Phone Numbers in the Session Initiation Protocol (SIP), March 2011.

24. CableLabs, PacketCable Enterprise SIP Gateway Specification, PKT-SP-ESG-I01-101103, November 2010.

25. Broadband Forum, TR-069 CPE WAN Management Protocol (CWMP), TR-069 Ammendment-5, November 2015.

26. Broadband Forum, TR-104 Release 2.0 Provisioning Parameters for VoIP CPE, April 2014.

OPERATIONS, ADMINISTRATION, MAINTENANCE, PROVISIONING, AND TROUBLESHOOTING FOR BUSINESS SERVICES

Victor Blake, Brian Hedstrom,
Sergio Gambro, Stuart Hoggan and Mehmet Toy

5.1 INTRODUCTION

In Chapter 3, we described operations, administration, maintenance, provisioning, and troubleshooting (OAMPT) for residential services. OAMPT is also crucial for business services. In fact, the OAMPT requirements for business services are tighter than those for residential services. This is due to the fact that degraded or lost services can have larger financial impact.

In order to process service orders, keep inventory, configure equipment, provision services, collect measurement and accounting data, secure network access, report service degradation and failures, troubleshoot the failures and fix them, and bill customers, multiple system operators (MSOs) have operations systems dedicated for business services.

Equipment manufacturers usually have their own element management system (EMS) to manage a network element (NE) such as a router and their own network management system (NMS) to manage a subnetwork of NEs. It is possible to combine these management systems with operators' operation systems to perform the OAMPT functions. In some cases, operators may choose to use only their operations systems to manage their network consisting of multiple vendor equipment.

Cable Networks, Services, and Management, First Edition. Edited by Mehmet Toy.
© 2015 The Institute of Electrical and Electronics Engineers, Inc. Published 2015 by John Wiley & Sons, Inc.

Every network requires computer systems to configure, operate, and maintain the network. While all networks require systems to configure and monitor equipment, typically NMS and EMS, service provider networks require additional systems to provision, test, control, monitor, and provide services such as security, accounting, and billing, among others. Collectively, these are operations systems or operations and management systems.

In the following sections, we will describe OAMPT including security for the business network and services described in Chapter 4.

5.2 OPERATIONS SYSTEMS AND MANAGEMENT ARCHITECTURES

5.2.1 Requirements for Business Services Operations

Service providers, as opposed to network operations in enterprises or other entities, have a need to tie together ordering, payment, and operation of systems. For example, a typical enterprise-run service once it is provisioned would typically operate it at its maximum capacity. By contrast, a service provider selling a similar or even the same service to an enterprise may sell a fixed bandwidth service—a burstable or a billable service that must measure and count traffic. The scale of service provider's networks also dictates more automation. Service providers must also build processes to accept changes from business customers such as a move in location, additions of new locations or services at an existing location, or other changes such as capacity upgrades or downgrades. These changes, so-called MAC (Move–Add–Change), must be handled in a timely and efficient manner without disrupting other services provided to the same customer or the services of other customers on the operator's network. These too must be automated to the greatest degree possible to reduce operating expenses and keep competitive prices for the business that needs the services.

Finally, unlike most enterprise networks, service providers must have means to control the traffic in their networks so as to avoid conflicts between the performance, quality, and reliability for both different services running on the same network and for different customers running on the same network. The various operational systems of service providers are used to provide automation, control, and oversight of the processes mentioned earlier.

We detail each of the typical subsystems and those subsystems functions for MSOs in the following text.

5.2.2 Operations Systems Architecture

Although the cable industry's history includes many formerly vertically integrated technologies such as traditional CATV broadcast, the industry now requires standards based that supports multivendor interoperability. With the combination of multiple vendor systems' constant push for new services, there is no Operations Support Systems (OSS) that does it all. Even within the cable operations, the "OSS" is really a collection of OSS used for various purposes. In addition to integrating vendor-specific NMS support, third-party OSS tools, and widely used NMS tools (i.e., event correlation and system management tools), most cable operators have custom built various portions of the OSS. These systems vary in functionality, scale, and complexity. Larger operators have custom-built distributed systems that provide massive scaling (i.e., tens of millions of managed customers and devices). This is especially the case for Data Over Cable Service Interface Specification (DOCSIS) that requires a very large-scale and tight-service integration with Business Support Systems (BSS) for service activation, security, accounting, billing, and most recently mobile services.

Custom-developed OSS may include scripts using languages such as PERL, TCL, and PYTHON. Compiled software may be written in C# or C++. Compiled OSS applications are typically only those that are very compute or resource-intensive. Supporting web systems may also use the above scripted languages along with Ruby, JavaScript, Java, and other web services applications. Finally, 3GPP IMS-based services include a much broader diversity of application infrastructure to support Value-added services (VAS) and unified communications services (UCS). These services make heavy use of existing Java Application Program Interface (API)'s for Integrated Networks (JAIN)[1], Parlay (and Parlay X) and the Open Mobile Alliance (OMA) [1] standards, and many other APIs. Some of those APIs in turn rely on other APIs or interfaces. For example, Parlay (now substantially replaced by OMA) relies on the Common Object Request Broker Architecture (CORBA)[2] defined by the Object Management Group (OMG) [2]. Recently, many of the organizations that develop these APIs have published or at least begun work on a new generation of REST or preferably RESTful APIs described in more detail in Section 5.4.

OSS systems typically include at least the required elements for DOCSIS such as Dynamic Host Configuration Protocol (DHCP), certificate servers, Domain Name System (DNS), Trivial File Transfer Protocol (TFTP) servers, Simple Network Management Protocol (SNMP) servers, and so on.

Although the telecom industry has developed comprehensive architectural models such as the Telecommunications Managed Network (TMN)[3, 4][3], these models do not reflect the infrastructure commonly used in telecom generally, but most specifically not the cable operator's OSS.

For example, although rarely used today, service delivery of pay cable channels was traditionally managed by the insertion and removal of filter taps that blocked or permitted specific channels. Besides this extreme example, more recent examples included Digital Rights Management (DRM) that goes from mobile devices to cloud-based infrastructure, bypassing the traditional distribution network control of content security.

---

[1] JAIN is a "open" standard that is not controlled by any one particular group or organization, but is managed by the Java Community Process (JCP), http://jcp.org.

[2] CORBA is a standard of the Object Management Group (OMG).

[3] The Telecommunications Management Network (TMN) "standards" are part of the International Telecommunications Union—Telecommunications (ITU-T), X.700 series of recommendations. The TMN model discussed here is from ITU-T M.3000 series (M.3010 Principles for a TMN).

5.2.3 Operations Subsystems

While each of these terms may have a very specific meaning within a given organization, the terms are unfortunately used more loosely across all of the telecom industry. There is no strict definition that makes it clear that specific elements are part of one subsystem or another. Protocols do not dictate what subsystem belongs to. For example, a server running SNMP may be used for configuring devices as part of the OSS, it may be used as part of an NMS, an EMS (which is usually vendor specific), or perhaps not even any of the those mentioned earlier and instead be used for accounting or other BSS functions. Let's examine each type of subsystem utilized by service providers.

Typical MSO systems include a wide variety of both multipurpose systems and some vertically integrated systems used only by specific services or even more narrowly only for specific NE:

- Element Management System (EMS);
- Network Management System (NMS);
- Intrusion Detection System (IDS);
- Operations Support System (OSS);
- Business Support System (BSS);
- Cable-specific systems (for DOCSIS, for example).

There may be significant overlap between these systems. For example, most DOCSIS-specific operations systems are considered to be part of the OSS. But not all OSS systems are DOCSIS related. Some vendor-specific EMS may be integrated with that vendor's NMS or even be a single EMS/NMS hybrid system. Some EMS or NMS may be considered a part of the OSS, especially if there are real-time functions in those systems. But in other cases, where for example the EMS is only used for troubleshooting, the EMS may not be considered part of the OSS, but is still part of the OAMPT systems.

Although some systems management standards or model have very formal and specific definitions for some of these terms, many of those definitions do not apply to either or both of legacy and contemporary MSO systems. For example, the typically custom order entry, management and BSS for DOCSIS do not fit neatly into classic or contemporary telecom models. These systems may operate executive dashboards (BSS), report generation, and order management—in a single system.

5.2.3.1 Element Management Systems

EMSs are typically systems utilized to support either a single-vendor network element (NE) such as a certain vendor's switch (or router) or sometimes various NEs, for example switches and routers, from a single vendor. Usually, the term "EMS" refers to a dedicated system or less likely software residing on a server for the express purpose of managing a single vendor's type of device or a class of devices. EMS is likely to be used for low-level diagnostic functions, initial configuration of the device—before provisioning or service activation—and other operator or engineering-level work. An EMS may, for example, run on a laptop connected directly to an NE or across a network. Some EMSs operate with textual interfaces, but most now use graphical user interfaces

(GUIs). In recent years, most EMSs have become web-based systems allowing for the installation of EMS application on shared web servers or at least on virtual machines (VMs) rather than dedicated hardware. Some EMSs are still vertically packaged as an appliance and may even be integral to a rack or other subsystem included with other NEs (i.e., switches, routers and optical equipment).

Some NEs do not support industry standard protocols for operations support, and must therefore use an EMS to translate proprietary commands to industry standard interfaces (i.e., syslog, SNMP, XML, and SOAP)

In the cable industry, EMS is used especially for diagnostics and for new technologies that may not be well supported by existing NMS and tools in the network. In these cases, the EMS and NMS systems may often by combined by the vendor and packaged in a way that forces operators to utilize the EMS.

Outside of testing and diagnostics, EMS is otherwise looked down upon by operators because, by definition, they offer functionality only for a specific vendor and often only for a specific subset of products, or in the worst case—just for one specific NE. This is unattractive to operators because the overhead of purchasing, installing, training for, operating, and maintaining EMS just for a single type of element (when a cable operator may have hundreds or even thousands of different types of NEs in their network) is inefficient. Organizations such as the ITU-T, Metro Ethernet Forum (MEF), and CableLabs have worked diligently to introduce industry standards for the initial configuration and low-level diagnostics and testing for devices in order to allow standards-based systems to replace individual vendor EMS. Exemplary protocols include CableLabs DOCSIS (see Section 2.2) and DPoE (see Section 4.3), IEEE 802.1ab [5], other IEEE 802.1 standards and many portions of IEEE 802.3 [6], ITU-T Y.1731 [7], and numerous MEF [8] specifications. Notably, many of those organizations (CableLabs, IEEE, IEEE-SA, and MEF) both work collaboratively with each other through industry liaisons and share many of the same participating member companies and participants. Chapter 4 gives a good overview of how, for example, the CableLabs DPoE specifications incorporate IEEE and MEF requirements. At the same time, IEEE SA 1904.1 SIEPON [8] includes the requirements from the cable industry identified by CableLabs and cable operators.

5.2.3.2 Network Management Systems

Note that sometimes (although rarely) the acronym NMS is used to refer to a Network Management Station which is a single station (computer or terminal) that is part of an NMS. A second even less used meaning for the acronym is Network Monitoring System referring to NMS or an independent system that is used to monitor network activity.

An NMS differs from an EMS because an NMS supports the management of multiple devices from either one vendor or from many vendors. It may manage more than one technology or aspect of the network or just one technology across multiple locations. For example, a commonly used "Network Management System" is HP OpenView[4]. In this example, the NMS is a system that allows for the addition of applications or tools, each with

[4] "HP OpenView" is Hewlett-Packard OpenView, a software product developed by Hewlett-Packard. The product continues to be sold and supported by Hewlett-Packard Development Company, L.P. which today includes the OpenView functionality as part of a product set called "HP OpenCall." Wikipedia.com (2014), "HP OpenView" includes a description of the history of the software, Wikimedia Foundations, Inc.

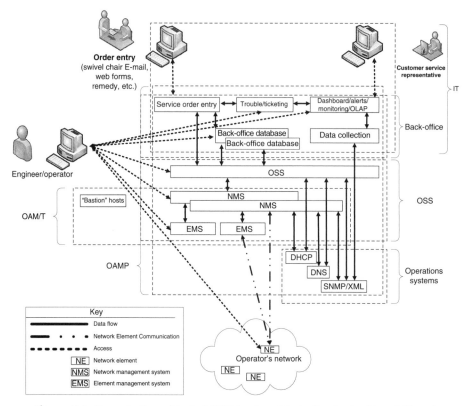

Figure 5.1. Organization of typical MSO "systems" including back-office, OSS, and Operations systems.

its own functionality. For example, OpenView supports SNMP polling and SNMP traps for configuration management. Since OpenView is a platform that acts like a "desktop" it can be used to invoke any other third-party applications providing additional functionality.

NMS may be a single-vendor solution or may be a multivendor solution. NMS may communicate directly with elements, through a set of distribution and collection of systems (or servers), or through EMS. All three variations are shown in Figure 5.1.

5.2.3.3 Intrusion Detection Systems

Intrusion detection systems (IDS) can be vertically integrated with a particular vendor's IDS such as an IDS appliance, may be a part of another NE such as IDS functionality for a router or switch with monitoring capabilities such as Deep Packet Inspection or sFlow, which is an industry standard technology for monitoring high-speed switched networks, or can be an off-the-shelf—but more likely—custom-integrated solution that aggregates monitoring data from multiple sources (and optionally correlates those sources) to "detect" intrusions or security breaches in a network or on networked systems such as computers, servers, and so on. Although IDS support industry standard reporting mechanisms such as SNMP traps and syslog event generation, many IDS functions are complex and unique and can only be operated by a vendor specific CLI, web interface, or EMS using a web interface or an EMS client application.

5.2.3.4 Operations Support Systems

Operations Support Systems or OSS is a broad term that usually refers to all of the systems required to operate a network for a service provider, excluding only BSS.

OSS means different things to different operators and most certainly to different vendors. Nearly all vendors of NEs offer (and many require) an EMS for their product. Many vendors offer (or require) an NMS. Very few vendors offer an "OSS." Of those that do, the term, when used by network equipment vendors, is typically misappropriated because vendor "OSS" rarely supports other vendors' products, therefore, cannot really be an OSS. Some "third"-party (i.e., neither operator nor equipment vendor) vendors offer purpose built OSS. These systems usually offer NMS and operational features such as end-to-end or "A to Z" provisioning, change control, monitoring, and other functions.

A full-featured OSS would be one that includes EMS, NMS, and other OSS functions such as service management (i.e., provisioning, configuration management, service inventory, turn-up, and testing); service assurance (monitoring, data collection, error reporting, etc.); and operations management (trouble ticketing, customer support, service escalation, etc). While there are a few vendors that offer a subset of these functions, there is no one product or vendor that offers all of the required capabilities of an OSS for any service provider. This is especially true for cable operators who have a diverse array of business and residential services including internet access, voice services, wireless services, video services, and so on.

OSS often include web systems for web services-aka cloud like-operations as well as interfaces with web services for administrative (i.e., NOC, call centers and customer service, self-service, and other applications) and BSS functions.

5.2.3.5 Business Support Systems

BSS, as the name suggests, are systems that tie together the business operations with the operations systems, network, and network systems. Exemplary BSS include functional systems such as the following:

- Accounting;
- Accounting;
- Billing;
- Process monitoring (process dashboards, executive dashboards, exception statistical data collection, etc.);
- Law Enforcement support (where applicable);
- Service assurance (SLA monitoring for contract based services);
- Billing (for consumption and service availability);
- DRM (where applicable);
- Order Entry and some aspects of service management (such as service creation).

BSS typically also includes administrative subsystems that include data entry, monitoring, and other user interfaces for the functionalities given earlier:

- Customer Service Representative (CSR) interfaces;
- Accounting and billing that may include APIs for third-party billing agencies and collections;

- Self-service;
- Reporting (dashboard, monitoring, error reporting, etc.).

Service creation is not a discrete function. Although it may operate as a unified service—a single administrative interface to create new products—service creation usually creates databases, templates, options, and other object and data structures required for use with other functions such as order entry, provisioning, billing, and so on.

Although BSS may use standards-based APIs (or more likely a combination of both standards based and custom APIs), the systems themselves are almost always custom built among the largest operators. Smaller MSOs are more likely to use a combination of turn-key systems and services.

Historically, the benefits of outsourcing billing have centered on the cost efficiencies of mailing and collections. But those large-scale benefits are diminished with increasingly capable self-service options and electronic payments. As MSOs move both up market and down market, respectively, increasing the complexity of services on the higher end, and the number of customers with simpler flat rate services on the bottom end push for automation and self-service has become stronger.

Still, today, BSS for business services lag behind the mature and highly scaled systems used for tens of millions of MSO residential customer.

5.2.3.6 Cable-Specific Systems

In addition to the earlier broadly used telecom terms, the cable industry has some of its own systems that are somewhat more vertically industry. Specifically, the cable industry widely uses the DOCSIS guidelines that include OAMPT specific protocol specifications from which both operators and vendors have built OSS functionality. DOCSIS systems are used to deliver cable business services including IP (VPNs and Internet access), Ethernet, and telephony (Voice over Internet Protocol or VoIP) services. Historically, DOCSIS was a vertically integrated set of protocols specific to the cable hybrid fiber coax plant. More recently, we developed a clever virtualized model called a virtual cable modem (vCM) that allows for the continued use of the DOCSIS back-office specifically including support for DOCSIS Operations and Support System Interface (OSSI) [9]. This allows cable operators, with minor extensions and tools, to use their existing OSS systems to provision IP, Ethernet, and VoIP services over EPON or more broadly any Ethernet technology.

5.2.4 Networks

One of the challenges of business services is the diversity of technologies required to delivery every possible service to customers. Business services encompass all of the same services as residential services plus additional voice, wireless, video, Ethernet, hosting, and other data services. The networks connect customers to the service providers' networks. Access networks use a wide range of technologies including the following:

- DOCSIS;
- Wireless (Wi-Fi and some point-to-point or point-to-multipoint wireless technologies);

- Ethernet;
- Baseband Ethernet (such as BX10, ZX, LH, LHA, et al.);
- CWDM and DWDM;
- PONs (RFoG, EPON, and GPON);
- New technologies such as Packet Optical Transport Systems (POTS).

Even for a single customer and a single service, that customer may have sites connected with different technologies. A single Ethernet customer, for example, might have multiple sites on DOCSIS, one or more sites on CWDM, one or more on EPON, and even a couple on DWDM or baseband utilized for 10 GigE.

5.2.5 Systems

For each operations system requirement, operators have developed solutions. Each solution involves both the selection of network technologies and systems accompanying the elements of that network technology. The systems are used to OAMPT services.

Sometimes, the network technologies dictate the systems and processes. For example, if a customer needs 40 or 100 Gbps access, with current technologies, that customer is likely to be served not with a shared access technology but with a dedicated service over DWDM or perhaps with baseband over fiber. By contrast, a customer that requires 100 Mbps might be served by any number of technologies including DOCSIS, DPoE, coarse wavelength division multiplexing (CWDM), wireless, or others. The technology that an operator selects will be based on many factors such as the performance needs of the customer and the location(s) for the customer (relative to existing network assets), and so on.

The processes and systems for business services in cable have traditionally been vertically integrated with specific technologies. DOCSIS has its own provisioning and operational processes. The cable industry has now adopted those technologies to some business services with DPoE (see Section 4.3). Still others remain with their own independent processes. Although it may be possible to build DPoE like automation for CWDM or even wireless services, to date the processes for business services built over CWDM, DWDM, wireless and baseband services, vary widely. They may use a mix of vendor specific EMS, NMS, off-the-shelf "OSS" systems, and home-grown tools and systems. Typically systems sold as off-the-shelf "OSS" are in fact just subsystems that handle a particular task such as inventory, provisioning, data collection, trouble reporting, or some other discrete function. Today, there is no off-the-shelf system that a cable operator could purchase that provides an "OSS" meeting all of the OAMPT needs for business services. The current and growing variety of technologies, architectures, and services make such a product not only a difficult target but also one with at best only a few customers. Given both the lack of available products for this need, and the growing demand (from customers) and growing importance (as the fastest growing revenue opportunity in the cable business), operators have aggressively grown their own in-house systems to integrate the discrete and manual process-driven operations into automated OAMPT systems for business services.

5.2.6 Protocols

Operations systems use a variety of protocols for communication between NEs (i.e., switches, routers, and transport elements) and the operations systems. Some of the same protocols are also used for communications between operations systems.

Historically, the most widely used operations system protocol is the IETF SNMP that includes functionality for both setting and retrieving parameters on devices and for logging (the so-called SNMP traps). While SNMP remains popular because of its widespread support and simplicity, its favor among operators is declining as the overhead of the SNMP model proves inefficient with increasingly denser devices. SNMP is based on a model of a discrete transaction for each parameter. Thus, as devices have grown in complexity to include more configurable parameters; collect more data, have increased in density, and have a larger number of ports; the operation of a transaction for each setting (when configuring with an SNMP "SET" command) or retrieving (with an SNMP "GET") command creates both unreasonable amounts of traffic and CPU requirements for network transactions.

For devices on the customer premise, operators have always favored either a configuration file (which may be any of a variety of formats) or SNMP settings (for simpler) devices. In the past, configuration files were typically transferred with the IETF Trivial File Transfer Protocol (TFTP) (as they still are with DOCSIS), or more commonly now with HTTP. With increasingly demand for security, the use of secure FTP, but even more commonly secure HTTP (HTTPS) have become more common and will likely replace other transfer mechanisms.

For internal NEs (in the operator's) network, SNMP has never been practical for configuration and "operations," although it is commonly used for data collection (SNMP GETs). Nearly all internal NEs are configured through the use of proprietary configuration files that are transferred with TFTP, FTP, SFTP, HTTP, or HTTPS. Some NEs support interactive or transaction (load, commit, etc.) configurations that are scriptable, or which support Application Programming Interfaces (APIs). Increasingly, APIs are becoming imperative because of the inefficiencies of scripting against custom CLIs and the push among operators for standards. Although many APIs use a web service-based infrastructure such as SOAP with Extensible Markup Language (XML) over HTTP, they are typically complex overlay models.

The network industry has yet to fully embrace true RESTful web services-based APIs (see Section 5.4.4.8), but this may well be forthcoming with the industry push for software-defined networks (SDNs) (see Section 6.4), software-driven networks (SDNs), and Network Functions Virtualization (NFV). All three efforts seek to standardize the operation of the network for which the first step is standardized APIs.

Well before we have SDNs and NFV, there is already a push for "cloud service" functionality within cable networks. In order to provide the level of automation (in breadth, depth, and speed), industry-standard APIs for NEs will need to be further standardized. Today, this is done by building configuration abstraction layers that unify the network functions across various infrastructure NEs, across different vendor products. Because the abstraction layers are often scripted, the number of vendors is still often limited to a small and manageable number for each type of function (switches, routers, DWDM, CWDM, load balancers, etc.).

5.2.7 Security

Security concerns for network operators include all the same concerns for enterprise networks and many additional challenges. In addition to protecting their own enterprise operations like any other enterprise, MSOs are also expected to protect their customers from each other and from outside threats (on the Internet, over the air for wireless, etc.). MSOs are legally required to assist law enforcement with the lawful collection of data for public telecommunications services such as Internet access and telephony. Although these obligations do not extend to private network services such as VPNs and Ethernet services, many business services operate over the Internet. Finally, MSOs have to protect against theft of services. Because many business services involve specific locations and deep service integration, they are not subject to the typical and somewhat easier theft of service as broadcast services. But with their customers' employees increasingly working from home and traveling, the challenges for network security for cable business services have grown more complex along with the complexity of the services.

We will not address Internet security here, because the requirements and implementation of security for Internet services for business is largely the same as that for customers including systems for monitoring, detection, and activate disruption of attempted breaches.

The first approach used to provide security for business customers, which differs from residential, is the segregation of business services, even Internet access, into separate IP address ranges and classes of services. By logically separating these services, even when on a common IP transport network, the monitoring of those services can be implemented with varying or higher quality monitoring and faster response times to detected or reported problems. All of these measures require more time, and so are typically included in higher levels of service assurance (SLAs) that justifies the expense of higher staffing costs to monitor and respond to threats. Private data services such as Ethernet are perhaps the most secure because they are both logically and practically separated. For a dedicated or virtual private Ethernet service (sometimes called a Layer 2 VPN), there is no way for an IP packet from the Internet to enter into a pseudowire and ingress to a customer's private Ethernet network. Although the transport equipment within the operator's network may be common to both IP and Ethernet forwarding, there is no logical way to forward traffic into or out of a L2VPN within the service providers' network. The access networks utilized for Ethernet services may be the same (DOCSIS for example), they could equally well be different (CWDM, EPON, et al.). The core transport of Ethernet may be on isolated Ethernet networks, but in most cases cable operators use converged IP based transport that utilize pseudowires to transport Ethernet. So while the services are by definition isolated and protected, the transport runs over IP routers. In order to protect the common transport equipment, it is common with all service providers to isolate their signaling and control protocols (routing) and all operations systems traffic into separate IP address space from IP transit operations. This separation ensures the integrity of operations and highly simplifies the process of protecting the networks at peering locations.

Metro Ethernet services are perhaps the fastest growing segment of business services. As Metro Ethernet networks grow beyond the scope of a single operator, there are increasingly complex challenges to provide an open Ethernet peering architecture [10]. While standards organizations such as the MEF[5] have been developing the protocols and methods for forwarding, much work remains to be done to build secure operations systems for order entry, management, and operations across service providers. Some private exchanges [11] have capitalized on this need and operate custom Ethernet exchanges that offer partial solutions to these problems. But between cable operators or for that matter between any service providers, there is no universally accepted system for wholesale Ethernet circuit management between networks.

5.2.8 Operations

Putting all of the requirements and needs for business customers together into a single operations systems is an ongoing challenge for all operators. When new systems and technologies are introduced, operators resort to the oldest system around in telecommunications, the dreaded "swivel chair." That is, an operator configuring one subsystem on one computer or system, and rotating to another system to configure another part of the system.

As long as there is innovation, and especially "disruptive" innovation [12], there will always be new systems that are, by definition, not a part of the old system. In these cases, operators that want to offer new services or the same services with new technologies or new delivery methods will always be using the swivel chair.

But the goal of operations systems is to anticipate or at worst react to those new needs and integrate all of the automation required to delivery and operate all telecom services.

5.2.9 Distributed Systems

It's safe to say that all of the services that cable operators provide are distributed systems. Even video broadcast, whether sold to residential or business customers, is not a centralized broadcast. In fact, ads are locally inserted, varying packages with different channels are bundled and sold to customers with different needs, and even language variations differ by locations for those operators serving multilingual communities. These services have highly distributed systems.

IP-based services such as video over IP, video conferencing, and VoIPare also distributed. The operations systems support then come from a variety of vendors' products serving specific functions. For example, voice mail may be provided by one vendor and email by another. Or, both may be combined with a unified communications solution. Along with the systems providing the services, the operations systems (in parallel with the applications) may likewise be separate or unified. You can see why there is such a strong push for unified applications, because when the services must be unified, building the operations systems for disparate applications grows in complexity as customers have come accustomed to a large suite of services for voice services, for example.

Apart from the parallels with the applications, the need for distributed systems is also driven by load distribution. Distributing the load across multiple operations systems allows for greater geographic diversity, and therefore reliability against failures

---

[5] Wikipedia.com, "Ethernet Exchange" entry lists active Ethernet exchanges included experio.com, neutraltandem. com, tlex.com, equinox.com, and cenx.com., Wikimedia Foundation, Inc., 2014.

within a service, within the network, as well as connectivity problems or failures outside of the network.

Access and transport services such as VPNs and Ethernet services are themselves distributed, but have more centralized control of operations systems.

5.2.10 Control and Centralization

Any organization running a network has to maintain a delicate balance between centralized control and distributed control. Centralized control of the network helps to assure uniform or consistent service, quality, and efficiency. Distributed control both of automated systems and of the processes themselves allows for both reliability (protection against failures) in systems and in the process. After all, the majority of failures in networks today remain human errors both in the operation of technology and in failed processes that were supposed to be underdistributed or centralized control, but were not.

Distributed control of engineering, process, and operations allows for the more rapid development of new ideas and experimentation. Centralized control makes experimentation more difficult to develop by consensus, slower, and discourages innovation. As much as innovation drives success in the long term, pressures on reducing costs are constant and the push to centralize is often a stronger force.

Cloud computing, which provides highly centralized operations, but allows for the distribution of people away from the physical resources, may ironically be the best compromise yet in the ongoing balance between control and distribution. Anyone, anywhere in an organization, can develop a new process, service, or application; layers are more independent, and control remains practically centralized.

It's safe to say that today, MSO operations systems are at least partially cloud computing platforms. They are virtualized, load balanced, and many parts of them are automated. A classic cable business service which is "cloud computing" is the so-called "hosted PBX" where. Today, cable operators do not host a PBX for a given customer. They run a virtual PBX for each customer on feature servers as part of their platform. We expect to see this trend continue because the drive to innovation in operations is no different from the drive to innovate with products. Whereas in the past testing, a new operations system element required budgets and systems, a cloud or platform model allows for the virtual instantiation of systems (servers), rapid turn up of standardized daemons (web servers, for example) and prototyping new operational systems. After development, the same cloud or platform allows for the rapid replication of those setups through the duplication of virtual machine instances or desktops that act as templates (sometimes called virtual machine images or VMIs). Finally, the dynamic capabilities of cloud or platform solutions allow for the flexible use (so-called elastic use) of CPU, disk space, and other resources over time.

Because business services are so complementary to residential services (they have opposing peak utilization times), cable operators have an enormous opportunity to take advantage of underutilized residential scale resources in the operational platforms, during the day, and then relinquish those resources back at the end of the business day into evening hours and weekends when residential services need them, but business services do not.

5.2.11 SDNs and OAMPT

SDN (see section 6.4) or whatever it evolves into is likely to have a significant effect on operations systems. At the very least, as the new movement called software-driven networking (still SDN) suggests, operations systems are likely to take a more prominent role. The factors driving this are the demand for faster development and rollout of new products and the economic pressures for standards-based interoperability. Although there are those that would argue that SDN seeks to do away with existing standards, such a view, in the author's opinion, is not realistic. Instead, what we expect to see is standards moving from protocol-driven interoperability deeper into functional standards and functional interoperability. That is, today we expect systems to interface to each other using standards protocols. But the function of the device (within the "black box" so to speak) may differ widely. NFV offers a path for defining the functions. But traditional standards bodies like the IETF, IEEE, ME, F, and CableLabs are likely to continue to define standards for interfaces because operators must have backward compatibility with existing interface protocol-driven standards (like routing) as well as similar functionality between networks. Even if the most optimistic visions of SDN come true, the SDN may still be limited in scale to a single operator. Given that, interface standards between operators (like the Border Gateway Protocol [12] that operates between all Internet service providers) will be with us indefinitely.

5.3 SERVICE ORDERS

A service is a defined function or a set of functions that a service provider, such as a cable operator, provides to a customer. A service order is a request to provide such a service. How service orders are created, communicated, and fulfilled varies widely by the type of service, the cable operator, the location of the customer relative to the customer's network, and many other factors.

5.3.1 Product Management

Well before service orders are handled, the service provider must clearly establish which functions are included (i.e., packaged) with what services, what the installation cost or non-recurring costs (NRC), regularly monthly recurring cost (MRC) that are fixed, what services are measured (by means called accounting), and what the rates are for metered services. This process is sometimes called "service design" (especially by vendors), but is more often called "product management."

Product managers' package functions into services and may bundle multiple services together into a single service or menu of services offered to customers.

Outside of these defined services, it is still common to create what are called custom or one-off services specific to an individual business customer, if the business is substantial enough in revenue to cover the additional overhead costs of developing a custom solution. Those costs involve identifying the customized functions and services, pricing them, designing, testing, building, delivering, and supporting the services. Unfortunately, most custom services do not also get customized OSS enhancements. What this means is that many aspects of OAMPT for customized services are not optimized. This makes the OAMPT for custom services labor-intensive and therefore costly. Over time, operators and customers alike realize the need for automation and data collection, for example to monitor and enforce SLAs, and these solutions still need to be developed. Unfortunately, if the solutions are not developed before selling a service, the cost of the OAMPT solutions is not included in the Independent Rate of Return (IRR) calculations used to decide if a sale of a proposed custom service is profitable. Nonetheless, almost all new services begin as custom innovations or special requests by valued or new customers. The result is that new services are often "launched" before they are ready. Over time, such services accumulate to a point where building a system around previously installed and operational customers itself represent a change and risk that is typically viewed as disruptive. This "catch 22," while not unique to business services, dominates the development of business services strategy and has become the focal point of when, where, and how innovation can be packaged and delivered for new business telecommunication services.

5.3.2 Service Order Management

Once a service order is submitted, it is "parsed" into orders for one or more functions or services that comprise the service that is ordered. We quote "parsed" because the order itself may not contain the information. For example, the order may simply be a reference order number, referring only to a key (entry) in a database. The order, order number, order code, and order key may contain information to be parsed or it may be a link or reference to a record that can contain additional data or links to other data elements. This process is sometimes referred to as "parsing" or "decomposition," but within MSOs is typically called service order management.

Typical functions to be performed include the creation of work orders, notifications, account records setup, and so on. Many of these functions must be performed in a particular order, especially if the order entry system has dependencies that are not automated as a fully integrated state machine. For example, many operators will have systems that can only create work orders for existing customers. Therefore, the customer's account must be created before a work order can be issued. In turn, it may be that an account can only be created once an installation or setup fee is collected, and this may involve charging functions with dependencies on other systems—some of which may be external to the operator such as a billing or payments vendor.

Order management systems (or service order management systems) typically include a function that can halt the order process and automatically create a trouble ticket if an error is returned from some function or, if quality controls are implemented, if some anomaly is detected. Classic examples of these anomalies are mismatched addresses. This is common where a street address in a geographic information system (GIS) does not match a billing address (and both addresses were not collected in the order entry process).

Order management systems are also typically tied into different types of resources that must be available to provide a service. These may involve logical inventories such as licenses required to use software or services—so-called "seats" in enterprise lingo, but more generally subscriber licenses—logical inventory such as IP address space,

CPU, memory, disk space, or other resources; network "inventory" such as wavelengths or ports on network equipment; and physical inventory such as set-top boxes, CMs, routers, gateways, or other tangible devices. A highly mature system might even inventory electrical power and space consumption. The benefits of a mature inventory management system include a better accounting for the real costs of the system and, obviously, better management of those resources for planning, execution, and management of expenses.

Finally, service order management systems should have the functionality to automatically create the objects necessary for the transition to operations. This includes creation of objects for EMS, NMS, fault management (FM), performance management (PM), and other functionality. The objects required differ widely, for example, between DOCSIS services and Metro Ethernet services. As MSOs expand beyond access and network services to include packaged infrastructure as a service (Iaas) to applications as a service (AaaS) and software as a service (Saas), platform as a service (Paas), or in general anything as a service (Xaas pronounced "zaas"), they increasing involve data moving between the MSO and third-party cloud service wholesalers.

One example of such a platform is Comcast's Xfinity platform. There is no reason to believe that it could not or would not be sold as a PaaS for other MSOs, especially smaller MSOs in the future. In such a future scenario, even the service order system itself could be "outsourced" to the PaaS provider which may be another operator or service provider!

Although many well-established services—such as video, voice, and Internet services—have reached that level of maturity, the customized needs of many businesses still leave many service orders dependent on the so-called one-off or custom engineering solutions. In these cases, the order management system typically creates a work order, which could be as simple as an email, that is a design request for additional engineering work.

5.3.3 Service Order Standards

Although there are standards for order management or service order management, for example, the Telecommunications Management Forum or TM Forum, outlines a process for "Service Management" which includes processes for [13]:

- Service configuration management;
- Service design/assign;
- Service inventory management;
- Service performance management;
- Service problem management;
- Service quality monitoring and impact analysis;
- Service rating/discounting management;
- Service specification management;
- SLA management;
- Service delivery;
- Retail store solutions.

While most aspects of these are, in practice, covered by some MSO process, the terminology and process of the TM Forum are not widely used. The greater the automation of the service order management system, the less likely it is to include the many variations that are possible with product sales that are contract-oriented, versus more packaged or retail services.

The TM Forum does offer the opportunity to standardize the language and processes used. With the growth in service complexity and interdependencies on both other service providers (operators) and vendors, the cable industry does actively participate in the TM Forum. CableLabs, Inc, Cogeco Cable, Cox Communications, Time Warner Cable, and other cable operators[i] are members of the TM Forum.

5.3.4 Service Management across Operators

The Cable Telecommunications and Marketing Association (CTAM), in conjunction with most of the cable operators in the United States, operates a service called "CableMover" [5]. Although it is limited currently to residential services, it highlights the value to consumers and operators alike to turn an existing customer into a service order for another MSO when that customer moves.

Although the physical relocation (moving) of large business customer is not common or frequent for businesses, providing services across multiple operators including both MSOs and other service providers is commonly needed with larger businesses. When businesses have locations that span across the geographic boundaries of two or more operators, they are often left with few options and end up piecing together networks themselves. Similarly, if businesses have needs that include a wide variety of services with access not only at their office locations but also for mobile and work-at-home staff, they too have had to piece together their own solutions. The challenge for operators including MSOs, telcos, cloud service providers is the lack of systems or integration that allows data to be shared between operators for provisioning across service providers. Despite all of the technology available, there is really no automation between different operators or service providers today.

Today, these processes are largely manual or at best may involve inventory and (wholesale) pricing available through web-based applications. These are slowly migrating from web-based applications to web-based services through the use of APIs. But today, there is no end to end service order management for most IaaS. PaaS and SaaS are far simpler for automation because they do not have geographic and access resource and inventory requirements. Both simply "assume" the underlying Infrastructure (the "I" in IaaS) as present and reliable.

5.3.5 Service Order Depending on Service Type

In today's cable networks, the methods and processes for creating, submitting, and fulfilling service orders vary based on the type of service. The broadest definition of a service is a group or set of functions that are collectively provided (sold) as a package. For example, what we commonly understand as "Internet Access" is a service that at least includes functions such as the access portion of the service, to connect the customers location, premise, or equipment to the service provider, transit through the service provider to other ISPs (the Internet at large), and DNS for domain name lookups. A service (called a product by Product Managers within the service providers) may have options. Continuing our example, an Internet Access service may have an option for fixed bandwidth and burstable bandwidth. It could have options for a redundant (protected) access circuit, filtering for one or more applications, BGP peering, and so on. Instead of or in addition to options, service providers may offer packages of service that "bundle" multiple services into a single fixed price or a fixed price plus additional measurable costs for consumed services (minutes, bits, lookups, routes, addresses, etc.). Services that are consumption based (which can be measured) must include means for measuring the consumption, which is called accounting (or metering). That is the same underlying assumption of all Internet services.

5.3.5.1 DOCSIS-Based Services

DOCSIS services are the foundation for many packaged business services and, when combined with added options, an expanded portfolio of business services. Most DOCSIS services are actually "designed" or identified as products, before the protocols are written. They are unique in this way because the DOCSIS specifications include both protocols and APIs required to perform OAMPT for the services. That said, DOCSIS specifications do not spell out the operation of the OSS. Although there are some commercially available DOCSIS OSS that include service order (and other OSS functionality), like the other OSS functions, DOCSIS service orders are typically handled by custom-developed order systems. These are among the most mature order systems not only in cable but also in all of the telecommunications industry. They include support for self-service, retail services, call center activation and service changes, and many other complex service order capabilities such as the CableMover [6] program.

When MSOs refer to DOCSIS, they are usually referring to both the technology (the protocols and specifications) and the services considered integral to DOCSIS. These are Internet access and voice or PacketCable services. Although PacketCable is a separate set of specifications, most references are collectively to both services. As a "transport" that can be used to carry IP packets or Ethernet frames, DOCSIS and DPoE are capable of supporting a much wider array of services. In fact, nearly any telecommunication service can be built on top of either DOCSIS or DPoE or both. Some very constrained requirements, such as those for storage area networks, may limit the support topologies (especially distances). Some applications may also have preferably large frame sizes than either DOCSIS or DPoE can handle, but these amount to an tiny fraction of business telecommunications services.

5.3.5.2 Internet Access

Some business service, such as Internet Access, are readily packaged with DOCSIS by developing business-specific variations of DOCSIS Internet access such as DOCSIS service with symmetric (matching downstream and upstream) rates, fixed IP addresses, and so on. Today, these are all delivered with existing DOCSIS systems operated by MSOs, requiring only variations in product packaging and pricing. Some variations such as burstable bandwidth are offered as options or add-ons. These service orders are entered through GUI-based order entry systems, usually by account salespersons or sometimes by customer service representative (for upgrades or service changes only).

5.3.5.3 Voice Services

For voice services, the PacketCable architecture supports multiline services for businesses. For smaller businesses, PacketCable offers a turnkey solution. More complex voice services typically require functions beyond FXS services and are offered through Session Initiation Protocol (SIP) solutions either integrated with PacketCable specification, or more commonly SIP solutions run "over the top" of PacketCable, but on dedicated service flows that are separated from Internet applications. That is, although they are IP applications, they run over what are separate virtual circuits (DOCSIS service flows) from other IP services such as Internet services. Although voice traffic (even for very large customers) is quite small compared to banwidth-intensive services such as Internet access, private VPNs, or private Ethernet services, voice services are very sensitive to variations in delay (jitter) and are (for that reason) logically and functionally separated in the access portion of the network. Such separations also allow for the subsequent logical marking of that and other IP traffic even in converged transport, for example, between the DOCSIS CMTS and the then separate paths for Internet or voice services into separate gateways in the service providers' network. Like Internet access services over DOCSIS networks, PacketCable specification-based voice service orders are typically entered through GUI-based order entry systems, usually by account salespersons or sometimes by customer service representative (for upgrades or service changes only).

DPoE can be utilized to transport voice with PacketCable specification support of optional Embedded Multimedia Terminal Adapter (eMTA), or Embedded Digital Multimedia Adaptor (eDVA) in a DPoE ONU (D-ONU), or in theory with a DPoE Bridge ONU with a port connected to a standalone DVA (sDVA). Customers with substantially more complex voice service needs would typically run SIP over a dedicated LLID on DPoE or EPON, without PacketCable. Such solutions are based on SIP or IMS/SIP service management instead of PacketCable. Since DPoE uses DOCSIS-based provisioning, the service orders can also be provided by the same order entry systems (although the products may vary because of the different capabilities).

5.3.5.4 Ethernet Services

Ethernet, like voice, can be delivered integrated with DOCSIS using the DOCSIS L2VPN standard. L2VPN was not widely used in part because there were few products (cable modems) that supported it, and in part because Ethernet service delivery continued to evolve with new functions and very complex service, for which L2VPN could only provide a piece of the solution. Having automation in DOCSIS provisioning for Ethernet in the access network (L2VPN) is not very attractive if it isn't integrated with OAMPT for the remainder of the solution required for MEF-specified services such as E-LINE and E-LAN. In theory, L2VPN service orders can be based on the same back-office systems supporting DOCSIS. But because these services always require connectivity beyond the CMTS, they typically involve "swivel chair" OAMPT—which reduces service orders to email requests to network engineers or technicians to configure the transport of Ethernet from the CMTS to another CMTS or other Ethernet systems as part of an Ethernet service. It is this limitation that has limited the adoption of L2VPN.

DPoE of course runs over a native Ethernet transport, opening the door to easy integration with CE devices and on the service provider's side with Ethernet transport in the access and transport networks. In particular, DPoE Version 1.0 and 2.0 offer options and specifications for end-to-end service integration to deliver MEF services including E-LINE, E-TREE, and E-LAN (including both private and virtual private variations). In addition, DPoE 2.0 offers a full suite of MEF-based service OAM (SOAM) capabilities that go beyond the "P" for provisioning in OAMPT. The challenge for DPoE vendors has been to provide the full suite of transport integration including IP routing, MPLS, VPWS, VPLS, and SOAM capabilities, while also trying to keep the system costs competitive with the low cost of DOCSIS-based solutions. With DOCSIS 3.1, the bar has moved above 1 Gbps (for a single customer location) as the minimum entry point where DPoE begins to make more sense (depending on many factors). Interestingly, there is no reason why a CMTS vendor could not adapt the DPoE 2.0 IP routing, MPLS, VPLS, VPWS (core transport), and SOAM functions to bring complete MEF service integration into the DOCSIS platform. As new CCAP systems are rolled out with both DOCSIS and DPoE capabilities, we are expecting to see exactly that integration. CCAP could turn out not only to be the inflection point for video and data integration but also for residential and commercial service integration—if the full suite of Ethernet OAMPT from DPoE 2.0 finds its way into the CCAP systems.

DPoE offers the potential for completely automated OAMPT. The automatic provisioning capabilities combined with an IP/MPLS core that can support VPWS or VPLS can provide complete automation for ELINE and ELAN services. Although this is possible, and has been technologically demonstrated, there are no publicized or wide-scale deployments of the technology. The primary obstacles are the lack of vendor support for the full DPoE System capabilities and organizational and process challenges to adopting a DOCSIS-based methodology for Ethernet services. The technology itself is not an obstacle. The combination of these challenges has meant a slow adoption despite the incredible demand and growth in Ethernet service delivery.

5.3.5.5 Inter-Carrier Ethernet and the Ethernet Network to Network Interface

The biggest gap in Ethernet services across the OSS, including service order entry and management, is the lack of protocols, tools, and systems necessary to provide Ethernet services between operators. Although some protocol work has been done by the MEF, notably OVCs and ENNI, a complete solution requires a marketplace and system for the wholesale offering of inventory and circuit ordering. The process is complicated by the complexity of possible service offerings, competition, and lack of mature standards for bandwidth and inventory management.

5.3.5.6 Fixed Wireless Services

MSOs have, for some time, used fixed wireless technologies as a last-mile substitute or backup. Some of these technologies support any IP or Ethernet transport-compatible product. Still, the cost, complexity, and reliability of these technologies typically result in the use of the technology as turn-up tool to get a customer online while construction crews build out permanent wired access. In these and other cases, the relatively small market share for fixed wireless services has created little demand for mature OSS systems including service order entry and management.

5.3.5.7 Mobile Wireless Services

As the MSOs have literally ventured (joint ventures and partnerships) into and out of relationships with wireless operators, there has been significant integration necessary for supporting wireless services. For example, a fundamental limitation of older cable systems was the "hard-wired" limitation to only sell services to customers within a franchise area. Given their legal obligations to conform to franchise agreements, having a billing system that prohibited sales out of franchise made and still does make perfect sense for those traditional services. But what about wireless services? Cable operators as a mobile virtual network operator (MVNOs) or otherwise, are under no contractual obligations regarding the sale of wireless services. To accommodate this, and other possibilities, has required changes in the service order systems for MSOs that continue to evolve.

5.3.5.8 Access and Beyond

Although the bulk of MSO services remain access driven, for both residential and business customers, the opportunities for non-access services including cloud storage, applications (SaaS, etc.), home and business automation, and energy management are compelling. Whether alone or in partnership with other MSOs or service providers, MSOs bring to the table local operations who know their customers, neighborhoods, and networks. Their capabilities include high-capacity networks, local installation and turn-up services, diverse networks, and extensive experience supporting local customers. As online service operators strive to go beyond early adopters, they are finding that partnerships with MSOs can ease the transition in terms of scaling and support. For each service, the MSOs have continued to expand their definitions of service far beyond "cable TV" and even beyond the "access network."

5.3.6 Defining Services

Beyond product management, MSOs are embracing the packaging of services that rely on underlying services from or with other partners. These range from other access networks such as wireless cellular networks, to cloud applications, monitoring services (for alarms, for example), and so on. Beyond these are the "services" that customers use, which are not defined by the MSO. In the industry, these are called "over-the-top" (OTT) services. While such services may not be defined by the MSOs, their value to MSOs customer may be important enough to understand them, and ultimately to partner with those OTT providers to improve the experience for the customer. It could be as simple as a combined bill or discount, or it could be more complex. Although OTTs often think of themselves as competing with access service providers such as MSOs, there is a long history of OTT applications that have driven tremendous growth in the telecom industry. And this is certainly the case with OTT providers who today have driven the adoption of MSO Internet access services to its current take rates.

Just as the shipping industry (UPS and FedEx) works closely with large distributors, so too are the cloud and OTT providers learning to work with the large distributors for the Internet, the MSOs, telephone companies, and wireless operators. We can always expect new services to emerge OTT; but mature services, especially as they proliferate in number, are likely to consolidate into managed services for consumers and business who do not want to manage individual billing and support relationships and more than the service providers themselves do not want to have to manage troubleshooting that many variations in delivery for their services.

Although these support processes are not often part of the OSS, it is very typical for the MSOs and large OTTs to have good working relationships between their network operations centers (NOCs) and even between engineering staff as they learn that working together to keep their mutual customers happy (quiet) is in their mutual interest.

5.3.7 Service Order Security

Service order entry system that have limited sources of orders (i.e., customer service representatives) have simpler order entry security criteria than those that support order entry from customers (self-service), through third parties (for wholesale), and even from other systems. The challenges for secure order entry are growing ever more complex as networks and systems themselves become services (IaaS). But the efficiencies gained by automation are driving down costs; and to be competitive, service providers must embrace automation. Security provides the checks and balances to such automation.

Service order entry, management, fulfillment, and delivery use cryptographic methods to identify systems, operators, and users. They further use IP-VPNs, firewalls, and a litany of system security measures (usernames, passwords, cookies, and other means) to provide many layers of security. Whether exchanges are between systems within a single operator's network, between operators, between an operator and their customer, or between an operator and one or more third parties, these systems are required to mitigate the risks of vulnerabilities to any number of service thefts, denial of service, or even human errors.

Although there is no industry-wide standard for each aspect of service order entry, management, fulfillment, and delivery, most operators employ a full mix of identity, system, network, and process (including electronic approvals processes) to limit, control, and monitor service orders. Data analysis and meta-data analysis of orders on a regular (usually daily, weekly, monthly, sometimes quarterly, and annual) basis also provide a comparative analysis for growth, anomaly detection, and expectation setting for orders of existing and new products. Some of the best security comes not from secure systems, but from the people who know the normal operation of the service flow process and who quick identify anomalies as suspect. Although many of these turn out to be "gaps" in the service order process rather than security problems, the watchful eyes of these experts always improves the processes.

5.3.8 Wholesale Markets

As long as there is no single telecom operator (i.e., monopoly), there will always be a need to provide services between operators. In wireline and wireless, there is a long (and sometimes, but not always, difficult) history of connectivity between networks. Today, the biggest gap in telecom services is for an automated system to provide for the exchange of Ethernet traffic and services. While both established and new players have built "Ethernet exchanges," these exchanges have been limited by one or more of automation, location, services, or the degree to which they are open to function as an open market.

Although the protocols have been developed for the description of network services based on OVCs [14], the signaling required to automate Ethernet services provisioning (and operations) between networks (on a peer-to-peer basis). Until both the technology and the processes are in place, service orders between operators for Ethernet remain largely based on lengthy email exchanges that are further limited by a lack of adequate but simply language to describe the limitless variations of Ethernet "peering." Starting, as the MEF has done, from a description of the service, and working down into the network is a rational approach to developing both the descriptive language, and ultimately the products and systems it will take to automate Ethernet service orders. We look forward to the continued development and maturation of that technology.

5.4 PROVISIONING

Provisioning is the process, as the name suggests, of making provisions with network and system resources required to provide a service. "Provisions" include the allocation, configuration, and enabling of resources. While many aspects of provisioning are virtual and can be automated, the most critical part of the provisioning process is, when required, the installation of devices and physical connections of devices to the network for services to work, except wireless devices that are provisioned, but not physically connected.

Provision is typically a process that involves the ordered execution of separate processes. Some processes may operate in parallel, while others are stateful and require the completion of one or more previous processes before execution. For example, if a provisioning process requires "picking" a device from a warehouse or storage and loading on a truck for installation, clearly the installation cannot occur until the picking and loading process is complete.

5.4.1 Service Orders

As explained in Section 5.3 the ordering process begins with the creation of a service order. After the receipt and processing of a service order, it may require further specification such as service design. After all aspects of a service order have been verified, approved if approvals are required, and resources are identified or allocated, the service order process transitions into the next process which is provisioning.

Although we make a clear distinction between the processes here, the automated systems involved don't stop and say "now the Service Order processes are done, provisioning begins." The end of service order processing and start of provisioning is an abstract concept. The delineation is best described as starting provisioning when the OSS systems begin the provisioning of NEs or network systems other than elements of the OSS (i.e., database entries, creation of objects, etc.).

Provisioning may include the configuration of either or both of network and systems resources. For example, provisioning an access service such as DOCSIS-based access requires configuring NEs such as the CMTS and CM. But it also requires configuration of DNS, DHCP, and other network systems. Many operators bundle email for example, with their access services. In such cases, provisioning extends beyond NEs and network systems to application systems such as email servers. We address each of these.

5.4.2 Provisioning Interfaces

There any various interfaces, supporting various protocols, available to operators for provisioning both their systems and NEs. Most systems and NEs support two or more interfaces. For example, it is common for equipment to support both a command-line interface (CLI) and SNMP. Web interfaces, both user web interfaces and machine-to-machine (M2M) web service interfaces are the fastest growing in terms of adoption of the technology and addition of new functionality. Over the Past few years XML (XML over HTTP or HTTPs)-based APIs have become popular because they allow scripting and automation. The development of structured models such as Network Configuration (NETCONF) Protocol and Yang has further promoted XML-based interfaces. More recently, the RESTful paradigm from web-service architectures has begun to find its way into network operations. We explore the options available for XML, SOAP, JavaScript Object Notation (JSON), REST, and RESTful APIs, alongside the legacy SNMP protocol later.

5.4.2.1 Command-Line Interface
The CLI remains the mainstay of service provider provisioning for business services. By definition, business services are often highly customized and present particular configurations that are unique to a customer. For operators with a small number of customers, the CLI is often the only provisioning tool. As business services grow within an operation, enterprise-like solutions usually take hold offering tools for some level of automated provisioning. Most of these tools, however, lack carrier-grade features and are not interoperable across multiple vendor products. Even a simple point-to-point Ethernet circuit such as an MEF EPL, can require the configuration of several or more NEs not including systems used for monitoring, accounting, or billing. The more complex the service, the more likely that it is provisioned manually with a CLI than automatically.

Manual provision involves provisioning each element, typically one at a time. The term "swivel chair" refers to the figurative description of an operator, technician, or engineer "rotating" their swivel chair, shifting from once screen to another configuring each different element. Figure 5.2 depicts an example where manual provisioning is used to build an OTT L2VPN for Ethernet where the service is transported over DOCSIS on one end (left) and over EPON on the other end (right) of the EPL. The network engineer depicted configures from their station, each element, one at a time, represented by a red line and arrow from the station to each and every NE.

An improvement upon manual provisioning is the use of an NMS. This is illustration with the network engineer accessing the NMS from their station (a single bidirectional yellow line) and then communication from the NMS to each and every NE (yellow lines with arrows).

A further improvement, and ultimately the end goal is complete automation where order entry, perhaps after review or approval by engineering (for custom business products), enters the OSS is and fully automated, as depicted by the bidirectional green arrows.

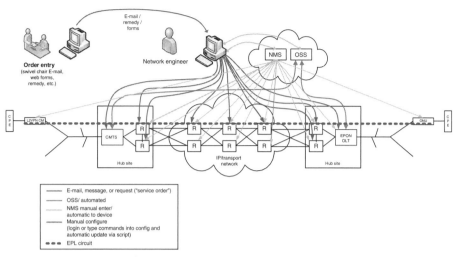

Figure 5.2. Manual service provisioning.

Many MSOs use a mix of automation for some systems, with manual provisioning for other functions. For example, a dedicated Internet access product may automate IP addresses with DHCP and helpers (proxies), DNS entries for A records, routing, and other provisioning, but still require manual provision for Ethernet ports (on a P2P solution or L2VPN overlay) because the specific port must be manually cabled, and there is not automated way to correlate the physical cabling to the provisioning of a port. While that particular problem can be automated with work orders, there is always the "next" challenge to offer a new service that cannot be fully automated. As soon as that challenge is tackled, the product and sales folks will always dream up new products for which the CLI will always come to the rescue.

The popularity of CLI is based on the efficiency of its use interactively as well as the ability to script the interface. Not all text-based interfaces are considered CLIs. For example, menu-driven systems are not considered CLIs. The classic example of a CLI is Unix CLI for the bourne shell (sh program in Unix). Systems with CLIs, like Unix itself, can easily be scripted and indeed often have integral tools for capturing input and output (script, history, etc.), scripting languages (PERL, TCL, Python, etc.), and tools for parsing (diff, grep, nroff, etc.) captured or real-time input and output.

A CLI is the only interface that can be used interactively and automated. The protocol interfaces are particular to automation for M2M communication (i.e., provisioning server to NE). The protocol interfaces are not designed to be used interactively by operators (although I've known colleagues that have memorized OIDs and other data structures to manually provision NEs, or for operator to NE (human-to-machine) communication (i.e., a GUI). The GUIs are designed for use by operators, but are not readily scripted. Although again, I know that has been tried, with little success.

The CLI must often be learned for interactive troubleshooting, testing, and other functions that occur after provisioning. Although the CLI may be seen as difficult or challenging, it remains the best "bang for the buck" in terms of time investments because you can do every aspect of OAMPT (operations, administration, maintenance, provisioning, and testing) for NEs. Many NEs operate with CLIs similar to each other or similar to Unix. For operators who otherwise have used a similar element or with Unix experience, there is almost no learning curve.

A close tool to the NE CLI is the Unix systems typically used by operators as their jumping off point for systems and network management. Although it is beyond the scope of this section to explain all of the uses, the use of a Unix (or Linux) host allows for ease of integration with Unix scripting (script, history, and shells), file storage, file transfer (FTP, SFTP, TFTP), SSH, logging (syslog), and other functions. The power of these tools should not be underestimated. With experience, a single operator can literally run an entire network and many services.

Despite the push for "user-friendly" interfaces such as the GUI, the CLI has remained the "least common denominator" for network operators. Many equipment designers, managers, novice operators, and novice engineers fail to realize that the complex systems that must be running to make GUIs work (web servers, web browsers, etc.) will not operate over a network when the network itself is not working. When "all else fails," operators continue to fall back to the CLI that can be operated on a cell phone with a USB to serial cable, from a terminal port with HyperTerm on a laptop, from a terminal server with SSH or Telnet, or even with an old-fashioned "crash cart" in the facility!

5.4.2.2 Graphical User Interface

Although GUIs are the preferred interface for configuring devices in the enterprise, they are not well appreciated among large-scale operators. Vendors often favor and promote GUIs because they simply ease the use of their products and reduce the demand for training and support that are seen as expenses that affect profitability.

While GUIs are advantageous for their ease of use, they present significant challenges to scaling OAMPT. In particular, GUIs cannot be easily automated. Although some GUIs include built-in automation for repetitive tasks within the GUI, the challenge for operators isn't provisioning a single device. It's provisioning devices in the proper sequence required to activate a service. Although some vendors have products that claim to manage other vendors NEs, the introduction of new products from other vendors always leaves such tools behind in functionality and support for new elements and for new functions even on currently supported elements. From an operational standpoint, GUIs are typically slower to perform most tasks than scripts or copy/paste CLI commands.

Some GUIs may be web (HTTP or HTTPS)-based. Some GUIs are executable applications. Executable applications that are compiled are obviously specific to a specific computer platform. This introduces the problem of relying on particular operating systems or configurations of workstations or servers. Some GUIs operate exclusively as a client on a workstation or server. Others operate in a distributed environment with some portions on a server and other portions (the GUI itself) executing on a server. This compounds the operating system and workstation or server-dependency problems.

Many newer systems are web-service based and rely on an application that resides on a web server and client execution of a scripted application (i.e., AJAX, JavaScript, Ruby, etc.). These sometimes suffer from performance problems (they are slow or require a lot of memory to execute). Although most web-based GUIs are not operating system specific, they are usually developed and tested to specific browsers (Internet Explorer, Mozilla/Firefox, Safari, etc.). All of these dependencies place operators at the mercy of workstation and server dependence that defeat their efforts at scaling, automation, and reliability.

GUIs offer great potential to visualize networks, systems, performance, and other analytical functions. For analytics, visualization is critical and the role of GUIs is well accepted. But for OAMPT, GUIs are a sore point for operators who would rather see open, published, and standardized APIs. If nothing else, the SDN movement is plea to move away from the proprietary and vertically integrated GUIs that vendors seem to love.

5.4.2.3 Scripting and Automation

As discussed in Section 5.4.2.1, the CLI is defined by its function to work interactively with a human operator or by scripting. But there are limits to CLI scripting. Scripting very large changes can introduce data flow control problems that force the developers of the script to introduce complex timing parameters to handle and regulate the flow of input and output text or data. These console messages and other irregularities present challenges to writing robust software that automates provisioning (and other OAMPT aspects) for NEs.

The purpose of scripting is to automate operations (including provisioning). As the needs of operators have grown more complex (to offer more services and greater variations on each service), operators have to use equipment from an increasing number of NE and systems (equipment) vendors. The addition of each new product requires support throughout the OSS and for OAMPT processes for provisioning and beyond to the end of the product life.

Here, the same EPL service as depicted in Figure 5.3 is provisioned automatically. The sequence of events begins with a service order placed using "Web forms"; back-office systems that create provisioning requests to the NMS and other OSS elements;

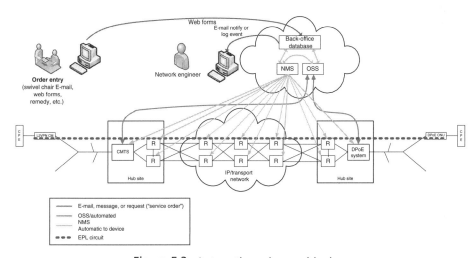

Figure 5.3. Automatic service provisioning.

and provisioning via APIs, protocols, or scripting to various NEs. The network engineer may still receive notification, or may be alerted to any exceptions that occur during the provisioning processor, which prevent the process from orderly fulfillment.

NE vendors are incented to vertically integrate management systems including provisioning to make their product more "sticky." Their goal is to make it as difficult as possible to replace their equipment with another vendor's equipment. Operators have exactly the opposite incentive. Their goal is to diversify their sources of equipment, including more vendors in order to explicit avoid that "stickiness." It is this tension and the race to efficiency that have driven standards for provisioning and other OAMPT.

Market forces between service providers and vendors are not the only factors that have affected the evolution of OAMPT and provisioning in particular. For much of the early history of IP and Ethernet networks, both technologies were seen by vendors and the technology community at large as "enterprise" or "information" technologies. Even today's major telecom equipment vendors or IP and Ethernet saw these technologies as suitable for the enterprise but not "carrier grade." As a result, enterprises management needs to take precedence over the needs of the comparatively smaller "service provider" market that was then in its infancy. Enterprises have traditionally favored GUIs and simple legacy protocols such as SNMP as sufficient to solve their problems. Most IP and Ethernet equipment (even today) are built with the assumption that devices should automatically connect and be self-provisioned to the greatest degree possible. As it turns out for service providers, this is often the last thing they want.

5.4.3 Pay for Play

Service providers are in the business of selling telecom services. What they really want is services that are gated or controlled based on what is paid for by the customer.

The underlying assumption in many network protocols, including OAMPT and provisioning protocols, is that NEs should always automatically connect. This assumption is based on the assumptions of protocol designs with historical roots in the enterprise (and later in consumer electronics) where the objective is to simplify and maximize connectivity. Protocols such as ARP, BOOTP, DHCP, NDP, and the many higher layer protocols built on top of these build on those assumptions. These protocols are just as important to service providers, but they must often be operated on systems with algorithms specifically designed to gate or control their functionality based on the services that customers have ordered or paid for.

Protocols describe the exchange of data between elements and systems. How each individual system implements its stateful process is beyond the scope of telecommunications protocols. It is there, within the state machines that operate protocols, that service providers insert the logic required to validate service orders and gate or control provisioning and fulfillment. These gating functions are control points for access to resources. Designing them to be able to handle operations at scale, while also providing absolute control, often requires both complex software and very capable hardware. The hardware utilized in these systems often includes very large dynamic memory and fast solid-state drives in order to maximize the memory available to hold numerous states while waiting for replies to requests from databases in the back-office or OSS. Often the worst-case scenario for a service provider is a complete shutdown of services, which is followed by rapid onslaught of login or reconnection attempts. DHCP systems, for example, are typically designed and distributed to handle very large-scale (millions or tens of millions) requests per second for such a worst-case scenario.

5.4.4 Provisioning Protocols

CLIs and GUIs use protocols to communicate directly with NEs or more typically with an intermediate system. For example, most CLIs operate over TCP/IP using the Telnet or SSH protocols. GUIs may use a proprietary or standard protocol for communication between the user and an intermediate system or directly to the NE. Today, it is common for GUIs to operate over HTTP or HTTPS. Within HTTP or HTTPS, they may use a property protocol, but are increasingly using standards-based protocols such as SOAP, XML, REST, NETCONF, or other protocols. Some systems may also use a combination of protocols.

Provisioning elements does not typically happen just once. It may involve an initial configuration to get a device onto a network before or after physical installation. Later as individual ports, interfaces, or services are configured, NEs and customer premise equipment (CPE) or customer-edge (CE) equipment may be configured regularly (with each service change).

The wide range of protocols and combinations makes it impossible to illustrate every possible path to provision elements. Even taking a single element such as a router, there may be dozens of ways in which it is configured. An initial configuration may be done by the manufacturer, according to the service provider's specifications, prior to shipping; upon arrival, it may be initially configured before inventory or before installation. Initial configurations may be by the insertion of a flash card or USB stick, via a serial port, or over a network. Each of these methods supports one or more protocols.

We illustrate a number of the possible paths:

- CLI directly from a workstation or host computer to an element;
- GUI directly from a workstation or host computer to an element;
- GUI via a proxy server that may be an NMS or EMS;
- Scripts and tools on servers (systems) that interface directly with an NE, EMS, NMS, or OSS;
- GUI to an order entry tool or system that in turn populates a service order that is passed on for fulfillment in the back-office. The back-office turns the order fulfillment into separate provisioning instructions or requests for each NE. Each element may be provisioned with one or more different protocols.

This list is only exemplary, as illustrated in the Figure 5.4.

In Figure 5.4, a more detailed view of OAMPT systems shows the relationship between service orders (orange); BSS and OSS internal protocols and communications (green); OSS (i.e., OSS, NMS, and EMS) protocols and communication with other systems (blue); and OSS (i.e., OSS, EMS, and NMS) protocols and communication with NEs (purple). Note that although data collection may use the same protocols as

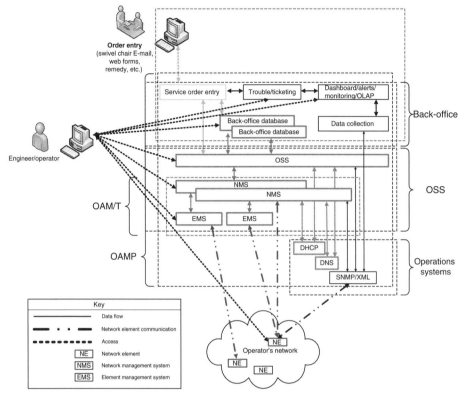

Figure 5.4. Detailed view of MPT systems.

provisioning (SNMP, XML, etc.) the systems are typically separate and in any event are not part of the provisioning process (excepting their possible use for validation of the configuration or later testing).

Now we will examine a number of both widely used provisioning protocols, some utilized only for specific types of NEs (i.e., Class 5 VoIP switches, IMS subsystems, Mobile devices, or DOCSIS and DPoE devices) and move on to examine protocols that are just coming into use.

5.4.4.1 Simple Network Management Protocol

Simple Network Management Protocol (SNMP) is the most widely used protocol for management of NEs. Many GUIs use SNMP for provisioning and data collection from NEs. SNMP is less often, but occasionally, used to manage some aspects of systems (servers or computers).

SNMP uses a data model called the management information base (MIB) that models parameters in a tree with instances for parameters for which there can be more than one instance (i.e., interfaces). SNMP uses three basic types of instructions to the NE from an agent or computer: "Get," "Set," and "Trap." As the names suggest, Get functions are used to retrieve data from an element. Set functions are used to set (and therefore provision) parameters. Unlike get and set functions that are initiated by a system (computer or server), the Trap function is initiated by the element itself to notify a system (an SNMP trap collector) of an event.

SNMP is specified by a set of IETF RFCs. The original version, just called SNMP, is now called SNMPv1 since the introduction of the current and second version, SNMPv2. Additional RFCs specify variations known as SNMPv2c and SNMPv2u. Of these, both SNMPv1 and SNMPv2c are widely used by service providers. The current version, SNMPv3, is more commonly used in newer equipment deployments. The principal reason to use SNMPv3 for most operators is for improvements in security supporting MD5 and SHA authentication.

The bulk of SNMP use is actually for data collection, as depicted in Figure 5.4. Traps are also widely used. But SNMP is rarely used for provisioning. The MIB data structure that SNMP uses includes both standardized (published) models and allows for vendor-specific additions. The model is difficult to use for provisioning because there is no strong correlation between the standard models and vendor specifics. For example, if you had to configure the Ethernet settings on an interface, you might use standard MIB models to do so, but then would have to load and use a vendor-specific model for configuring routing protocols on that same interface. The model does easily allow for the arbitrary formation of groups of parameters, such as all of the setting for an interface, or all routing settings for a router across all interfaces. In summary, it is a rigid, inefficient, and largely insecure protocol.

Using SNMP for data collection only limits the security vulnerabilities because SNMP settings are not used or if possible blocked. Even then, SNMP is so wildly inefficient that the demand to replace it is mounting. Large-scale service providers networks have been known to have more SNMP traffic (as a percentage of packets and bits) then production VoIP. When a network management protocol consumes more bandwidth than one of the principal (albeit smallest) applications in the network, it is a sign of gross inefficiency.

As to provisioning, it is rare, outside of vendor-specific EMS that does so, to see SNMP used for provisioning.

5.4.4.2 Extensible Markup Language APIs

Extensible Markup Language (XML), like HTML, a structured markup language. Like HTML, it features a human readable (textual) format. It is also similar to HTML in syntax because it uses tags (Figure 5.5).

XML uses tags syntactically similar to HTML. Beyond tags, the syntax also support attributes such as sources, classes, or other parameters, again, in the same way as HTML.

Although HTML is limited to markup for displays, XML could be used for such applications, but can be used for any type of structured data exchange and formatting. XML is so powerful that a new markup language, XHTML is based on XML. XML supports various models called schemes for data models. These include the Document-Type Definition (DTD), XML Schema, Document Schema Description Languages (DSDL), and so on. These are more formal definitions of data modeling, which offer defined or standardized structures and language.

<start-tag>data</end-tag>

Figure 5.5. XML example.

XML is very appealing for network operators because it does not have a strict (standardized) object or data structure. It allows vendors and operators to develop any possible data structure quickly and easily with a text editor. The resulting objects and the applications (programs) that utilize them are easy to develop with because they are human readable and portable. Portable means creating a format and readily sharing it. A related protocol, Extensible Messaging and Presence Protocol (XMPP) extend XML beyond file exchanges into the role of shorter form messages.

XML is widely supported in web services development environments including all of the typical scripting languages and web development languages. XML-based syntax is also used by other protocols such as SOAP and NETCONF. XML or XML and SOAP have been growing in use for provisioning systems. They are widely used through the telecom industry including in telephony, cable, Ethernet, and mobile systems.

REST and RESTful APIs are beginning to display XML and SOAP. Although XML and SOAP are very capable, extensible, and portable, REST and RESTful APIs offer an even further improvement because they place data directly in visible URIs instead of being hidden in data structures in files or messages. This ability to more simply form communications is particularly attractive in the network. REST models that also use SOAP offer a hybrid approach with URI signaling, but payloads that are SOAP objects (which may contain XML data structures within). RESTful APIs, by definition, never contain such objects, but can be burdensome and inefficient for lengthy transactions. For example, imagine sending an initial configuration for a router, which may be thousands of lines of commands, with RESTful HTTP messages!

Although RESTful APIs are likely to take a domain position for short-message provisioning and other OAMPT functions, REST APIs that utilize XML/SOAP or other object and data models are likely to be used for more complex data exchanges such as initial provisioning.

5.4.4.3 Simple Object Access Protocol

Simple Object Access Protocol (SOAP) is a protocol designed to communicate XML data. SOAP itself is usually carried over HTTP or HTTPS. It is widely used in web services because it offers a simple method for sending and receiving objects. The simplicity of the protocol hides the underlying complexity of objects and data they may contain. This places SOAP in the position of doing the opposite of RESTful exchanges. But the simplicity for complex objects, despite the appeal of RESTful notions, is likely to remain in use because of the very complex data and object models used, at least in telecommunications services. Like HTTP and XML, SOAP[6] is defined by the W3C [15].

5.4.4.4 JavaScript Object Notation

JSON[7] is a structured data object model, originally derived from JavaScript (although not dependent on Java or the use of Java in any way) similar in capability to XML. The primary distinction of JSON (Figure 5.6) is that it omits the markup language syntax required in XML (Figure 5.7) and SOAP, making JSON far easier to read than XML.

---

[6] http://www.w3.org/TR/soap/.

[7] The JSON-RPC is an open standard collaboratively developed outside of any standards organizations. JSON-RPC home is at http://www.jsonrpc.org/, JSON-RPC 2.0 Specification 2.0 (2010) is available there.

<object-name>object-data</object-name>

Figure 5.6. JSON object example.

object-name: object-data

Figure 5.7. XML object example.

JSON is specified by the Internet Engineering Task Force (IETF). There is a MIME type [16] for JSON, and a remote procedure call (RPC)[8] has been developed for JSON. JSON is widely used in web services and telecommunications services.

JSON is widely used in JavaScript, Ajax, and other web-based languages. JSON encoding and decoding has been integrated into most web browsers. It is also supported by Microsoft's .NET Framework[9].

5.4.4.5 Network Configuration Protocol

The NETCONF Protocol [17, 18] was developed by the IETF for the explicit purpose of providing a next-generation object-oriented transaction framework for the management of NEs. NETCONF is based on an RCP transaction model that uses XML data models (encoding). As discussed earlier, SNMP has a tree-based object model for devices, but each data parameter must be handled with individual objects. SNMP does not offer the ability to organize data into related objects (i.e., a subscriber or customer) and then set or get (using the transaction terminology of SNMP) the object with a single transaction. NETCONF overcomes this limitation offering a much more efficient model for moves, adds, and changes typical in provisioning. NETCONF operates over a variety of possible transports including the operation of SSH of both automated interfaces as well as an interactive (what NETCONF calls "console") interface. M2M implementations typically utilize SSH, TLS (and SSL), or BEEP for very lightweight implementations on simple devices.

NETCONF's transaction model supports a very simple model for functions typical in provisioning. The transaction model supports stateful transactions during an established session.

5.4.4.6 YANG

YANG [19] is an XML tree model that defines specific hierarchies for NETCONF called "YANG models." Each model describes a particular type of network service, function, or subsystem to be managed. In particular, a YANG model identifies the data types, lengths, or other characteristics (parameters) for each model. The obvious challenge is creating or requiring the consistent use of the same model across multiple vendors. To date standardizing those models has been a challenge across the telecom industry as a whole. But some efforts like the CableLabs CCAP (see Section 3.2) effort, driven by the MSOs, have created YANG models that are not vendor specific.

Although not required by NETCONF, YANG is the only standardized model for use with NETCONF.

---

[8] See footnote 7.

[9] http://en.wikipedia.org/wiki/Jayrock.

5.4.4.7 Common Object Request Broker Architecture

The CORBA [20] is a somewhat dated, object model system defined by the OMG[10] for systems and network configuration and management. Although widely accepted in traditional telephony systems management, CORBA is rarely used by MSOs because most NMS that utilize CORBA are legacy telephony systems. MSOs more recent entry into voice services has not burdened them with this legacy technology. Although efforts were made to update CORBA by adding support for HTTP- and XML-based transport and modeling (respectively), its use is largely deprecated in favor of XML-based APIs. In telephony, the Parlay and Parlay X APIs replaced CORBA functionality with SOAP (XML/SOAP). The current OMA/OneAPI [21] uses the OMA API that is a RESTful API [22]. Some other (typically older) OMA APIs use SOAP. As MSOs move to provide more mobile services, the OMA APIs will come into use for providing device and location information, software updates, and other mature OMA technology functions. Notably, while client telephony and mobile functions of CORBA were replaced with XML, SOAP, and REST, the infrastructure devices (not managed by OMA) have largely been replaced with IP-based core services[11] that utilize IP-based managed protocols and more recently web services-based architectures such as REST.

5.4.4.8 Representational State Transfer (REST and RESTful) Architecture APIs

RESTful architectures are thus distinguished as those (again using URIs transported over HTTP/S) that use only data contained in the URI (and which do not rely on extrinsic data or objects). Freedom from state and external objects is an important characteristic of RESTful architectures. By not keeping state, it makes it possible for humans, or machines capturing URI messages, to understand what the input or output (message) is and what action or actions (what are called HTTP verbs) are. Reply messages, for example, would contain both the action and a result code indicating the outcome of a requested action. By avoiding state, RESTful architectures can easily support complex environments where there are multiple peers, without synchronization and other transaction complexities. The software is much simpler, easier for programmers and operators to develop, use, understand, and improve upon. It also encourages interoperability because without "unknown" states and objects, everything in a transaction is "self-contained" in the URI (message).

The challenge for RESTful implementations, such as CLI implementations and SNMP, is that such simple representations of actions (verbs) and data are typically limited to one action (verb) for each message. As a result, like the CLI and SNMP before RESTful, RESTful architectures require a very high number of messages to accomplish large-scale provisioning tasks. By contrast, protocols that do not reveal the state and which carry data in object sets that are externally defined or not self-contained, are far more bandwidth and transaction efficient.

5.4.5 SDN Provisioning

SDN (see Section 6.4) can utilize any provisioning protocol. Most current efforts are focused on the use of NETCONF [17, 18] and Yang [19] or other XML-based APIs. Some approaches use REST alongside NETCONF and YANG. Other approaches use

---

[10] CORBA is a standard of the Object Management Group (OMG).

[11] See the Open Mobile Alliance (OMA) API inventory at http://technical.openmobilealliance. org/API/APIsInventory. aspx (accessed August 26, 2014).

REST with XML. Yet others offer a simpler RESTful approach. The latter of these, at least in part because of their simplicity, have been the first to market. Whether adding RESTful APIs and support really constitutes "SDN" or is just a new management interface, is a matter of marketing terms and what SDN is or is not.

The challenge for SDN and NFV is that both efforts have an objective of providing real-time or near real-time control of the forwarding with an external control source. This places a significant burden on provisioning not just to be efficient, but to be fast enough to support real-time or near-real time exchanges in order to support dynamic changes to forwarding in the network. In practice, the delivery of SDN and NFV systems has been limited to those which, like traditional systems, are provisioned in advance of services. With any external (provisioning) system, it will never be possible to have a control loop that operates in real-time as effectively (as fast) as a locally controlled embedded system. Thus, some "real-time" especially stateful security behaviors may never be moved up to an SDN controller. Other functions such as the addition of new customers, for which the time scale to implement is much slower (whole seconds is more than fast enough) are likely to be the strength of SDN's contribution to provisioning.

While SDN may be poster child for commoditization, standards-based OAMPT are not the real objective of SDN. SDN aims deeper to remove the forwarding logic from distributed systems and centralize not only provisioning, but also forwarding (see Section 6.4 for details of SDN).

5.4.6 Network Systems Provisioning

Network systems are the systems required to operate a network service. Widely understood examples of network services for IP networks are DHCP and DNS that are typically required to provide IP addresses and provide IP address to name and name to IP address lookups, respectively. Other examples many includes proxies, caches, load balancers, or other systems.

Because many of these systems operate on servers, some operators, OSS vendors, and some of the literature describe this as "server provisioning." Although arguably true in some cases, not all of these processes actually operate on servers. For example, most load balancing is done on NEs (sometimes called appliances) that are embedded systems and are definitely not servers. The same is true for caching, proxies, and other functions. Calling network systems "servers" also opens the possibility of confusing services provided by servers (which we call applications) with network services. Again, it might be true from some perspectives that even DNS and DHCP are applications; but more typically when we use the word "application," we mean an application that a subscriber or customer is utilizing. The term "server provisioning" may also refer to the process of configuring the storage, memory, operating system, and other functions on a real or virtual server. We use the term for this latter meaning (see Section 5.4.7).

Hereafter, we use the term "network systems" to refer to systems within the network that are required to operate the network and network-based services and which communicate directly with NEs. DNS and DHCP are network systems. If LDAP is used for DNS or DHCP, it operates behind the DNS, DHCP, and other systems and is not here considered a network system because it LDAP does not communicate directly with the NEs. Some systems such as SNMP collectors or relays could be considered network systems (because they do communicate directly with NEs) or they could be considered part of the OSS. Which portion of the architecture they belong to is often a function of a particular operator's organization (who runs or "owns" the box), then the particular functionality.

5.4.7 Server Provisioning

The term "server provisioning" refers here, to the process of configuring the storage, memory, operating system, and other functions on a real or virtual server. Server provisioning is rarely a part of the process for MSO services because most MSO services are focused on access and telecommunication services, not computing services. Computing services, hosting, and other server-based applications, software, and infrastructure require server resources. Historically, hosting providers are the service providers that had server provisioning. Some hosting providers, such as Rackspace, have transitioned from a traditional hosting model (of both dedicated and virtual hosting) to a cloud service model that is entirely virtualized.

As MSOs move increasingly not only to create applications for consumers but also to build multimedia and communication platforms that serve industries, businesses, and consumers, the number of servers within MSO networks has grown dramatically and will continue to grow.

Products like HP Opsware offer provisioning, configuration management, and automation for servers. Competing software from vendors such as Oracle vertically integrate some of that functionality both for standalone and blade servers.

5.4.8 NE Provisioning

NEs usually refer to elements in the network that are in the service provider's premise and are owned and operated by the service provider. Typical elements include transport equipment such as DWDM and (some legacy SONET) equipment, IP routers, Ethernet switches, load balancers, access equipment (CMTS, CWDM, Ethernet, EPON or DPoE Systems, or SONET), wireless access points, and so on.

NEs are typically provisioned with CLI or vendor-specific GUIs that are usually incorporated in vendors EMS. Among MSOs, CLI still dominate most provisioning. GUIs are however used for optical transport equipment such as DWDM. The rate of provisioning of these is much slower than the more rapid changes in the IP network and for higher layer services.

CMTS are provisioned by CLI, although scripts are regularly used to update and maintain systems. The same is true of IP routers and switches. Some CMTS and newer CCAP systems feature NETCONF, YANG, and XML provisioning and operations support. These systems are just coming to market now, but they and the OSS to support them are not yet widely deployed. Because CCAP systems are backward compatible with existing OAMPT systems, the availability of new APIs and systems does not force MSOs to support them. We can expect the OSS to develop over time to support new features more than to support new CCAP equipment.

SNMP is typically used for NE data collection, but not provisioning. Some Ethernet services utilizing CPEs that are Network Interface Devices (NIDs) are provisioned with SNMP.

5.4.9 Customer Premise Equipment Provisioning

Customer premises equipment (CPE) is equipment on the customers' premise that belongs to the service provider. The service provider is responsible for installing, provisioning, and operating CPE. The methods used to provision CPE are the focus of most MSO provisioning. This is, by count, CPE's outnumber all other equipment in the MSOs network. For every single Ethernet switch, there may be dozens or even hundreds of commercial service customers. For every single CMTS or CCAP system, there are typically hundreds or thousands of customers and CPE.

It is no surprise then that the bulk of OAMPT efforts are focused on managing and automating these devices. Although each device typically only supports one customer, their location at the end of the access network makes them inaccessible if they experience problems. They are also "single points of failure" (SPOFs). Unlike most NEs that can be operated in a redundant configuration, there is no redundant CPE model.

DOCSIS- and DPoE-based services use the same back-office and OSS systems that support residential DOCSIS provisioning. This model uses DHCP and TFTP to dynamically assign and address and retrieve (in some cases) or, more often, dynamically generate a configuration file for each CM or DPoE ONU. CM and DPoE ONU configuration files use Type–Length–Value (TLV) encoding. TVLs are common in many telecommunications protocol data units (PDUs); but in DOCSIS, the TLVs are initially communicated to the CPE in a configuration file called a bin file because it is in binary format. The bin (binary) file contains a binary representation of the TLVs that are typically notated in hexadecimal format when decoded in order to facilitate human readability when troubleshooting or testing.

A DPoE ONU configuration file (a CM file would be very similar) converted from binary to hexadecimal (and some fields further decoded to ASCII). As shown in Figure 5.8, the hexadecimal codes can be more readily understood. The Type codes are well known in DOCSIS.

The DOCSIS TLVs are nested using a universal numbering system. This means that, for example, for Type 22 shown later, the length 13 indicates that there will be a total of 13 bytes for all of Type 22. This includes 22, 22.1, and 22.14 in the example given. The Length (L) of 13 includes not only the data fields but also the Length (L) of the subtypes such as the Length 03 for 22.3, Length 01 for 22.1, and Length 14 of 22.14. If you add up the bytes 03, 02, 00 01, 01, 01 01, 14, 4 (04), 02, 02, and finally 00 11, you find 13 bytes as indicated in the Length for 22.

Following the TLVs, you ultimately find a subtype such as 22.14, which is a parameter called the S-VLAN ID. In this given example, the S-VLAN ID to S-Tag for short is 0x011, which is S-VLAN 17 (converted from hexadecimal to decimal). Other notable

```
03 01 01
03 02 0001 01 01 01 14 4 02 02 0x0011
23 13 03 02 0001 01 01 02 14 4 2 2 0x0011
24 44 06 01 07 10 04 5000000 08 04 5000000 15 01 04 17 04 5000 01 02 0001 43 14 08 03 FFFFFF 05 7 01 05 EPL1
25 19 06 01 07 01 02 0001 10 04 10000000 08 04 10000000
43 23 08 03FFFFFFFF 05 16 01 05 EPL1 04 04 0x40 13 01 01
255
```

Figure 5.8. Hexadecimal representation of TLV.

TABLE 5.1.   The Full De-coded Type Length Value Data for a Typical DPoE ONU Configuration

Required	Type	Length	Value or Sub-Type	Length	Value	Length	Value
	03	01	01				
22	22	13					
22.3			03	02	00 01		
22.1			01	01	01		
22.14			14	4	02	02	0×0011
23.13	23	13					
23.13.3			03	02	00 01		
23.13.1			01	01	02		
23.13.14			14	4	02	02	0×0011
24	24	44					
24.6			06	01	07		
24.10			10	04	5 00 00 00		
24.8			08	04	5 00 00 00		
24.15			15	01	04		
24.17			17	04	5 0 0 0		
24.1			01	02	00 01		
24.43			43	14			
24.43.8					08	03	FF FF FF
24.43.5					05	07	
24.43.1					01	05	EPL1
25	25	19					
25.6			06	01	07		
25.1			01	02	00 01		
25.10			10	04	10 00 00 00		
25.8			08	04	10 00 00 00		
43	43	23	08 03FFFFFFFF				
43.5			05	16			
43.5.1					01	05	EPL1
43.5.4					04	04	0×40
43.5.13					13	01	01
255	255						

parameters are the upstream data rate of 5 Mbps indicated in the values for 24.10 and downstream data rate in 25.10.

Type 255 indicates the end of the bin file.

A full decode of a typical DPoE ONU (a CM file would be very similar) is shown in Tables 5.1, 5.2, and 5.3. This particular configuration is for an MEF Ethernet private line (EPL) with a data rate of 5 Mbps in one direction and 10 Mbps in the other direction over SVLAN ID (S-VID) 17. DPoE supports provider bridging (utilized in this case) as well as provider backbone bridging (not depicted here).

TABLE 5.2.   The Full Coded Type Length Value Data for a Typical DPoE ONU
Configuration

Required	Complete TLV
	03 01 01
22	03 02 0001 01 01 01 14 4 02 02 0×0011
22.3	
22.1	
22.14	
23.13	23 13 03 02 0001 01 01 02 14 4 2 2 0×0011
23.13.3	
23.13.1	
23.13.14	
24	24 44 06 01 07 10 04 500000 08 04 5000000 15 01 04 17 04 5000 01 02 0001 43 14 08 03 FFFFFF 05 7 01 05 EPL1
24.6	
24.10	example shown above is 5Mbps or 5,000,000kbps
24.8	higher data rates such as 100Mbps will still have the LEN of the parameter at 04 for
24.15	example data rates such as 1Gbps or 1,000,000,000kbps will extend the LEN of the
24.17	parameter to 05
24.1	
24.43	
24.43.8	
24.43.5	
24.43.1	
25	25 19 06 01 07 01 02 0001 10 04 10000000 08 04 10000000
25.6	
25.1	example shown above is 10Mbps or 10,000,000kbps
25.10	
25.8	
43	43 23 08 03FFFFFFFF 05 16 01 05 EPL1 04 04 0×40 13 01 01
43.5	
43.5.1	
43.5.4	
43.5.13	
255	255

5.4.10 Customer Equipment Provisioning

Customer equipment (CE) is a term used by the MEF[12] to describe the equipment on the
customer premise that belongs to the customer (not to the service provider). More tech-
nically, CE is the equipment that "attaches to the network at the User-Network Interface
(UNI) using a standard 10 Mbps, 100 Mbps, 1 Gbps, or 10 Gbps Ethernet interface".[13]

[12] See footnote 6.

[13] Such as 3GPP Long Term Evolution (LTE) or LTE-Advanced.

TABLE 5.3. Description of Each TLV and sub-TLV

Required	Description	DPoE Default for full automation
	NACO	Y
22	Upstream Classifier	
22.3	SF Reference	N
22.1	Classifier Reference	N
22.14	SVLAN ID	N
23.13	Downstream Classifier	
23.13.3	SF Reference	N
23.13.1	Classifier Reference	N
23.13.14	SVLAN ID	N
24	Upstream Service Flow Encoding	
24.6	QoS Param Set Type Subtype	Y
24.10	Minimum Reserved Traffic Rate = US DATA RATE (IN KBPS)	US DATA RATE
24.8	Maximum Sustained Traffic Rate = US DATA RATE (IN KBPS)	US DATA RATE
24.15	Scheduling Type = RTPS	Y
24.17	Nominal Polling Interval	Y
24.1	Service Flow Reference = 0001	N
24.43	Identify service flow as classifier from TLV 43 above	N
24.43.8	Vendor ID for GEI	N
24.43.5	GEI 43.5 for L2VPN Encoding	N
24.43.1	VPNID Subtype (aka 43.5.1)	N
25	Downstream Service Flow Encoding	
25.6	QoS Param Set Type Subtype	N
25.1	Service Flow Reference = 0001	N
25.10	Minimum Reserved Traffic Rate = DS DATA RATE (IN KBPS)	DS DATA RATE
25.8	Maximum Sustained Traffic Rate = DS DATA RATE (IN KBPS)	DS DATA RATE
43	L2VPN Encap Mode	N
43.5	GEI .5 Subtype for L2VPN encoding (aka 43.5)	
43.5.1	VPNID Subtype (aka 43.5.1)	
43.5.4	CMIM(aka 43.5.4) = default	
43.5.13	L2VPN Mode (aka 43.5.12) = dpoe transport mode	
255	End of File	

CE differs from CPE in that CPE belongs to the service provider, but CE belongs to the customer. CE is almost always a term specific to Metro Ethernet services as described by the MEF.

In theory, the customer provisions their own CE. In practice, operators may recommend specific equipment (or even sell it to customers) and assist the customer in provisioning. Because the CE is outside of the service provider's network and on the other side of the UNI, there is no means to provision or manage the CE. It is, by definition and decidedly, outside of the scope of a Metro Ethernet service provider's network and responsibility to provision or manage CE.

5.4.11 Application Systems Provisioning

Many business services are focused on infrastructure such as access, VPNs, and so on, that do not require application systems. Those that do include voice, email, videoconferencing, messaging, and related media or multimedia services. Nearly all of these are SIP based and use SIP- or IMS-based application systems. Provisioning for these systems is typically vertically integrated into IMS systems by the IMS system vendors. The most significant exception is email and voice mail. In both cases, many MSOs have created the provisioning processes to manage these apart from other IMS elements. This is in part because of the need to support both IMS compatible and non-IMS compatible voice services for both business and residential customers.

Over time, we can expect MSOs to increasingly deliver either integrated applications, or partner with cloud service providers to provide these applications. In both cases, these are likely to use web service-based provisioning models that are widely uses in the SaaS industry.

5.4.12 Cloud Services Provisioning

As service providers increasing bundle more services and more complex services, they are either partnering with cloud services providers or operating their own services as a cloud service, where a separate team and separate systems operate the applications (AaaS), software (SaaS), or infrastructure (IaaS). Whether hosted in house or externally (in the cloud), provisioning processes now commonly interface with other service providers or equivalently place service orders (see Section 5.3 for a more detailed discussion of service orders) through order submittal processes or more preferably APIs.

Although in its infancy, this trend is likely to dominate the service order and provisioning processes in coming years.

5.4.13 Self-Service Provisioning

Self-service provisioning is a provisioning process that allows customers to provision their own services. A widely known examples is Amazon Web Services (AWS) that use either their web-based self-service provisioning or their APIs to allow customers to provision network infrastructure (i.e., Amazon Direct Connect, a VPN service), Content Delivery Network services, IP address, bandwidth, load balancing, servers (so-called AWS Elastic Computing or ECS), storage (elastic bandwidth service or EBS), and so on. Notably, AWS supports self-service provisioning not only for its own services but also for services available through third parties, so-called AWS marketplace services.

Among MSOs, the most popular self-service provisioning functions are for voice service setup (i.e., the calling party name used for caller-ID, voice mail setup, etc.) and for email. Although self-service provisioning has also been used for residential CMs (through partnerships with retailers), its use has been limited because the perquisite connectivity to the network must either be in place or must still be handled through field service.

In business service, self-service provisioning is typically limited by the need for security and reliability. Technologies such as the MEF E-LMI [23] allow for the possibility of customer self-service provisioning, but the processes required for security, contractual obligations, and other requirements have not yet been developed to offer such functionality to business services customers.

5.5 FAULT MANAGEMENT

Fault management for business services in the cable network is based on the fundamental principles of detecting, isolating and correcting abnormal operation of these services that general have SLAs. Often, such abnormal operation is detected by proactive performance management functions or from autonomous events and alarms from the network layer, typically via an NE in the service delivery chain. Once a failure is identified, a service trouble report is generated in the back-office so that on-demand fault isolation and corrective actions can be taken.

Faults at the link level can be identified by sending periodic IEEE 802.3ah [24, 25] Link OAM PDU messages between the link end points. The link end points, or OAM peers, establish their connectivity through a discovery process where they also establish their Link OAM capabilities. Figure 5.9 illustrates a common Link OAM use case. The protocols support:

- Remote loopback
  - Fault localization and link performance testing
- Remote failure indication
  - Dying gasp
  - Link fault
  - Critical event
  - Link events

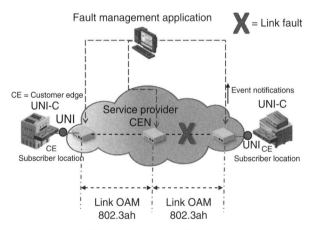

Figure 5.9. Link OAM use case.

- Link monitoring
  - Event logging and autonomous notifications of link failures
- Loopback control
  - Enabling and disabling remote loopback in the peer OAM client
- Discovery
  - Discovery of OAM peers and their capabilities
- Unidirectional mode

When the NEs detect remote failure indications, such as dying gasp or link fault, autonomous alarms and events can be sent to back-office network monitoring applications, such as SNMP notification receivers. At this stage, MSOs can enable IEEE 802.3ah remote loopback to the OAM peer and send Ethernet frames to the OAM peer to be looped back to test the link for performance and packet loss. This type of testing can be useful for fault isolation purposes.

Faults at the service level can be identified by sending continuity check messages (CCMs) [26, 27] at periodic intervals between configured adjacent MEPs within an MEG. If the CCM messages do not arrive at a receiving MEP, the MEP declares a loss of connectivity defect and may autonomously notify the back-office if provisioned to do so. CCMs are one-way messages similar to keep alive or heartbeat messages. CCMs are also capable of detecting inadvertently cross-connected EVCs where one MEP receives a CCM from an MEP associated with a different EVC. If the CCM interval is configured, MEPs are required to generate CCMs on a periodic basis as defined by the periodic interval. Figure 5.10 below illustrates a common CCM use case where MEP A and MEP B are exchanging CCM messages. A service fault occurs between the two MEPs causing MEP B to lose three consecutive CCM messages from MEP A. MEP B declares a loss of connectivity alarm and sets the RDI bit in the CCM messages sent toward MEP A.

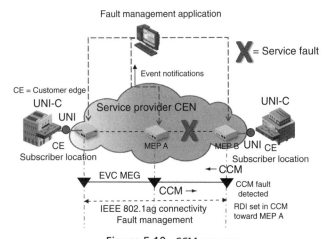

Figure 5.10. CCM use case.

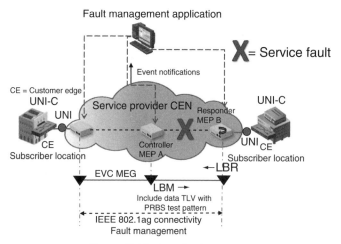

Figure 5.11. Loopback use case.

An alarm indication signal (AIS) [27, 28] is another SOAM function that can be used to suppress alarms following detection of defect conditions such as signal fail conditions in the case that CCMs are enabled. In addition, the SOAM function remote defect indication (RDI) can be used by an MEP to communicate to its remote peer MEPs that a defection condition has been encountered. RDI is a 1-bit field carried in the CCM frames and is not a separate message like AIS.

Once a defect condition is present, the SOAM function loopback [26, 27] can be utilized to trigger transmission of a configured number of periodic loopback messages (LBMs) from a source MEP to a destination MEP. When the destination MEP receives the LBM, a loopback reply message (LBR) is returned to the source MEP. The loopback function is similar to an IP layer ping function. MEPs are capable of generating LBMs upon on-demand request via management interfaces. Figure 5.11 illustrates a common SOAM loopback use case where Controller MEP A initiates a loopback with Responder MEP B. MEP A includes a data TLV in the LBM message include a test pattern. If the resulting LBR message received from MEP B contains test pattern errors (or MEP A fails to receive the LBR), additional troubleshooting can be performed on the link between MEP A and B.

In order to isolate detected faults, the SOAM function LinkTrace [26, 27] can be utilized to initiate transmission of a Linktrace Message (LTM) from a source MEP to a destination MEP. However, unlike the Loopback, intermediate MIPs and MEPs (referred to generally as maintenance points) reply a Linktrace Reply Message (LTR) along with the destination MEP. The intermediate nodes are required to forward the LTM message on to the destination MEP. In this scenario, the source MEP receives multiple LTR responses and learns the path taken by the LTM and the last reachable point in the link or EVC/OVC. The Linktrace function is similar to an IP layer traceroute function. MEPs are capable of generating LTMs upon on-demand request via management interfaces. Figure 5.12 illustrates a common SOAM Linktrace use case where Controller MEP A initiates a linktrace path discovery with Responder MEP B. The intermediate MP C responds with an LTR and forwards the LTM message on to MEP B.

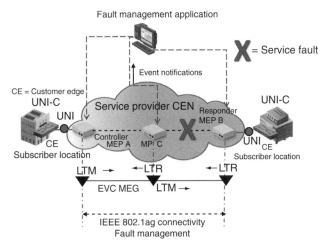

Figure 5.12. Linktrace use case.

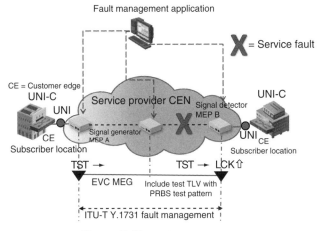

Figure 5.13. Lock/Test use case.

To perform additional diagnostic testing, the SOAM functions Lock and Test are provided [27, 28, 29, 30]. The Lock Message (LCK) is a signal function used to control and signal the administrative locking state of an MEP The Lock function is generally used in conjunction with the Test function. Locking an MEP prevents data traffic forwarding toward MEPs expecting this traffic plus signals the locking state to these MEPs. Once an MEP is administratively locked, the Test Message (TST) is used between peer MEPs to perform one-way in-service and out-of-service testing. This type of testing can assess and measure throughput, frame-loss and bit-errors. The data streams can contain pseudo random bit streams, all zeros, and other test patterns. Figure 5.13 illustrates a common SOAM Test and Lock use case where MEP A initiates a testing with MEP B. MEP A contains a test signal generator function and sends Test messages containing a Test TLV including a PRBS test pattern. MEP B contains a test signal detector function that checks for bit errors in the received test message, reporting any detected errors. It should be noted that for out-of-service testing, the MEPs are administratively locked, at which points the MEPs send LCK messages to higher layer MEGs, such as the Subscriber MEG in the use case described in Figure 5.13. The LCK message notifies MEPs on the higher layer MEG that customer traffic will be affected and to suppress defect notification.

Once the defect condition has been isolated and corrected, the service trouble report should be reported on and closed.

5.5.1 IEEE 802.1ag SOAM Framework

The fault management and performance management model is based on the IEEE 802.1ag [26] SOAM architectural framework. This framework is based on 8 hierarchical maintenance domains that bind OAM flows and assign OAM responsibilities. The hierarchical OAM domain concept is shown in Figure 5.14. Each OAM domain can be independently managed within the SOAM framework.

Figure 5.14 illustrates four different OAM domains. The customer domain extends from the UNI-C to UNI-C and is available to the customer. The provider domain extends from the UNI-N to UNI-N, or the full EVC, and is available to the service rovider. Each operator domain extends within each carrier Ethernet network (CEN) or across each OVC and is available to each of the access providers/CEN operators. Higher maintenance domains are transparent to lower domain levels. Note that maintenance domains and OAM domains are equivalent terms.

The OAM framework defines functional entities for OAM PDU message transport. A maintenance entity group end point (MEP) is an instantiation of a management interface where PDUs are sourced and sinked. It is associated with the interface of an NE such as a network interface device that is part of a data service path. MEPs are indicated by means of a triangle in Figure 5.1. MEPs are associated with peer MEPs and are located at

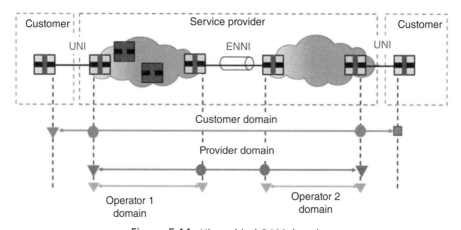

Figure 5.14. Hierarchical OAM domains.

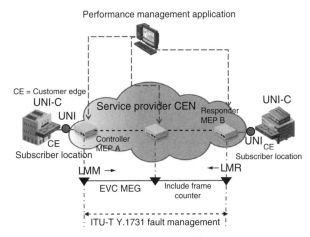

Figure 5.15. Frame loss ratio use case using DMM/DMR.

TABLE 5.4. MEG Levels

MEG	Usage	Default Maintenance Domain Level
Subscriber MEG	Subscriber monitoring of an Ethernet service	6
Test MEG	Service provider isolation of subscriber reported problems	5
EVC MEG	Service provider monitoring of provided service	4
Service provider MEG	Service provider monitoring of service provider network	3
UTA SP MEG	Service provider monitoring of UNI tunnel access	3
Operator MEG	MEN operator monitoring of their portion of a network	2
UNI MEG	Service provider monitoring of a UNI	1
ENNI MEG	Network operators' monitoring of an ENNI	1

the end of the data service path or EVC to bind the service path. A maintenance entity group intermediate point (MIP) is a simplified instantiation of an MEP with fewer functionality. MIPs generally only respond to OAM PDUs and do not source them. MIPs are also associated with the interface of an NE, such as an Ethernet switch, between two MEPs. MIPs are represented by means of a circle in Figure 5.15. The OAM PDUs that are exchanged between MEPs and MIPs are referred to as service OAM frames.

A maintenance entity group is a set of MEPs where each MEP is configured with the same maintenance domain level to establish and verify the integrity of a single service instance. The maintenance domain level is the unique integer identifier for the MEG, also known as the MEG level. Table 5.4 defines the key maintenance domain levels along with their usages [31].

5.6 PERFORMANCE MANAGEMENT

Performance management for business services in the cable network is based on the fundamental principles of gathering and analyzing statistical data for the purpose of monitoring and correcting the behavior and effectiveness of the services, network, network resources, and other equipment and to aid in planning, provisioning, maintenance, and the measurement of quality. Typically, this involves the recurring collection of counters and measurement data from the network resources in the service delivery chain. Therefore, much of the performance management functionality is a proactive management function. Such functions entail performing near real-time analytics and historical trending on the counters, measurement data, and Threshold Crossing Alert notification monitoring to evaluate the service quality and availability. Comparative analysis can also be performed against the SLA performance objectives to ensure that the service meets the quality guarantees promised to the end subscriber.

Performance management for Metro Ethernet services relies on a well-standardized SOAM framework that uses the ITU-T Y.1731 protocol [28, 30]. The MEF has extended this protocol with additional functionality in MEF 35 Service OAM Performance Monitoring Implementation Agreement [32, 33, 34]. The SOAM protocol provides the ability to monitor end-to-end service performance, for point-to-point and multipoint connections, across OAM domains as well as provides visibility into SLA compliance based on service-level specification performance objectives.

Performance attributes that are measured include the following:

- Frame loss ratio (FLR)
- Frame delay (FD)
- Inter-frame delay variation (IFDV)
- Availability performance
- Resiliency performance

FLR is a characterization of the number of lost frames between the ingress UNI and the egress UNI, for a point-to-point EVC, expressed as a percentage. FLR measurements use loss measurement messages (LMMs) and loss measurement replies (LMRs) that are sent between MEPs with recent nonmanagement traffic counter information. The counters [35] are used to estimate the FLR of the nonmanagement traffic. This measurement is only useful in point-to-point topologies; flooding makes the measurement useless in other topologies.

MEPs are capable of generating and receiving LMM and LMR messages for E-Line service types. MEPs are also capable of calculating and reporting the one-way FLR based on frame counter information received from LMM and LMR messages. Figure 5.15 illustrates a common SOAM one-way FLR use case where Controller MEP A initiates measurements with Responder MEP B using the LMM and LMR messages. The LMM message contains a counter of the frames sent to MEP B. MEP A receives the LMR response messages and calculates the near-end and far-end FL measurements.

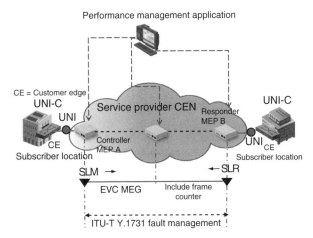

Figure 5.16. Frame loss ratio use case using SLM/SLR.

This scenario is referred to as a single-ended session since MEP A performs the measurement calculations.

Figure 5.16 illustrates the SOAM PM function using synthetic frames instead of data frames as shown in Figure 5.15. In this instance, the MEPs are using the SLM/ SLR protocol messages to measure and calculate the FLR PM metric. The SLM message is created and sent from the controller MEP A. This SLM message contains a Test Id to differentiate the session from other measurement sessions, the source MEP Id, and the value of the local counter containing the number of SLM messages as sent by MEP A. Once MEP B receives the SLM, MEP B copies the MEP A local counter from the SLM into the SLR that MEP B creates and also places it's local counter of received SLM messages and sends the SLR back to MEP A. The FLR metric is then calculated at MEP A and made available to the EMS. This use case supports the multipoint service definition, whereas Figure 5.15 only supports the point-to-point service definition.

Figure 5.17 illustrates one-way FLR use case where Controller MEP A initiates [34] measurements with Sink MEP B using the 1SL message. The 1SL is similar to the SLM message. MEP B receives the 1SL messages and calculates the near-end forward direction FL measurements. This scenario is referred to as a dual-ended session since MEP A initiates the measurement and MEP B performs the measurement calculations.

FD is a measure of the time required to transmit a service frame from the ingress UNI to the egress UNI, for a point-to-point EVC, where the performance is expressed as a measure of the delays experienced by the service frames belonging to the same class of service (CoS) instance.

One-way FD measurements use periodic 1DM messages sent by the source MEP to a destination MEP. The message contains a timestamp from the source MEP. The receiving MEP computes the one-way delay by subtracting the transmit timestamp from the received timestamp, which is created upon receipt of the 1DM. The 1DM protocol

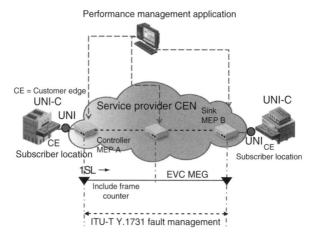

Figure 5.17. Frame loss ratio use case using 1SL.

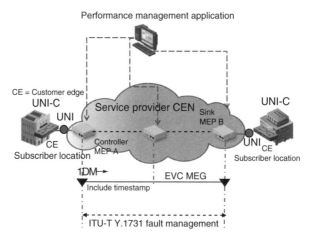

Figure 5.18. Frame delay use case using 1DM.

assumes that MEPs are all operating with synchronized clocks. MEPs are capable of computing one-way delay end-to-end across E-Line service types using successive 1DM messages. Figure 5.18 illustrates a one-way FD use case where Controller MEP A initiates measurements with Sink MEP B using the 1DM message. MEP B receives the 1DM messages and calculates the near-end forward direction FD measurements. This scenario is referred to as a dual-ended session since MEP A initiates the measurement and MEP B performs the measurement calculations.

Two-way FD measurements use two message types, the delay measurement message (DMM) and the delay measurement response (DMR). The DMM message is sent by the source MEP to a destination MEP. The message contains a timestamp from

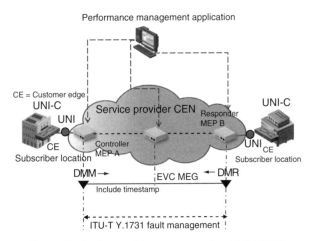

Figure 5.19. Frame delay use case using DMM/DMR.

the source MEP. The receiving MEP replies with a DMR message containing a copy of the received DMM timestamp, and optionally, a receive timestamp taken at DMM reception, and a transmit timestamp created at the time of DMR message transmission. The two-way delay can then be computed by the source MEP as the difference between its current timestamp and its original timestamp as received back in the DMR message. The addition of transmit and receive timestamps allows for the calculation of the destination MEP processing time. Two-way delay calculations do not require clock synchronization, and therefore MEPs are required to support two-way delay measurements using successive DMM/DMR messages. The MEF 35 standard does not require MEPs to support one-way delay measurements using the 1DM message. Figure 5.19 illustrates a SOAM two-way FD use case where Controller MEP A initiates measurements with Responder MEP B using the DMM and DMR messages. The DMM message contains a transmit timestamp sent to MEP B. MEP A receives the DMR response messages and calculates the two-way FD measurements. This scenario is referred to as a single-ended session since MEP A performs the measurement calculations.

IFDV is a measure of the difference in the FD experienced by a pair of service frames belonging to the same CoS instance. The same protocol and messages used for measuring FD are used for measuring IFDV.

Availability performance is defined as the percentage of time within a specified time interval T during which the service FLR is small. It is used to indicate the percentage of time that the Ethernet service is usable. Resiliency performance is defined as the number of high loss intervals (HLIs) and/or consecutive high loss intervals (CHLIs) in a time period T. HLI is a small interval contained in T with a high FLR. CHLI is a sequence of small time intervals contained in T, each with a high FLR. In order to measure and calculate the availability and resiliency performance metrics, the MEPs are capable of supporting the synthetic loss message (SLM) and synthetics loss reply (SLR) message.

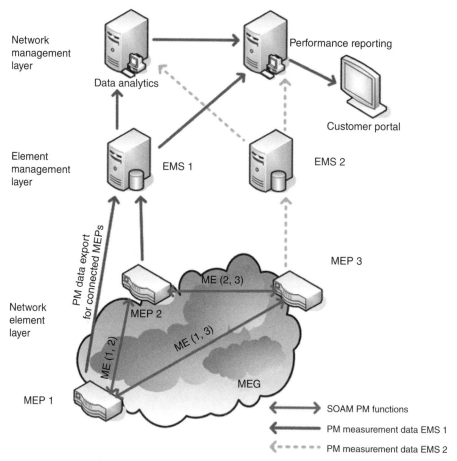

Figure 5.20. MEPs reporting PM metrics into back-office.

5.6.1 Performance Monitoring Solution Framework

The performance monitoring framework is based on collecting metrics, as discussed in Section 5.6, from the MEPs residing in the network layer. The network layer contains the resources implementing the various protocols to carry out the necessary quality and availability measurements. The performance monitoring service OAM framework utilizes the same OAM domain hierarchy as described in the fault management service OAM framework. Figure 5.20 illustrates three MEPs residing in the MEG all at the same MEG level. The MEPs are performing the DMM/DMR and SLM/SLR measurements as shown in the diagram with red arrows referred to as PM functions.

The EMS collect the resultant measurement data from the MEPs in the network, along the service delivery chain (e.g., along the EVC for a point-to-point E-Line service). Each EMS may perform processing on the data to determine if the service is performing within normal boundaries as set within the service-level specification or SLA. Any autonomous Performance Threshold Crossing Alert notification may likely be destined for an EMS. Each EMS also may forward the processed/filtered performance management data northbound to OSS systems for further data analytics and storage. In addition, this is where the customer portal presentation layer may be accessible, allowing the Ethernet service subscriber online access to a Web-based portal for performance-related statistics on the historical and near-real time quality of their service, as benchmarked against their SLA.

5.7 BILLING SYSTEMS AND FORMATS

The billing space for commercial services has been developing over the past 7–8 years within the cable industry. There are many solution providers that have complex service billing platforms with a great deal of flexibility in how bills are delivered to customers. Generally though, MSOs have faced the challenge of providing services to customers that have a variety of products that are delivered through complex provisioning as well as standard DOCSIS-based provisioning. For example, there might be a lawyers' office who has seven locations across a provider's footprint, and the firm wants to also pay for their employees Internet service at home. The provider could be asked to provide the bill for the seven business locations to a single accounts payable department, while all the residences are to receive their Internet service bill separately. The same firm might choose to have all the charges for all services included on a single bill sent to their accounts payable. The alternatives are simply too many to list, and this is what makes the business services ordering process so much more complex than residential services.

The industry has generally responded by deploying a second back-office with specific applications used to deliver the complex business services as described in earlier sections.

Below is a high-level diagram that illustrates how the business services systems were developed to capture the complexities of customers order and their installation particulars where appropriate and to leverage the existing "residential" platform to support small- and medium-sized business orders as they align more closely to "residential"-type products and services that don't require complex ordering.

Selling to a business is not the same as selling to an individual. The sales process can be significantly longer and requires a very different sales force than marketing-driven sales in the residential space. Because of this fact, the business services back-office leverages territory and lead management tools that ensure that multiple sales representatives don't sell to the same customer. When the lead becomes a customer, the order is generally taken in a business services complex order management platform because often the actual delivery of the services require solution design and build out leveraging potentially nonintegrated tools. In many occasions, the sale of a customer circuit to a location may require building out new fiber to the premise that requires outside plant construction and may require town planning approval and

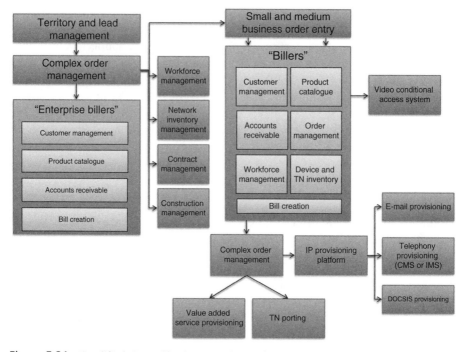

Figure 5.21. Simplified view of business services order management & billing vs residential.

permits. Like in the case of residential services, operators have purchased or developed many solutions to meet their needs.

Many of these tools leverage standard web services to integrate like SOAP or RESTful services to integrate. In some cases, the business services complex order management platform is the system that records an orders current status, and it requires team members to check in and update the status of the order in the platform manually (Figure 5.21).

From a billing viewpoint, customers still receive paper or pdf versions of their bills. There has been a growing call from larger customers to adopt some electronic data interchange (EDI) standard for providing their billing data. The number of customers requesting EDI-specific standards is very limited; and generally speaking, the enterprise billing platforms have made it possible to share the billing data in a structured format that can be electronically consumed by the customer and does not require the full implementation of an EDI standard.

5.8 SECURITY

The security threats that exist for residential services also apply for business services (see Section 3.8). All of the business service network architectures discussed in Chapter 4 have security functions and controls to help prevent those threats. The following sections explain what security mechanism exists for each architecture and recommendations on how to manage them.

5.8.1 DPoE Security

DPoE security provides the same features as DOCSIS security: preventing theft of service and information disclosure (loss of privacy). The main features are device authentication, traffic encryption, and secure software download. DPoE security details and requirements can be found in the DPoE 1.0 [36] and 2.0 [37] security specifications. SNMP MIB requirements are found in RFC 4131 [38], and the DPoE OSSI specification. While the features are similar the protocols used are quite different.

5.8.1.1 Device Authentication and Traffic Encryption

DPoE security has two different traffic encryption modes: downstream-only and bidirectional. The device authentication process and key exchange messaging is different between these modes.

For downstream-only mode, key exchange, and downstream encryption starts first before ONU device authentication. The DPoE system does not allow the ONU to complete registration and become operational unless both downstream encryption and device authentication are performed successfully.

Like DOCSIS security, X.509 digital certificates are used to authenticate the ONU. A unique device certificate and private key are installed in each ONU along with the manufacturer certification authority (CA) certificate that issued/signed the device certificate. The root CA certificate is installed in the DPoE system and acts as a trust anchor for validation purposes. CableLabs manages the DPoE certificate public key infrastructure (PKI), shown in Figure 5.22, for the cable industry.

The DPoE system authenticates each ONU before allowing it access the network. Authentication begins with the ONU sending its device certificate and manufacturer. CA certificate to the DPoE system using a protocol based on EAP-TLS. The DPoE system validates these certificates by verifying that they chain to a trusted root CA certificate. Validation includes other checks such as certificate expiration, revocation status, and MAC address verification.

Figure 5.22. DPoE certificate PKI hierarchy.

TABLE 5.5.   SNMP MIBS for Certificate Management [30]

Name	Description
docsBpi2CmtsDefaultSelfSignedManufCertTrust	Sets trust status of self-signed mfr CA certificates.
docsBpi2CmtsCheckCertValidityPeriods	Enables/disables certificate validity period checking.
docsBpi2CmtsProvisionedCmCertTable	Used to override CM device certificate validation status.
docsBpi2CmtsCACertTable	Table of known certificate authority certificates. Supports trust status configuration.

TABLE 5.6.   SNMP MIB for DPoE Key Exchange

Name	CM or CMTS MIB	Description
docsBpi2CmtsDefaultTEKLifetime	CMTS	Is used in DPoE Networks to configure the encryption key exchange timeout. Default is 600 s

There are a number of DPoE System SNMP MIBs for managing the DPOE certificate validation process (Table 5.5).

Once an ONU's device certificate has been validated, it is considered trusted and registration is allowed to complete.

The protocol used for key exchange is defined in the DPoE security specification. For downstream-only mode, the ONU generates an encryption key and sends it to the DPoE system. The DPoE system uses this key to encrypt downstream traffic. It is important to note that keys are sent upstream in the clear for downstream-only encryption mode. Those applications where there is a threat of upstream key compromise should use bidirectional encryption mode where encryption keys are never sent in the clear.

Bidirectional encryption mode starts with certificate device authentication based on EAP-TLS. Encryption keys are exchanged and setup securely using EAP-TLS. After the key exchange process, traffic encryption is applied in both downstream and upstream directions.

An SNMP MIB is used for DPoE key exchange (see Table 5.6):

Encryption is applied to the entire frame except the preamble. The preamble contains the LLID that is used by the ONU to determine which frame belongs to it and should be decrypted. The preamble also contains a security octet that has bit flags that indicate if the frame is encrypted and what the key identifier value is.

DPoE security supports early authentication and encryption (EAE) similar to DOCSIS security. EAE secures provisioning and ONU registration traffic that occurs before the configuration file is downloaded. The SNMP MIB used to configure EAE is named docsIf3MdCfgEarlyAuthEncrCtrl.

5.8.1.2 Secure Software Download

The integrity of software images downloaded to the ONU for update is verified using the secure software download (SSD) mechanism that is similar to DOCSIS SSD except that the signed image file is validated by the vCM in the DPoE System instead of the ONU. If validation is successful the vCM removes the signature and forwards the image to the ONU for installation. In DPoE 2.0 security image file validation will occur at both the vCM and the ONU.

A digital code verification certificate (CVC) is used to digitally sign the image. The format of the digital signature is PKCS #7, and it contains the CVC and the signing time. These items are used by the vCM during the validation process. The digital signature is appended to software image file. When the image file is downloaded to the vCM, it verifies that the CVC was issued/signed by the trusted CableLabs CVC root CA public key, that the organization name in the CVC matches the organization of the manufacturer of the ONU, that the CVC has the correct extensions, and it also checks the validity period of the CVC and the signing time. The image file must be signed by the manufacture and may be co-signed by the cable operator. CableLabs manages the certificate PKI for issuing CVCs to ONU manufacturers and cable operators.

DPoE SSD has a built in method for revoking CVCs and signed images. The ONU maintains the start time of the CVC validity period and the signing time of the last image update. These values are called time-varying control (TVC) values. They are sent to the vCM during registration. When a signature of an image is validated, the vC checks that the CVC validity start time and the signing time are equal to or greater than the TVC values. If they are greater, the TVC values are updated after successful completion of the other validation checks. If they are less than the TVC values the image file is rejected. This method automatically revokes the old CVC when a new CVC is used and revokes images with signatures having older signatures. Cable operators can control the signing time (keep it the same) on images to make it easier to switch back and forth between image download versions if needed.

Downloading an image involves first enabling SSD and then triggering the download. SSD is enabled by including the CVC in the configuration file. Once the vCM has validated the configuration file CVC, it is ready for triggering. Triggering the download is accomplished using configuration file TLV encoding settings or SNMP MIB sets.

TLV encoding 21 is used set the server URL and TLV encoding 9 is used to set the filename. If the filename is different than what the vCM already has it will trigger a TFTP download of the file from the server. When using SNMP, the docsDevSwServer MIB is used to specify the server URL, docsDevSwFilename MIB is used for the filename, and docsDevSwAdminStatus MIB is used to start the download. It is possible to disable the configuration file TLV encoding trigger when using SNMP.

5.8.1.3 Secure Provisioning

The configuration file provisioning process in the DPoE architecture is different than in the DOCSIS architecture. Instead of sending the configuration file to the ONU first and then having it forwarded to the DPoE System, the file is first sent to the DPoE system and then ONU-specific configuration settings are sent to the ONU. This avoids sending DPoE system service settings through an untrusted device that could be hacked by an attacker.

TABLE 5.7.   SNMP MIBs for Managing SAV [30]

Name	CM or CMTS MIB	Description
docsIf3MdCfgSrcAddrVerifEnabled	CMTS	Enables/disables SAV for the MAC domain
docsSecCmtsSavControlCmAuthEnable	CMTS	Enables/disables SAV static IP address policy settings
docsSecSavCmAuthTable	CMTS	Provides a read-only set of SAV policies based on configuration file settings
docsSecSavCfgListTable	CMTS	Defines the CMTS configured subnet prefix extension
docsSecSavStaticListTable	CMTS	Defines a subnet prefix extension based on CM statically signaled subnet prefixes to the CMTS.
docsSecCmtsCmSavStatsSavDiscards	CMTS	Provides the information about number of dropped upstream packets due to SAV failure.

Since the interface between the configuration file provisioning server and the DPoE system is expected to be secure the configuration file message integrity check TLV values are not used.

To help prevent unauthorized use of IP addresses, DPoE security has a feature called source address verification (SAV) like DOCSIS security. SAV only allows traffic to be forwarded from ONUs that are using IP address assigned by the cable operator. That can be dynamic IP addresses assigned using a DHCP server or static IP addresses that have been allocated to the subscriber (Table 5.7).

5.8.1.4 ONU Cloning

The DPoE architecture has the same ONU cloning threats as the DOCSIS architecture. The best way to prevent ONU cloning is to enable security using the configuration file TLV 29 parameter. This causes both certificate device authentication and traffic encryption to be performed.

5.8.2 EPoC Security

EPoC security is based on DPoE security. Downstream encryption mode is not allowed due to the increased threat of upstream snooping on the coax cable. Only bidirectional encryption mode is used.

When a Repeater-FCU is used, authentication and encryption traffic is forwarded between CNU and the DPoE system. When Bridge-FCU is used, it acts like a proxy. A bidirectional encryption connection is first established between the B-FCU and the

DPoE system. This connection aggregates traffic from multiple CNUs. Bidirectional encryption connections are also established between the B-FCU and each CNU. As traffic is forwarded between the CNU and the DPoE system, the B-FCU decrypts it from one connection and encrypts it for another.

When traffic is decrypted and encrypted as it passes from one connection to the next, there is a period of time when it is sent in the clear that is a potential vulnerability in B-FCU. Therefore, it is important that the B-FCU is located in a secure environment and that it only allows authorized access. It should also be hardened to help prevent any network-based attacks.

5.8.3 CCAP Security

CCAP security depends on the network supported. A DOCSIS implementation uses DOCSIS security (see section 3.8.1). A DPoE implementation uses DPoE security (see Section 5.8.1).

5.8.4 IMS Security

IMS performs mutual authentication between subscriber and the network before allowing them to receive service. The protocol used is called IMS Authentication and Key Agreement (AKA). It is based on HTTP digest. The network challenges the UE device by appending security headers to SIP registration messaging. The UE uses credential data stored in the USIM module to provide a challenge response.

Keys are derived from challenge information and credential data. On the network side, the S-CSCF (see Chapter 3) derives the keys after authenticating the UE and then sends them to the P-CSCF. The UE and the P-CSCF use these keys to encrypt signaling traffic using the IPSec protocol.

5.9 CONCLUSION

In this chapter, we have described OAMPT for MSO commercial networks and services.

Perhaps the greatest challenge for OAMPT is not the ongoing addition of new technologies to operate and support, but to maintain backward compatibility with often aging technologies. Some technologies outlive even the staff that promoted, developed, and supported the technology. These technologies often gain a kind of mythical status so that only the "chosen few" can put their hands upon them. Shutting down legacy services is sometimes more difficult even when building new services. Getting each last customer and service moved to a new platform can be a tedious and lengthy process.

Service orders are an important concept for service providers. It is often the technical process of implementing an order entry and order fulfillment system, for a particular product or service, that really identifies what is included in the product, how it is offered, and what variations are supported.

Telecommunications services for small and small-to-medium business can typically be provided by the same infrastructure utilized for residential customers. These are highly automated using the DOCSIS infrastructure. Additions or overlays to those services are sometimes provided by providing functionality with VPNs such as L2VPN or L3VPNs for Ethernet or IP services (respectively). Voice services may be provided by PacketCable over DOCSIS. For larger voice services customers where the needs go beyond primary line services, MSOs use SIP-based services that operate over DOCSIS or other access networks. DOCSIS and PacketCable provisioning are completely automated. Some aspects of SIP based services are automated. As the needs grow into complexity with larger and more demanding customers, those needs almost always push beyond the "canned" service offerings and require some level of custom engineering. In these cases, provisioning may use a mix of automation for some elements and manual provisioning for others. In access technologies, the MSOs have, through CableLabs, developed the DPoE specifications to expand their OSS automation beyond DOCSIS to include EPON which can provide services up to 10 Gbps. The basic concept of DOCSIS OSS service abstraction away from DOCSIS MAC and lower layers, applied in the DPoE example, could easily extended to provide similar functionality over CWDM or other access mediums. Such development could be prompted by any mass adoption of new access technologies for business customers. But DOCSIS continues to expand bandwidth capabilities staying well ahead of most small- to medium-sized businesses.

The systems used to support commercial customers can be the same used for residential platforms or can be extremely different applications depending on the market being targeted. Most of the DOCSIS-type services leverage the same components as residential however due to the very complex nature of the services that medium and large enterprises purchase, the tools used in today's operators are more complex. From the presales to the actual customer bill and e-Care (self-service) platforms, the commercial customer has many more variables than residential customers. Additional tools for lead management, complex order decomposition, and order provisioning (sometimes manual) are required to deliver these services. More and more automation is being added by operators as they scale to meet demand and will continue to ensure manual tasks are reduced and deployment timelines are reduced and billing options will continue to increase to meet customer demands.

Having good network security is even more important for business services. Businesses rely on networks that can provide authenticated connectivity with traffic confidentiality. The business services leverage the security features used for residential networks, enhances them when needed, and applies them in a business setting. For example, DPoE supports the same security features as DOCSIS, but using different MAC layer protocols for EPoN networks. These security features address the most significant threats and help protect high-speed Internet and VoIP telephony business services.

Fault management for business services relies on the IEEE 802.3ah Link OAM protocol for identifying faults at the point-to-point Ethernet link level. However, service-level faults can be detected, verified, isolated, and tested using the IEEE 802.1ag/ITU-T Y.1731/MEF 30.1 suite of specifications. This set of specifications defines an SOAM

Performance Monitoring protocol, including Continuity Check Messages, Remote Defect Indication, Alarm Indication Signal, Loopback, Linktrace, Lock, and Test. Along with this set of messages is an OAM domain architecture for administration of different MEG levels for different operators and the subscriber, enabling independent and isolated monitoring of segments of the end-to-end service.

Performance management for business services relies on the Service OAM ITU-T Y.1731/MEF 35 suite of specifications. This set of specifications defines an SOAM Performance Monitoring protocol, including LMM, LMR, DMM, DMR, 1DM, SLM, SLR, and 1SL. Along with this set of messages is an OAM domain architecture described in Section 5.5. The OAM architecture and message set provide a solution for performing Loss, Delay and Availability performance measurements on point-to-point and multi-point services, both using a single-ended (i.e. controller MEP initiates and calculates measurements) and dual-ended (i.e. controller MEP initiates measurements, sink MEP calculates measurements) approach. Typically, these types of measurements are performed on a proactive, measurement interval driven basis, but can be initiated on-demand by the MSO if desired.

REFERENCES

1. Open Mobile Alliance, http://openmobilealliance.org (accessed August 26, 2014).
2. Object Management Group, http://omg.org (accessed August 26, 2014).
3. Aidarous, S. and T. Plevyak, "Telecommunications Network Management into the 21st Century," IEEE Press, 1995.
4. Raman L.G., "Fundamentals of Telecommunications Network Management," IEEE Press, 1999.
5. Institute of Electrical and Electronics Engineers (IEEE) 802.1AB-2009, "IEEE Standard for Local and Metropolitan Area Networks—Station and Mediac Access Control Connectivity Discovery," 2009.
6. Institute of Electrical and Electronics Engineers (IEEE) 802.3–2012, "IEEE Standard for Ethernet," 2012.
7. International Telecommunications Union (ITU), Telecommunications Recommendation G.8013/Y.1731 OAM Functions and Mechanisms for Ethernet based networks, 2014.
8. Institute of Electrical and Electronics Engineers Standards Association (IEEE-SA), P1904.1 Standard for Service Interoperability in Ethernet Passive Optical Networks (SIEPON), (Draft D2.5), 2014.
9. CableLabs, DPoE-SP-OSSIv2.0, "DOCSIS Provisioning of EPON Operations and Support System Interface Specification," 2013.
10. Metro Ethernet Forum Technical Specification MEF 26.1, External Network to Network Interface (ENNI)—Phase 2, Metro Ethernet Forum, 2012. http://metroethernetforum.org/Assets/Technical_Specifications/PDF/MEF_26.1.pdf (accessed August 26, 2014).
11. Christensen, Clayton M., The Innovator's Dilemma: When New Technologies Cause Great Firms to Fail, Harvard Business Press, 1997.
12. Rekhter, Y., T. Li, and S. Hares (Eds.), "A Border Gateway Protocol 4 (BGP-4)," Internet Engineering Task Force, 2006, http://tools.ietf.org/html/rfc4271 (accessed August 26, 2014).

13. ITU-T Recommendation M.3400 (02/2000), TMN management function.

14. Me tro Ethernet Forum, http://metroethernetforum.org (accessed August 26, 2014).

15. World Wide Web Consortium, http://www.w3.org/Consortium/ (accessed August 26, 2014).

16. Crockford, D., IETF RFC 4627, "The application/json Media Type for JavaScript Object Notation (JSON)," The Internet Society, 2006.

17. Enns, R., M. Bjorklund, J. Schoenwaelder, and A. Bierman (Eds.), IETF RFC 6241, "Network Configuration Protocol," Internet Society, 2011.

18. Lear, E. and K. Crozier, IETF RFC 4744, "Using the NETCONF Protocol over the Blocks Extensible Exchange Protocol," The Internet Society, 2006.

19. Bjorklund, M. (Ed.), IETF RFC 6020, "YANG—A Data Modeling Language for the Network Configuration Protocol (NETCONF)," Internet Society, 2010.

20. See Section 2.7 "CCAP Architectures and Services."

21. Object Management Group, http://omg.org (accessed August 26, 2014).

22. GSM Association (GSMA), Open API. http://www.gsma.com/oneapi (accessed August 26, 2014).

23. Santitoro, Ralph, "Metro Ethernet Services—A Technical Overview," Metro Ethernet Forum, 2003–2006, http://metroethernetforum.org/Assets/White_Papers/Metro-Ethernet-Services. pdf (accessed August 26, 2014).

24. Institute of Electrical and Electronics Engineers (IEEE) 802.3–2012, "IEEE Standard for Ethernet," 2012.

25. IEEE Std 802.3ah™-2004: Part 3: Carrier Sense Multiple Access withCollision Detection (CSMA/CD) Access Methodand Physical Layer Specifications.

26. Virtual Bridged Local Area Network Amendment5: Connectivity Fault Management, IEEE 802.1ag, 2007.

27. Service OAM Fault Management Implementation Agreement: Phase 2, MEF 30.1, Metro Ethernet Forum, April 2013.

28. OAM functions and mechanisms for Ethernet based networks, Y.1731, ITU-T, February 2011.

29. Service OAM Fault Management Definition of Managed Objects, MEF 31, Metro Ethernet Forum, January 2011.

30. OAM functions and mechanisms for Ethernet based networks, Y.1731 Amendment 1, July 2012.

31. Metro Ethernet Forum, "Technical Specification MEF 16: Ethernet Local Management Interface (E-LMI). Metro Ethernet Forum, 2006, http://metroethernetforum.org/Assets/ Technical_Specifications/PDF/MEF16.pdf (accessed August 26, 2014).

32. Service OAM Performance Monitoring Implementation Agreement, MEF 35, Metro Ethernet Forum, April 2012.

33. Service OAM Performance Monitoring Implementation Agreement Amendment 1, MEF 35.0.1, Metro Ethernet Forum, October 2013.

34. Service OAM Performance Monitoring Implementation Agreement Amendment 2, MEF 35.0.2, Metro Ethernet Forum, February 2014.

35. Service OAM SNMP MIB for Performance Monitoring, MEF 36, Metro Ethernet Forum, January 2012.

36. DPoE 1.0 Security and Certificate Specification, DPoE-SP-SECv1.0, Cable Television Laboratories, Inc.

37. DPoE 2.0 Security and Certificate Specification, DPoE-SP-SECv2.0, Cable Television Laboratories, Inc.

38. IETF RFC 4131—Management Information Base for Data Over Cable Service Interface Specification (DOCSIS) Cable Modems and Cable Modem Termination Systems for Baseline Privacy Plus.

6

FUTURE DIRECTIONS IN CABLE NETWORKS, SERVICES AND MANAGEMENT

Mehmet Toy

6.1 INTRODUCTION

In the previous chapters, we have described highly complex MSO residential and commercial networks and services and their management. These networks and management systems have proven to be scalable and allow MSOs to create and offer new competitive services to their customers. However, MSOs can further reduce CAPEX, OPEX, and the amount of time to create new services. The concepts of cloud-based services, virtualization, software-defined networks (SDN),[1] and self-managed will help MSOs to achieve these goals.

Both cloud and virtualization will help MSOs to more efficiently utilize their resources resulting in lower CAPEX. SDN will help MSOs not only to reduce CAPEX due to simplified routers and switches but also to reduce OPEX by further automating provisioning of equipment and services. In fact, there are substantial efforts in MSOs to take advantage of these new technologies. Some MSOs do provide cloud-based services today. Self-managed networks will be an ultimate goal for the industry and require substantial changes in how equipment and management systems are built and operated.

In this chapter, we will describe cloud, virtualization, SDN, and self-managed concepts and provide examples.

---

[1]SDN is used to abbreviate "software-defined networks" as well as "software-defined networking" in the literature. This book will do the same.

Cable Networks, Services, and Management, First Edition. Edited by Mehmet Toy.
© 2015 The Institute of Electrical and Electronics Engineers, Inc. Published 2015 by John Wiley & Sons, Inc.

6.2 CLOUD SERVICES

Substantial growths in high-speed personal devices such as phones, laptops, and iPAD and high-definition (HD) IP video and HD IPTV applications are driving huge bandwidth demand in networks. Applications such as storage networking, video streaming, collaborative computing, and online gaming and video sharing are driving not only bandwidth demand in networks but also resources of various data centers (DCs) connected with these networks. The concepts of cloud computing and cloud-based services are expected to help service providers to deal with these challenges.

Cloud computing technologies are emerging as infrastructure services for provisioning computing and storage resources on demand in a simple and uniform way. Multiprovider and multidomain resources and integration with the legacy services and infrastructures are involved. Current cloud technology development is targeted to developing intercloud models, architectures, and integration tools that could allow integrating cloud-based infrastructure services into existing enterprise and campus infrastructures. These developments also provide common/interoperable environment for moving existing infrastructures and infrastructure services to virtualized cloud environment.

Cloud-based virtualization allows for easy upgrade and migration of enterprise application, including also the whole IT infrastructure segments, which brings significant cost saving compared to traditional infrastructure development and management requiring good amount of manual work.

Cloud-based applications operate as regular applications in particular using web services platforms for service and application integration; however, their composition and integration into distributed cloud-based infrastructure require a number of functionalities and services that can be jointly defined as Intercloud Architecture.

In 2010, NIST launched cloud computing program to support the federal government effort to incorporate cloud computing as a replacement for traditional information system and application models where appropriate.

Cloud computing [1] is a model for enabling convenient, on-demand network access to a shared pool of configurable computing resources (e.g., networks, servers, storage, applications, and services) that can be rapidly provisioned and released with minimal management effort or service provider interaction. The characteristics of cloud computing are:

- **On-demand Self-service:** A consumer can unilaterally provision computing capabilities, such as server time and network storage, as needed automatically without requiring human interaction with each service provider.
- **Broad Network Access:** Capabilities are available over the network and accessed through standard mechanisms that promote use by heterogeneous thin or thick client platforms (e.g., mobile phones including smartphones, laptops, tablets, and PDAs).
- **Resource Pooling:** The provider's computing resources are pooled to serve multiple consumers using a multitenant model, with different physical and virtual resources dynamically assigned and reassigned according to consumer demand.

There is a sense of location independence in that the customer generally has no control or knowledge over the exact location of the provided resources but may be able to specify location at a higher level of abstraction (e.g., country, state, or DC). Examples of resources include storage, processing, memory, network bandwidth, and virtual machines (VMs).

- **Rapid Elasticity:** Capabilities can be rapidly and elastically provisioned, in some cases automatically, to quickly scale out and rapidly released to quickly scale in. To the consumer, the capabilities available for provisioning often appear to be unlimited and can be purchased in any quantity at any time.
- **Measured Service:** Cloud systems automatically control and optimize resource use by leveraging a metering capability at some level of abstraction appropriate to the type of service such as storage, processing, bandwidth, and active user accounts. Resource usage can be monitored, controlled, and reported providing transparency for both the provider and consumer of the utilized service.

From an enterprise IT perspective, the overwhelming benefit with cloud computing is flexible on-demand access to IT resources without the usual purchasing, deployment, and management overhead. With cloud computing, terabytes of storage can be available instantly, with the swipe of a credit card. Customers pay for time and capacity.

The network acts as the foundation for cloud computing. Network optimization will play a major role in the transition to the cloud. It needs to become a virtual service; support mobile users, innovative applications, and protocols; and provide network visibility at a granular level. Metro Ethernet Forum (MEF) introduced Network as a Service (NaaS) concept to realize this vision and beyond.

The provider's computing resources are pooled to serve multiple consumers using a multitenant model, with different physical and virtual resources dynamically assigned and reassigned according to consumer demand. There is a sense of location independence in that the customer generally has no control or knowledge over the exact location of the provided resources, but may be able to specify location at a higher level of abstraction such as country, state, or DC. Examples of resources include storage, processing, memory, network bandwidth, and VMs.

Three service delivery models are defined:

- **Cloud Software as a Service (SaaS):** The capability provided to the consumer is to use the provider's applications running on a cloud infrastructure. The applications are accessible from various client devices through a thin client interface such as a web browser (e.g., web-based email). The consumer does not manage or control the underlying cloud infrastructure including network, servers, operating systems (OSs), storage, or even individual application capabilities, with the possible exception of limited user-specific application configuration settings.

In this model, an entire business or set of IT applications runs in the cloud. Enterprise consumers outsource the entire underlying technology infrastructure to a SaaS provider and thus have no responsibility or management oversight for SaaS-based IT components.

SaaS examples include Gmail from Google, Microsoft *live* offerings, and salesforce. com. There is no hierarchy in these service offerings.

- **Cloud Platform as a Service (PaaS):** The capability provided to the consumer is to deploy onto the cloud infrastructure consumer-created or acquired applications created using programming languages and tools supported by the provider. The consumer does not manage or control the underlying cloud infrastructure including network, servers, OSs, or storage, but has control over the deployed applications and possibly application hosting environment configurations.

PaaS provides the capability to build or deploy applications on top of Infrastructure as a Service (IaaS). Typically, a cloud computing provider offers multiple application components that align with specific development models and programming tools. For the most part, PaaS offerings are built upon either a Microsoft-based stack (i.e., Windows, .NET, IIS, SQL Server, etc.) or an open-source-based stack (i.e., the *LAMP* stack containing Linux, Apache, MySQL, and PHP).

- **Cloud IaaS:** The capability provided to the consumer is to provision processing, storage, networks, and other fundamental computing resources where the consumer is able to deploy and run arbitrary software. The software can include OSs and applications. The consumer does not manage or control the underlying cloud infrastructure, but has control over OSs, storage, and deployed applications and possibly limited control of selected networking components with firewalls.

This is the most basic cloud service model (CSM), aligning the on-demand resources of the cloud with tactical IT needs. IaaS is similar to managed service offerings such as hosting services. The primary difference is that cloud resources are virtual rather than physical and can be consumed on an as-needed basis. Enterprise consumers pay for VMs, storage capacity, and network bandwidth for a variable amount of time rather than servers, storage arrays, and switches/routers on a contractual basis. IaaS prices are based upon IaaS resource consumption and the duration of use.

Similar to the CSMs, cloud computing can be deployed in a number of ways depending upon factors like security requirements, IT skills, and network access. The IT industry has outlined four cloud computing deployment models that are described by NIST [2]:

- **Private Cloud:** The cloud infrastructure is operated solely for an organization. It may be managed by the organization or a third party and may exist on premise or off premise. The cloud infrastructure is operated within a single organization. The resources and services provided by an internal IT department or external cloud computing provider are consumed by internal groups.
- **Community Cloud:** The cloud infrastructure is shared by several organizations and supports a specific community that has shared concerns such as security requirements, policy, and compliance considerations. It may be managed by the organizations or a third party and may exist on premise or off premise. A community

cloud is a superset of a private cloud. The cloud supports the needs of several or an extended community of organizations. Again, community clouds can be built and operated by members of the community or third-party providers.

- **Public Cloud:** The cloud infrastructure is made available to the general public or a large industry group and is owned by an organization selling cloud services. The cloud infrastructure and services are available to the general public. Examples of public clouds include Amazon Elastic Compute Cloud (EC2), Google App Engine, Microsoft Azure, or Terremark cloud computing services.

- **Hybrid Cloud:** The cloud infrastructure is a composition of two or more clouds (private, community, or public) that remain unique entities but are bound together by standardized or proprietary technology that enables data and application portability such as load balancing between clouds.

The cloud infrastructure amalgamates private or community clouds with public clouds. In this case, private or community cloud services have the capability to extend or *burst* to consume public cloud resources.

Cloud computing builds upon current IT trends like DC consolidation and server virtualization. Enterprises might adopt new technologies, transitioning from physical to virtual IT assets and adapting existing IT best practices to a new dynamic world [3].

Public clouds allow fast-moving companies to outsource compute and storage on a pay-as-you-go basis. Private clouds describe an architecture in which the servers and networks in the DC can rapidly respond to changing demands, by quickly scaling compute capacity and connecting that server capacity where it is needed. The technology underlying this fluidity is server virtualization.

Cloud computing architectures are being massively deployed in DCs since they offer great flexibility, scalability, and cost-effectiveness. These advantages are important to financial firms in Wall Street, which have implemented cloud computing to help them innovate and compete more successfully. Fortune 500 enterprises are also using cloud computing to quickly scale application capacity in response to changing business conditions.

Smaller businesses are turning to cloud computing to accomplish more work with limited budgets, using virtualization to squeeze out optimal efficiency from their server and network investments.

In building out a DC infrastructure, the compute elements, the fabric that connects those elements, and the tools that orchestrate the entire environment need to be considered:

- **Compute Elements:** VM mobility is used to quickly establish and move VMs around the network that allows DCs quickly add application capacity and support Dynamic business requirements.

- **Connectivity Fabric:** A feature-rich fabric connecting compute elements with high bandwidth and low latency is needed to cloud computing in supporting applications such as disaster recovery, provisioning, and load balancing.

- **Orchestration:** As the DC becomes more complex and takes advantage of more technology, integrated management tools become important, especially when

DCs scale to tens of thousands of servers using cloud computing. Supporting this environment requires DC administrators to manage policies, service-level agreements (SLAs), and network, application, and storage traffic from any location in the cloud and between clouds.

Cloud-enabled DCs place higher technical demands on the network in areas such as speed, flexibility, cost-effective operation, and scalability. From a business perspective, a viable network architecture for today's cloud computing applications should also support incremental deployment that can employ existing facilities. Multivendor implementation and freedom to select best-in-class hardware and software components are necessary. To meet the technical and business requirements of cloud computing, the networking layer of a cloud needs to offer:

- **High Bandwidth with Low Latency:** Low-latency 10 GbE (or higher rate) switches at a lower port cost make cloud computing a realistic alternative.
- **Converged Communications and Storage:** One of the main advantages of cloud computing is the ability to carry massive amounts of data. Managing and maintaining separate local area network (LAN) and storage area network (SAN) infrastructures for such vast quantities of data do not make sense when converged networks can provide sufficient performance and scalability. With network-attached storage (NAS), iSCSI, and Fibre Channel over Ethernet (FCoE), Ethernet-based network is feasible. FCoE needs to be deployed with lossless characteristics to ensure that storage traffic is delivered reliably and in a timely manner.
- **Agile Networks for Mobile VMs:** One of the main advantages of cloud computing is on-demand access to resources, and virtualization plays a key role in providing those resources. The advantage of on-demand resources can be greatly magnified with the ability to move VMs between physical servers while applications continue to run. An infrastructure with mobile, active VMs can respond to new requirements much more quickly and very cost-effectively. Cloud computing users can gain even greater advantages from mobile VMs when they can be moved not only within a cloud but over greater distances to connect multiple clouds. Movement between clouds enables applications such as disaster recovery and data replication.

For example, elastic cloud infrastructure built on OpenStack enables any IT group to deploy cloud services that can rapidly scale resources, achieve new levels of agility, and improve market responsiveness, with full control and governance in the privacy of the user's on-premise DC. The cloud capabilities are unlikely to be delivered from the same environment that supports traditional or legacy applications. Applications or workloads best suited for elastic cloud infrastructure include:

- Scale-out web applications such as social, mobile, and collaboration
- PaaS running packaged and custom business applications

- Distributed applications
- Development/test
- Big data

A recent report from Gartner Research [4, 5] predicted that the public cloud services market total to $131 billion worldwide. Infrastructure as a service (IaaS), including cloud compute, storage and print services is to be the fastest-growing segment of the market. That is solid business built upon increasing demand for these services. However, a majority of cloud services available now don't meet all the reliability, security, and performance needs of the enterprise. This is because these clouds are built using an older server-centric approach that works well for low-cost, commodity-type physical computing but not for running VMs in a cloud.

A server-centric approach for an on-premise IT infrastructure is difficult to emulate in the cloud, so enterprises have been forced to architect around the cloud environment instead of integrating with the systems and security already in place. A network-centric cloud architecture enables enterprises to use their on-premise enterprise architecture to power their cloud environments and then manage networking hardware at the networking layer. This improves overall cloud performance and provides higher levels of security and control.

Recently Cloud Ethernet Forum (CEF) defined Cloud Services Architectures and Cloud Services [5] to standardize user, network and application interfaces as well as Cloud Services.

6.2.1 Standards

Standards and industry groups that are working on cloud computing, infrastructure, and services are listed in Table 6.1.

6.2.2 Architecture

The NIST cloud computing reference architecture [2] is a generic high-level conceptual model that defines a set of actors, activities, and functions that can be used in the process of developing cloud computing architectures and relates to a companion cloud computing taxonomy. It is described in Figure 6.1.

The architecture defines five major actors: cloud consumer, cloud provider, cloud carrier, cloud auditor, and cloud broker. Each actor is an entity (a person or an organization) that participates in a transaction or process and/or performs tasks in cloud computing. The actors are listed in Table 6.2.

A cloud consumer may request cloud services from a cloud provider directly or via a cloud broker. A cloud auditor conducts independent audits and may contact the others to collect necessary information.

In general, a cloud broker can provide service intermediation, aggregation, and arbitrage. Service intermediation enhances a given service by improving some specific capabilities and providing value-added services to cloud consumers. The improvement can be managing access to cloud services, identity management, performance reporting, enhanced security, and so on. Service aggregation combines and integrates multiple

TABLE 6.1. Standards Organizations for Cloud

ARTS—Association for Retail Technology Standards
ATIS—Alliance for Telecommunications Industry Standards
CCIF—Cloud Computing Interoperability Forum
CEF—Cloud Ethernet Forum
CSA—Cloud Security Alliance
CSCC—Cloud Standards Customer Council
DMTF—Distributed Management Task Force
ETSI—European Telecommunications Standards Institute
itSMF—IT Service Management Forum
OASIS—Organization for the Advancement of Structured Information Standards
ODCA—Open Data Center Alliance
OpenStack
OGF—Open Grid Forum
TM Forum—Telecommunications Management Forum
IEEE Intercloud Working Group (IEEE P2302)
IETF—Internet Engineering Task Force
ITU-T Focus Group on Cloud Computing (FG-Cloud)
ISO—International Organization for Standardization
NIST—National Institute of Standards and Technology
    Cloud Computing Target Business Use Cases Working Group
    Cloud Computing Reference Architecture and Taxonomy Working Group
    Cloud Computing Standards Roadmap Working Group
    Cloud Computing SAJACC Working Group
    Cloud Computing Security Working Group
SNIA—Storage Network Industry Association
    Cloud Standards Customer Council
    Distributed Management Task Force (DMTF)
    Open Virtualization Format (OVF)
    Open Cloud Standards Incubator
    Cloud Management Working Group (CMWG)
CADF—Cloud Auditing Data Federation Working Group
ETSI—The European Telecommunications Standards Institute
TC CLOUD
CSC—Cloud Standards Coordination
GICTF—Global InterCloud Technology Forum
ISO/IEC JTC 1 SC38 Distributed Application Platforms and Services (DAPS)
ITU-T SG13: Future networks including cloud computing, mobile, and next-generation
  networks
JCA—Joint Coordination Activity on Cloud Computing
NIST—National Institute of Standards and Technology
NIST Working Definition of Cloud Computing
SAJACC (is it a group under NIST?)—Standards Acceleration to Jumpstart Adoption of
  Cloud Computing
OGF—Open Grid Forum
OCCI Working Group Open Cloud Computing Interface Working Group
OMG—Object Management Group

(Continued)

TABLE 6.1. *(cont'd)*

OCC—Open Cloud Consortium
OASIS Organization for the Advancement of Structured Information Standards OASIS
  Cloud-Specific or Extended Technical Committees (TC)
OASIS Cloud Application Management for Platforms (CAMP) TC
OASIS Identity in the Cloud (IDCloud) TC
OASIS Symptoms Automation Framework (SAF) TC
OASIS Topology and Orchestration Specification for Cloud Applications (TOSCA) TC
OASIS Cloud Authorization (CloudAuthZ) TC
OASIS Public Administration Cloud Requirements (PACR) TC
SNIA
SNIA Cloud Data Management Interface (CDMI)
The Open Group
Cloud Work Group
ARTS—Association for Retail Technology Standards
TM Forum
  Cloud Services Initiative
  TM Forum's Cloud Services Initiative Vision
  Barriers to Success
  Enterprise Cloud Leadership Council Goals (ECLC)
  Future Collaborative Programs
  About the TM Forum
    TM Forum's Frameworx

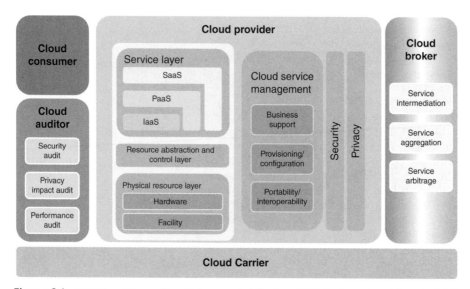

Figure 6.1. NIST Cloud Computing Reference Architecture (CCRA). Reproduced from Ref. [2].

TABLE 6.2. Actors in Cloud Computing

Actor	Definition
Cloud consumer	A person or organization that maintains a business relationship with, and uses service from, *Cloud Providers*. The cloud consumer maintains a business relationship with and uses the service from a cloud provider, browses the service catalog from a cloud provider, requests the appropriate service, sets up service contracts with the cloud provider, and uses the service. The cloud consumer may be billed for the service provisioned and needs to arrange payments accordingly
Cloud provider	A person, organization, or entity responsible for making a service available to interested parties. A Cloud Provider acquires and manages the computing infrastructure required for providing the services, runs the cloud software that provides the services, and makes arrangement to deliver the cloud services to the Cloud Consumers through network access
Cloud auditor	A party that can conduct independent assessment of cloud services, information system operations, performance, and security of the cloud implementation
Cloud broker	An entity that manages the use, performance, and delivery of cloud services and negotiates relationships between *Cloud Providers* and *Cloud Consumers*
Cloud carrier	An intermediary that provides connectivity and transport of cloud services from *Cloud Providers* to *Cloud Consumers*. A cloud carrier provides connectivity and transport of cloud services between cloud consumers and cloud providers. Cloud carriers provide access to consumers through network, telecommunications, and other access devices. For example, cloud consumers can obtain cloud services through network access devices, such as computers, laptops, mobile phones, mobile Internet devices (MIDs), and so on [1]. The distribution of cloud services is normally provided by network and telecommunications carriers or a transport agent [6], where a transport agent refers to a business organization that provides physical transport of storage media such as high-capacity hard drives. Note that a cloud provider will set up SLAs with a cloud carrier to provide services consistent with the level of SLAs offered to cloud consumers and may require the cloud carrier to provide dedicated and secure connections between cloud consumers and cloud providers

services into one or more new services. The broker provides data integration and ensures the secure data movement between the cloud consumer and multiple cloud providers. Service arbitrage is similar to service aggregation except that the services being aggregated are not fixed. Service arbitrage means a broker has the flexibility to choose services from multiple agencies. The cloud broker, for example, can use a credit-scoring service to measure and select an agency with the best score.

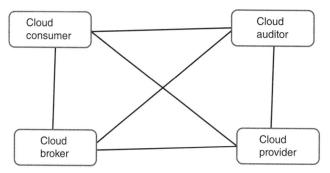

Figure 6.2. Interactions between the actors in cloud computing. Reproduced from Ref. [2].

Figure 6.3. Usage scenario for cloud carriers.

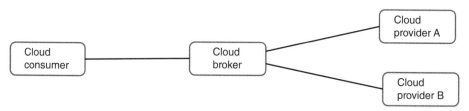

Figure 6.4. Usage scenario for cloud brokers.

The proposed architecture is suitable for purposes where network performance is not critical but needs to be extended with explicit network service provisioning and management when the cloud applications are critical to network latency like in the case of enterprise applications, business transactions, crisis management, and so on.

The interactions among the actors are depicted in Figure 6.2. A cloud consumer may request cloud services from a cloud provider directly or via a cloud broker. A cloud auditor conducts independent audits and may contact the others to collect necessary information.

Cloud carriers provide the connectivity and transport of cloud services from cloud providers to cloud consumers (Figure 6.3). A cloud provider may have unique SLAs with a cloud carrier and a cloud consumer.

A cloud consumer and Cloud Provider may interface to a Cloud Broker for coordination instead of interfacing directly with each other as depicted in Figure 6.4.

For a cloud service, a cloud auditor may conduct independent assessments of interactions with both the Cloud Consumer and the Cloud Provider including the operation and security of the cloud service implementation (Figure 6.5).

Figure 6.6 presents some examples of cloud services available to a cloud consumer.

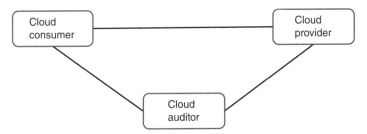

Figure 6.5. Usage scenario for cloud auditors.

For SaaS, the cloud provider deploys, configures, maintains, and updates the operation of the software applications on a cloud infrastructure so that the services are provisioned at the expected service levels to cloud consumers. The provider of SaaS assumes most of the responsibilities in managing and controlling the applications and the infrastructure, while the cloud consumers have limited administrative control of the applications. The consumers of SaaS can be organizations providing their members with access to software applications, end users who directly use software applications, or software application administrators who configure applications for end users. SaaS consumers can be billed based on the number of end users, the time of use, the network bandwidth consumed, the amount of data stored, or the duration of stored data.

For PaaS, the Cloud Provider manages the computing infrastructure for the platform and runs the cloud software that provides the components of the platform, such as run-time software execution stack, databases (DBs), and other middleware components. The PaaS Cloud Provider typically also supports the development, deployment, and management process of the PaaS Cloud Consumer by providing tools such as integrated development environments (IDEs), development version of cloud software, software development kits (SDKs), and deployment and management tools. PaaS consumers can be application developers who design and implement application software, application testers who run and test applications in cloud-based environments, application deployers who publish applications into the cloud, and application administrators who configure and monitor application performance on a platform. The billing of PaaS consumers can be according to processing resources, DB storage, and network resources consumed by the PaaS application and the duration of the platform usage.

PaaS Cloud Consumer has control over the applications and possibly some of the hosting environment settings, but has no or limited access to the infrastructure under-lying the platform such as network, servers, OSs, or storage.

For IaaS, the Cloud Provider acquires the physical computing resources under-lying the service, including the servers, networks, storage, and hosting infrastructure. The Cloud Provider runs the cloud software necessary to make computing resources available to the IaaS Cloud Consumer through a set of service interfaces and com-puting resource abstractions, such as VMs and virtual network interfaces (VNIs)[2].

---

[2]VNI is used to indicate Virtual Network Interface as well as Virtual Network Instance in the literature. We will do the same here.

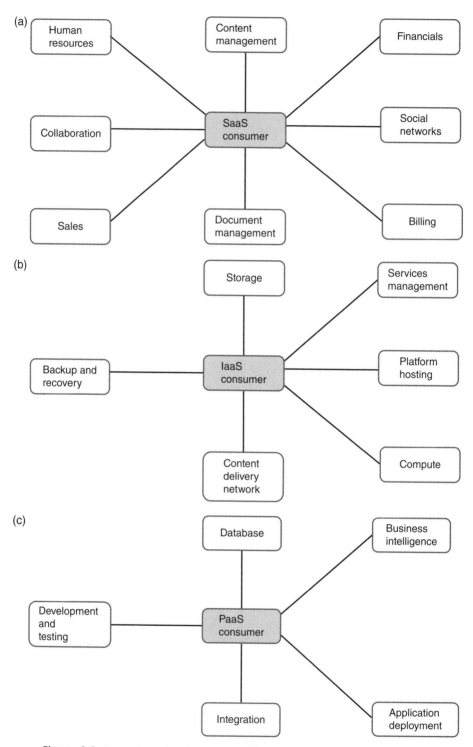

Figure 6.6. Example services for (a) SaaS, (b) IaaS, and (c) PaaS cloud consumer.

The consumers of IaaS can be system developers, system administrators, and IT managers who are interested in creating, installing, managing, and monitoring services for IT infrastructure operations. The billing for IaaS consumers is according to the amount or duration of the resources consumed, such as CPU hours used by virtual computers, volume and duration of data stored, network bandwidth consumed, and the number of IP addresses used for certain intervals.

The IaaS Cloud Consumer in turn uses these computing resources, such as a virtual computer, for their fundamental computing needs. Compared to SaaS and PaaS Cloud Consumers, an IaaS Cloud Consumer has access to more fundamental forms of computing resources and thus has more control over the more software components in an application stack, including the OS and network. The IaaS Cloud Provider, on the other hand, has control over the physical hardware and cloud software that makes the provisioning of these infrastructure services possible, for example, the physical servers, network equipment, storage devices, host OS, and hypervisors for virtualization.

A Cloud Provider conducts its activities in the areas of service deployment, service orchestration, cloud service management, security, and privacy.

The Cloud Provider and Cloud Consumer share the control of resources in a cloud system as depicted in Figure 6.7.

In application layer, software applications can be used by SaaS consumers or installed/managed/maintained by PaaS consumers, IaaS consumers, and SaaS providers.

The middleware layer is used by PaaS consumers and installed/managed/maintained by IaaS consumers or PaaS providers and provides software building blocks (e.g., libraries, DB, and Java VM) for developing application software in the cloud. The middleware is hidden from SaaS consumers.

The OS layer is controlled by IaaS of Cloud Provider and IaaS of Cloud Consumer. It includes OS and drivers and is hidden from SaaS consumers and PaaS consumers.

An IaaS cloud allows one or multiple guest OSs to run virtualized on a single physical host where consumers are free to choose an OS among all the OSs supported by the cloud provider. The IaaS consumers should assume full responsibility for the guest OSs, while the IaaS provider controls the host OS.

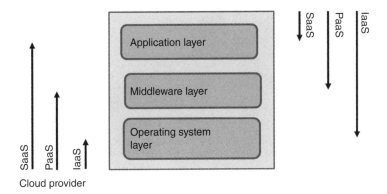

Figure 6.7. Scope of controls between provider and consumer.

6.2.3 Service Deployment Architecture

A cloud infrastructure may be operated as public cloud, private cloud, community cloud, or hybrid cloud based on how exclusive the computing resources are made to a Cloud Consumer.

A public cloud (Figure 6.8) is owned by an organization selling cloud services and serves a diverse pool of clients where the cloud infrastructure and computing resources are made available to the general public over a public network.

A private cloud (Figure 6.9) may be managed either by the Cloud Consumer organization or by a third party and may be hosted on the organization premises (i.e., on-site private clouds) or outsourced to a hosting company (i.e., outsourced private clouds) where a Cloud Consumer organization has the exclusive access to and usage of the infrastructure and computational resources.

A community cloud (Figure 6.10) may be managed by the organizations or by a third party and may be implemented on customer premise (i.e., on-site community cloud) or outsourced to a hosting company (i.e., outsourced community cloud) and serves a group of Cloud Consumers that have shared concerns such as mission objectives, security, privacy, and compliance policy, rather than serving a single organization as does a private cloud.

A cloud consumer can access the local cloud resources and also the resources of other participating organizations through the connections between the associated

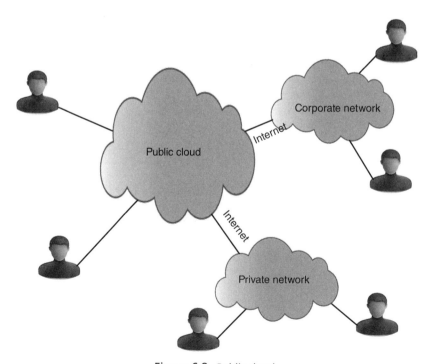

Figure 6.8. Public cloud.

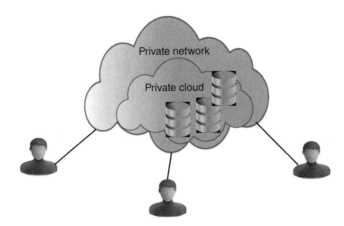

Figure 6.9. Private cloud.

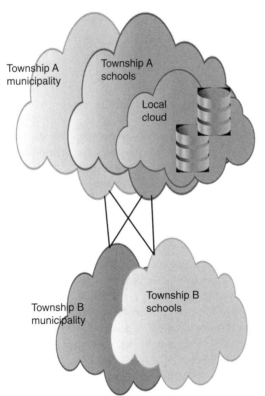

Figure 6.10. On-site community cloud.

organizations. The server of private cloud can be outsourced to a hosting company that can build its infrastructure off premise.

A hybrid cloud is a composition of two or more clouds (on-site private, on-site community, off-site private, off-site community, or public) that remain as distinct

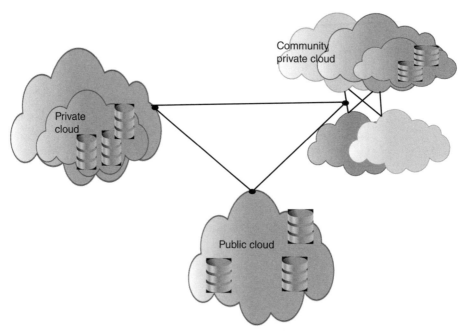

Figure 6.11. Hybrid cloud.

entities but are bound together by standardized or proprietary technology that enables data and application portability. Figure 6.11 presents a simple view of a hybrid cloud that could be built with a set of clouds.

System components need to be orchestrated, service orchestration, to support the Cloud Provider activities in arrangement, coordination, and management of computing resources in order to provide cloud services to Cloud Consumers. Figure 6.12 depicts a generic stack diagram of service orchestration that underlies the provisioning of cloud services.

In Figure 6.12, system components are grouped into three layers. The *service layer* is where Cloud Providers define interfaces for Cloud Consumers to access the computing services. Access interfaces of each of the three service models are provided in this layer. SaaS applications can be built on top of PaaS components, and PaaS components can be built on top of IaaS components. For example, a SaaS application can be implemented and hosted on VMs from an IaaS cloud, or it can be implemented directly on top of cloud resources without using IaaS VMs.

The *resource abstraction and control layer* contains the system components that Cloud Providers use to provide and manage access to the physical computing resources through software abstraction. The *resource abstraction* components include software elements such as hypervisors, VMs, virtual data storage, and other computing resource abstractions. It needs to ensure efficient, secure, and reliable usage of the underlying physical resources. The *control* function in this layer refers to the software components that are responsible for resource allocation, access control, and usage monitoring. This

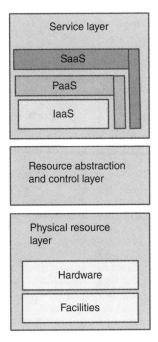

Figure 6.12. Cloud system layers.

is the software fabric that ties together the numerous underlying physical resources and their software abstractions to enable resource pooling, dynamic allocation, and measured service. The resource abstraction and control layer exposes virtual cloud resources on top of the physical resource layer and supports the service layer where cloud service interfaces are exposed to Cloud Consumers, while Cloud Consumers do not have direct access to the physical resources.

The *physical resource layer* includes all the physical computing resources such as computers (CPU and memory), networks (routers, firewalls, switches, network links, and interfaces), storage components (hard disks), facility resources (such as heating, ventilation, and air conditioning (HVAC), power, communications, and other aspects of the physical plant.

6.2.4 Cloud Service Management

Cloud service management is the management and operations of all services for cloud consumers. As illustrated in Figure 6.13, cloud service management can be categorized as *business support, provisioning and configuration,* and *portability and interoperability.*

Business support entails business services and supporting processes. It includes the components used to run business operations that are client facing. The business services and supporting processes can be categorized as customer management, inventory management, accounting and billing, reporting and auditing, and pricing and rating.

Customer management is to manage customer accounts, open/close/terminate accounts, manage user profiles, manage customer relationships by providing points of

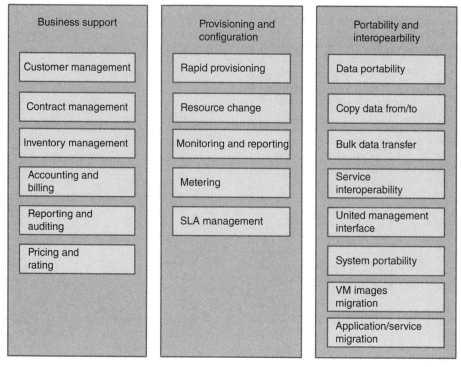

Figure 6.13. Cloud service management components.

contact and resolving customer issues and problems, and so on. *Contract management is to m*anage service contracts and set up/negotiate/close/terminate contract. *Inventory management is to s*et up and manage service catalogs. *Accounting and billing is to m*anage customer billing information, send billing statements, process received payments and track invoices. *Reporting and auditing is to m*onitor user operations and generate reports. *Pricing and rating is to e*valuate cloud services and determine prices, handle promotions and pricing rules based on a user's profile.

Provisioning and configuration can be categorized as rapid provisioning, resource changing, monitoring and reporting, metering, and SLA management. *Rapid provisioning is* automatically deploying cloud systems based on the requested service/ resources/capabilities. *Resource changing is a*djusting configuration/resource assignment for repairs, upgrades, and joining new nodes into the cloud.

*Monitoring and reporting is d*iscovering and monitoring virtual resources, monitoring cloud operations and events, and generating performance reports. *Metering is* providing a metering capability at some level of abstraction appropriate to the type of service (e.g., storage, processing, bandwidth, and active user accounts). *SLA management* is defining Quality-of-Service (QoS) parameters for the SLA contract, SLA monitoring, and SLA enforcement according to defined policies.

The adoption of cloud computing depends greatly on how the cloud can address security, portability, and interoperability. Customers want to move their data or applications across multiple cloud environments at low cost and minimal disruption. Cloud providers should provide mechanisms to copy data objects into or out of a cloud or to use a disk for bulk data transfer, to use their data and services across multiple cloud providers with a unified management interface, to migrate a fully stopped VM instance or a machine image from one provider to another provider, or to migrate applications and services and their contents from one service provider to another. For IaaS, it is necessary to capture VM images and migrate to new cloud providers that may use different virtualization technologies. Any provider-specific extensions to the VM images need to be removed or recorded upon being ported. For SaaS, given the focus is on data portability, data extractions and backups in a standard format are essential.

6.2.5 Security

Security in cloud computing architecture concerns is not solely under the purview of the Cloud Providers but also Cloud Consumers and other relevant actors. Cloud-based systems need to address authentication, authorization, availability, confidentiality, identity management, integrity, audit, security monitoring, incident response, and security policy management.

The three service models, SaaS, PaaS, and IaaS, present consumers with different types of service management operations and expose different entry points into cloud systems. For example, SaaS provides users with accessibility of cloud offerings using a network connection, normally over the Internet and through a web browser. Therefore, web browser security is important for SaaS. IaaS is provided with VMs that are executed on hypervisors on the hosts. Therefore hypervisor security for achieving VM isolation is important for IaaS.

Cloud deployment models have security implications. A private cloud is dedicated to one consumer organization. On the other hand, a public cloud is shared by various tenants; therefore, isolation of tenants is a security concern in a public cloud.

Cloud Provider and the Cloud Consumer have differing degrees of control over the computing resources in a cloud system. They collaboratively design, build, deploy, and operate cloud-based systems. Both parties now share the responsibilities in providing adequate protections to the cloud-based systems. Security controls need to be analyzed to determine which party is in a better position to implement. For example, account management controls for initial system privileged users in IaaS scenarios are typically performed by the IaaS provider, whereas application user account management for the application deployed in an IaaS environment is typically not the provider's responsibility.

Though cloud computing provides a flexible solution for shared resources, it also brings additional privacy challenges to consumers using the clouds. Cloud providers should protect the assured, proper, and consistent collection, processing, communication, use, and disposition of personal information (PI) and personally identifiable information (PII) in the cloud [7]. PII can be used to distinguish or trace an individual's identity, such as their name, social security number, and biometric records, combined with other personal or identifying information that is linked or linkable to a specific individual, such as date and place of birth, mother's maiden name, and so on [8].

6.2.6 Carrier Ethernet for Cloud Services

The NIST cloud model promotes availability and is composed of five essential characteristics: on-demand self-service, broad network access, resource pooling, rapid elasticity, and measured service.

The MEF uses the NIST models and augments the terminology and definitions as they apply to Carrier Ethernet Networks and Services [9].

MEF introduces Ethernet Cloud Carrier and extends Cloud Broker definition. The Ethernet Cloud Carrier uses Carrier Ethernet as the wide area networking technology. Carrier Ethernet can be used for creating private or virtual private network.

The MEF augments NIST's Cloud Broker definition by adding the Ethernet Cloud Carrier actor role to the definition. Cloud Brokers could also act as the broker between Cloud Consumers, Cloud Service Providers, and Cloud Carriers for private WAN connections between Cloud Consumers and Cloud Service Providers. An example of this would be an Ethernet Exchange Provider. A Cloud Broker could also be an Ethernet Cloud Carrier or Cloud Service Provider and broker between the three actors while delivering the functions of the Cloud Carrier or Cloud Service Provider.

For example, the Ethernet Cloud Carrier may have the business relationship with the Cloud Consumer and wholesales cloud services from one or more Cloud Service Providers (Figure 6.14). The Cloud Consumer has no business relationship, for example, billing, service agreement, and so on, with the Cloud Service Provider. The Ethernet Cloud Carrier in the model is acting in two different roles, namely, Cloud Broker and Ethernet Cloud Carrier.

The Cloud Service Provider may have the business relationship with the Cloud Consumer and wholesales network connectivity and transport services from one or more Ethernet Cloud Carriers (Figure 6.15). The Cloud Consumer has no business relationship, for example, billing, service agreement, and so on, with the Ethernet Cloud Carrier.

The Cloud Service Provider in the model acts in two different roles, namely, Cloud Broker and Cloud Service Provider. In the telecoms industry, this type of Cloud Broker is analogous to an ISP delivering Internet access service to their customers and

Figure 6.14. Cloud carrier acting as a cloud broker [9].

Figure 6.15. Cloud service provider acting as a cloud broker. Reproduced from Ref. [9].

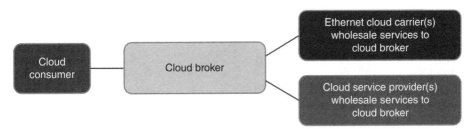

Figure 6.16. Cloud broker providing broker service for Ethernet Cloud Carrier and cloud service provider [9].

wholesaling the access network connection from the local incumbent telecoms service provider who owns the access lines to the customer premises. In effect, the ISP is delivering and managing Internet access and providing network connectivity and transport to the customer premises.

The Cloud Broker may have the business relationship with the Cloud Consumer and wholesales network connectivity and transport services from one or more Ethernet Cloud Carriers and cloud services from one or more Cloud Service Providers (Figure 6.16). The Cloud Consumer has no business relationship, for example, billing, service agreement, and so on, with either the Ethernet Cloud Carrier or Cloud Service Provider. In the telecoms industry, this type of Cloud Broker is analogous to a mobile virtual network operator (MVNO) who sells cellular wireless voice and data services to their customers even though they do not own or operate a cellular wireless network. However, from the customer's perspective, they are buying cellular wireless voice and data services from a cellular wireless network operator.

6.2.6.1 Ethernet Private Line and Ethernet Virtual Private Line

MEF Ethernet Private Line (EPL) services are applicable to applications where a Cloud Consumer requires a dedicated Ethernet WAN connection to a Cloud Service Provider at a DC. Each UNI at the Cloud Consumer and Cloud Service Provider sites supports a single Ethernet Virtual Connection (EVC). An EPL service does not require any VLAN ID coordination with the Ethernet Cloud Carrier since all VLAN IDs are passed transparently across the network. In Figure 6.17, two Cloud Consumers (A and B) each use an EPL service to connect to the Cloud Service Provider's primary DC. A third EPL service is used to interconnect the Cloud Service Provider's primary and secondary DCs.

When a Cloud Consumer or Cloud Service Provider wants to multiplex EVCs (services) onto the same UNI, MEF Ethernet Virtual Private Line (EVPL) service is needed. This enables them to save the cost of an extra Ethernet port on their attaching equipment but often saves service cost because the Ethernet Cloud Carrier doesn't need to install a separate port for the additional EVC as would be the case with an EPL service. However, the Cloud Service Provider will need to coordinate customer VLAN IDs (C-VLAN IDs) to ensure that Cloud Consumers A and B do not use overlapping C-VLAN IDs.

In Figure 6.18, two Cloud Consumers (A and B) each use an EVPL service to connect to the DC in which their Cloud Service Provider is located. At the DC, the EVCs

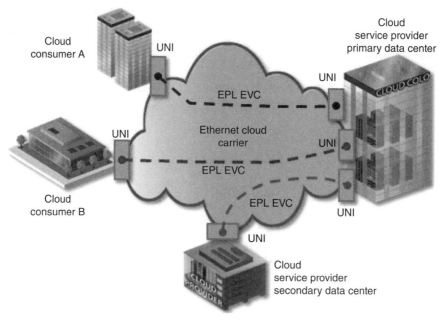

Figure 6.17. EPL connecting cloud consumers to cloud service provider. Reproduced from Ref. [9].

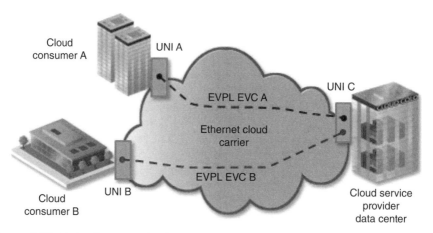

Figure 6.18. EVPLs from two cloud consumers multiplexed at cloud service provider's UNI. Reproduced from Ref. [9].

from each EVPL service are multiplexed onto the same UNI (UNI C). Because the EVPL service supports EVC multiplexing at a UNI, additional EVCs can be added to UNI C at the DC to connect to additional Cloud Consumer sites, connect to the Internet, or even connect to another Cloud Service Provider DC.

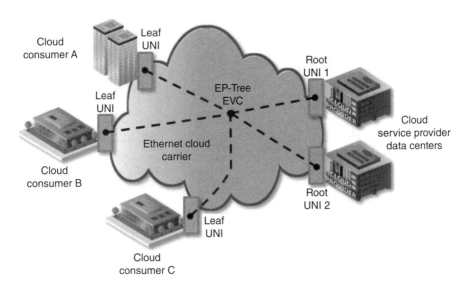

Figure 6.19. EP-tree to interconnect cloud consumers with cloud service provider data centers. Reproduced from Ref. [9].

6.2.6.2 E-Tree Services

MEF EP-Tree services can be used to connect different Cloud Consumer locations to a common Cloud Service Provider DC. In Figure 6.19, Cloud Consumers A, B, and C connect to two Cloud Service Provider DCs. Leaf UNIs are used at the Cloud Consumer sites and Root UNIs are used at the DC sites.

The EP-Tree provides the traffic separation and isolation between Cloud Consumers A, B, and C and enables the Cloud Service Provider to add additional Cloud Consumers to the same EP-Tree. If C-VLAN IDs are used, then they must be coordinated with the different Cloud Consumers to ensure no overlapping values are used. Cloud Consumers A, B, and C have access to either of the Cloud Service Provider's DCs connected by Root UNIs one and two. Since Root UNIs enable intercommunication, each Root UNI can be used to interconnect a different DC facilitating data replication, disaster recovery, and cloud bursting applications.

MEF EVP-Tree services can address the same applications as EP-Tree services and support multiple services on the same UNI.

In Figure 6.20, Cloud Consumer A has an additional EVPL EVC providing a connection to the Internet on the same UNI as the EVP-Tree EVC connecting to the Cloud Service Provider DCs. The Cloud Service Provider also has an EVPL EVC on Root UNI 1 providing a connection to the Internet that could be used to deliver public cloud services or support cloud bursting over the Internet to another DC when local compute and storage resources are fully consumed.

6.2.6.3 E-LAN Services

E-LAN services facilitate data replication, disaster recovery, and cloud bursting applications between Cloud Service Provider DCs connected to the E-LAN service EVC.

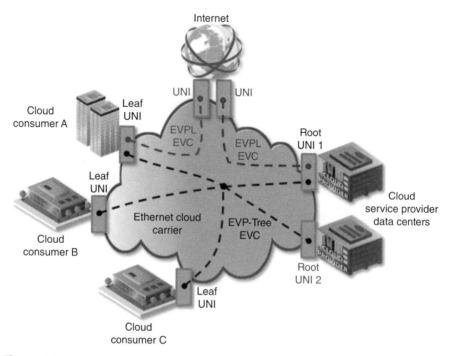

Figure 6.20. Example EVP-tree use case with EVPLs used to provide Internet connectivity. Reproduced from Ref. [9].

Furthermore, Cloud Consumers connected to the same E-LAN service EVC as the DCs have full access to both primary and backup DCs that could simplify the Cloud Consumer's disaster recovery business requirements. Finally, E-LAN services enable the Cloud Consumer organization to add additional sites (UNIs) to the same E-LAN EVC. The newly added sites will get the same level of connectivity and data exchange capabilities as all of the other connected sites.

MEF EP-LAN services can be used to connect different Cloud Consumer locations to one or more Cloud Service Provider DCs. EP-LAN services can only support a single service at each UNI. In Figure 6.21, a Cloud Consumer has three sites that connect to a Cloud Service Provider's primary and secondary DCs. The secondary DC could be used for cloud bursting applications to provide additional VM or data storage capacity to the primary DC. It could also be used as a disaster recovery DC.

6.2.7 Intercloud Architecture

Enterprise IT infrastructure migration to cloud from legacy infrastructure requires the integration of the legacy infrastructure with cloud-based components initially. Later on, this migration can become progressive transfer from general cloud infrastructure services to specialized private cloud platform services [8, 10].

Communications between cloud applications and services belonging to different service layers (vertical integration) and between cloud domains and heterogeneous

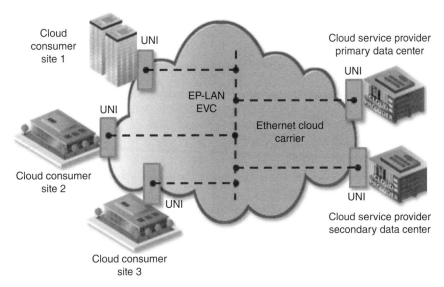

Figure 6.21. EP-LAN interconnecting cloud consumer sites with cloud service provider data centers.

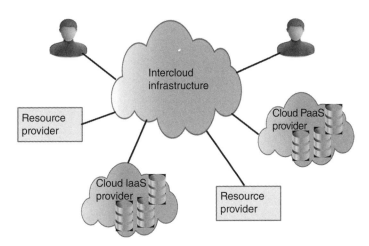

Figure 6.22. Network interconnecting infrastructure.

platforms (horizontal integration) are necessary to intercloud control and optimize resource usages are necessary.

For example, IT infrastructure disaster recovery requires not only data backup but also the whole supporting infrastructure restoration/setup on possibly new computer/cloud software or hardware platform. The whole infrastructure of computers, storage, network, and other utilities is to be provisioned on demand, independent from physical platform. The local persistent utilities need to be integrated with legacy services and applications.

Figure 6.22 illustrates a typical case when two different types of cloud services such as IaaS and PaaS based need to interoperate to allow consistent hybrid cloud infrastructure control and management.

In summary, an intercloud architecture should:

- Provide communication between cloud applications and services belonging to different service layers (vertical integration) and between cloud domains and heterogeneous platforms (horizontal integration).
- Compatible and provide multilayer integration of existing CSMs—IaaS, PaaS, SaaS, and Apps clouds.
- Provide a possibility that applications could control infrastructure and related supporting services at different service layers to achieve run-time optimization (Intercloud control and management functions).
- Have common Intercloud Control Plane and signaling for better cloud services and network integration.
- Support cloud services/infrastructures provisioning on demand and their life cycle management, including composition, deployment, operation, and monitoring, involving resources and services from multiple providers.
- Provide a framework for heterogeneous intercloud federation.
- Facilitate interoperable and measurable intraprovider infrastructures.
- Provide guaranteed intra- and intercloud network infrastructure provisioning (as NaaS service model).
- Support existing Cloud Provider operational and business models and provide a basis for new forms of infrastructure service provisioning and operation.

In order to support the aforementioned capabilities , the following intercloud architecture components are proposed [8]:

- Multilayer CSM for vertical cloud service interaction, integration, and compatibility that defines both relations between CSMs (such as IaaS, PaaS, SaaS) and other required functional layers and components of the general cloud-based service infrastructure.
- Intercloud Control and Management Plane (ICCMP) for Intercloud applications/ infrastructure control and management, including interapplication signaling, synchronization and session management, configuration, monitoring, run-time infrastructure optimization including VM migration, resources scaling, and jobs/objects routing.
- Intercloud Federation Framework (ICFF) to allow independent clouds and related infrastructure components federation of independently managed cloud-based infrastructure components belonging to different cloud providers and/or administrative domains; this should support federation at the level of services, business applications, semantics, and namespaces, assuming necessary gateway or federation services.
- Intercloud Operation Framework (ICOF) that includes functionalities to support multiprovider infrastructure operation including business workflow, SLA management, and accounting. ICOF defines the basic roles, actors, and their

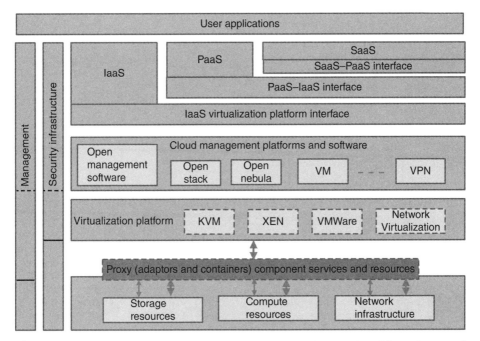

Figure 6.23. Reference multilayer cloud service model (CSM). Reproduced from Figure 3 of Ref. [10].

relations in sense of resources operation, management, and ownership. ICOF requires support from and interacts with both ICCMP and ICFF.

Figure 6.23 shows the basic CSMs IaaS, PaaS, and SaaS that expose in most cases standards-based interface to user services or applications but actually use a proprietary interface to the physical provider platform. In this Intercloud layered service model, the following layers are defined:

- **Cloud Layer 7:** User client or application
- **Cloud Layer 6:** SaaS (or cloud applications) as a top cloud layer that represents cloud applications
- **Cloud Layer 5:** PaaS provided as a service or used as a platform for hosting cloud applications
- **Cloud Layer 4:** IaaS provided as infrastructure or used for hosting cloud platforms or applications
- **Cloud Layer 3:** Cloud virtual resources composition and orchestration layer that is represented by the Cloud Management Software (such as OpenNebula, OpenStack, or others)
- **Cloud Layer 2:** Cloud virtualization layer (e.g., represented by VMware, Xen, or KVM as virtualization platforms)
- **Cloud Layer 1:** Physical platform (PC hardware, network, and network infrastructure)

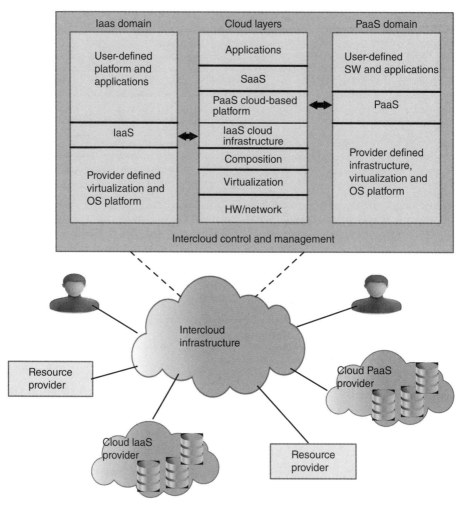

Figure 6.24. Intercloud Control and Management Plane providing single control and management domain to heterogeneous intercloud infrastructure. Reproduced from Ref. [10].

Upper layer applications and processes are expected to control underlying layers of the cloud infrastructure or platform. Figure 6.24 illustrates a scenario where IaaS and PaaS cloud domains communicate via standard and proprietary interfaces.

6.3 VIRTUALIZATION

Cost-effective utilization of IT infrastructure to meet budget constraints, responsiveness in supporting new business initiatives, and flexibility in adapting to organizational changes are necessary to accommodate business challenges. Virtualization is expected to bring solutions to such business challenges.

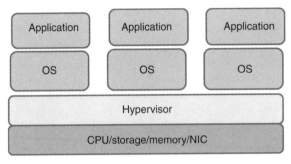

Figure 6.25. Virtualized server with hypervisor layer [11].

Virtualization separates a resource or request for a service from the underlying physical delivery of that service. With virtual memory, for example, computer software gains access to more memory than is physically installed, via the background swapping of data to disk storage. Similarly, virtualization techniques can be applied to networks, storage, server hardware, OSs, applications, and so on.

Virtual infrastructure provides a layer of abstraction between computing, storage, and networking hardware and the applications running on it (see Figure 6.25). Virtual infrastructure gives administrators the advantage of managing pooled resources across the enterprise, allowing IT managers to be more responsive to dynamic organizational needs and to better leverage infrastructure investments.

Virtualization can be in various areas [6, 12]:

- **Management Virtualization:** User and group policies in Microsoft Windows XP, 2003, and Vista are examples of virtualized administration rights. A user may read access to all the files that the user needs to back up, but doesn't have rights to install new files or software.
- **Network Virtualization:** An Ethernet port may support multiple virtual connections from multiple IP addresses and networks, but they are virtually segmented using VLAN tags. Each virtual IP connection over this single physical port is independent and unaware of others' existence, but the switch is aware of each unique connection and manages each one independently.
- **Hardware Virtualization:** Hardware virtualization breaks up pieces and locations of physical hardware into independent segments and manages those segments as separate, individual components. Both symmetric and asymmetric multiprocessing are examples of hardware virtualization. In both instances, the process requesting CPU time isn't aware which processor it's going to run on; it just requests CPU time from the OS scheduler and the scheduler takes the responsibility of allocating processor time. The process could be spread across any number of CPUs and any part of RAM, so long as it's able to run unaffected.
- **Storage Virtualization:** Storage virtualization can be broken up into block virtualization and file virtualization. An OS or application can be mapped to a virtual block device, such as a mounted drive, to a local network adapter instead of a

physical drive controller, as in iSCSI. The iSCSI network adapter translates block calls from the application to network packets and then back again, essentially providing a virtual hard drive.

File virtualization abstracts the physical location of a file from the end user requesting it similar to DNS. The file virtualization appliance receives all file requests from users and routes them to the storage devices that currently hold each file.

- **OS Virtualization:** Each OS instance is unaware that it's virtual and that other virtual OSs are (or may be) running at the same time. VMs are typically full implementations of standard OSs, such as Windows Vista or Red Hat Enterprise Linux, running simultaneously on the same physical hardware. A hypervisor or virtual machine monitor (VMM), which is a piece of computer software, firmware, or hardware, creates and runs each VM individually. VMs are highly portable and can be moved or copied to any industry standard (x86-based) hardware platform, regardless of the make or model.
- OS virtualization allows enterprises to reduce the number of physical machines in their DCs without reducing the number of underlying applications. This ultimately saves enterprises money on hardware, colocation fees, rack space, power, and cable management.
- **Application Server Virtualization:** An appliance or service provides access to many different application services transparently where one server is presented to the world, hiding the availability of multiple servers behind a reverse proxy appliance. In a typical deployment, a reverse proxy will host a virtual interface accessible to the end user on the *front end*, while on the *back end*, the reverse proxy will load balance a number of different servers and applications such as a web server.
- **Application Virtualization:** Virtual application runs locally, but its application logic runs remotely. Also, it is managed remotely. For example, Microsoft Word 2007 may reside in a laptop providing the CPU and RAM required to run the software, but nothing installed in the machine. The binaries, PI, and state information are all stored on and managed and delivered by SoftGrid.
- **Service Virtualization:** Service virtualization connects all of the components utilized in delivering an application over the network and includes the process of making all pieces of an application work together regardless of where those pieces physically reside. For example, a web application typically has many parts: the user-facing HTML; the application server that processes user input; the service-oriented architecture (SOA) gears that coordinate service and data availability between each component; the DB back end for user, application, and SOA data; the network that delivers the application components; and the storage network that stores the application code and data. Service virtualization allows each component to function independently and be *called up* as needed for the entire application to function properly. The web server is load balanced between multiple VM OSs, the SOA requests are pushed through any number of XML gateways on the wire, the DB servers are located in one of multiple global DCs,

and so on. Service virtualization combines these independent pieces and presents them together to the user as a single, complete application.

Without virtualization, single OS image per machine, where software and hardware tightly coupled, running multiple applications on the same machine often creates conflict. Resources are underutilized. The infrastructure is costly and inflexible.

With virtualization, OS and applications are hardware independent. VMs can be provisioned to any system. OS and application can be managed as a single unit by encapsulating them into VMs. Deployment of systems as VMs that can run safely and move transparently across shared hardware increases server utilization rates from 5–15% to 60–80%.

The hypervisor sits between hardware and the OS. Virtualization allows multiple OSs and applications to cohabitate on a physical computing platform. Server virtualization especially when coupled with blade technology increases computing and storage density.

Virtual infrastructure initiatives often spring from DC server consolidation, which reduces existing infrastructure equipment count and retires older hardware or legacy applications. Server consolidation benefits result from a reduction in the overall number of systems and related recurring costs (power, cooling, rack space, etc.)

While server consolidation addresses the reduction of existing infrastructure, server containment leads to a goal of infrastructure unification. Server containment uses an incremental approach to workload virtualization, using VMs rather than physical servers. Partitioning alone does not deliver server consolidation or containment. Beyond partitioning and basic component-level resource management, comprehensive system resource monitoring of CPU activity, disk access, memory utilization, and network bandwidth; automated provisioning; high availability; and workload migration support are required.

The adoption of smaller form-factor *blade servers* is growing dramatically. Virtualization is an ideal complement for blade servers, delivering benefits such as resource optimization, operational efficiency, and rapid provisioning. The latest generation of x86-based systems feature processors with *64-bit extensions* supporting very large memory capacities are able to host large and memory-intensive applications as well as allow many more VMs to be hosted by a physical server deployed within a virtual infrastructure.

Likewise, *dual-core* processor technology dramatically reduces the costs of increased performance. Compared to traditional single-core systems, systems utilizing dual-core processors will be less expensive, since only half the number of sockets will be required for the same number of CPUs. By significantly lowering the cost of multiprocessor systems, dual-core technology will accelerate DC consolidation and virtual infrastructure development.

6.3.1 Architecture

A DC, which is a physical complex housing physical servers, network switches and routers, network service appliances, and networked storage, provides applications, computation, or storage services. A DC may consist of multiple virtualized DCs where each provides virtualized compute, storage, and network services. Furthermore, each Virtual Data Center (VDC) managed by one tenant can contain multiple VNs and multiple

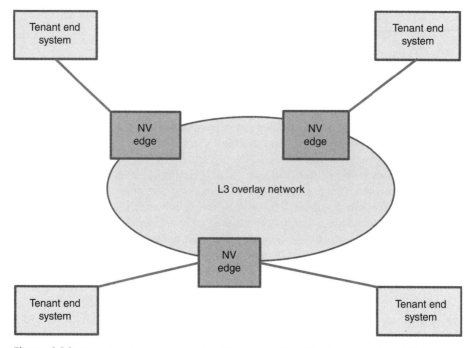

Figure 6.26. Generic reference model for DC network virtualization over a layer 3 infrastructure. Reproduced from Ref. [13].

Tenant End Systems (TESs) that are connected to one or more of these VNs [13]. Virtualized DC services are also known as IaaS.

Several VMs can share the resources of a single physical computer server using the services of a hypervisor that is a server virtualization software running on a physical compute server that hosts VMs. The hypervisor that often embeds a virtual switch provides shared compute/memory/storage and network connectivity to the VMs that it hosts.

The virtual switch switches Ethernet frames between VMs, providing similar services of a physical Ethernet switch. A customer who consumes virtualized DC services offered by a cloud service provider is called Tenant. A single tenant may consume one or more VDCs hosted by the same cloud service provider. An end system of a particular tenant is called TES, which can be a VM, a nonvirtualized server, or a physical appliance.

A virtual network (VN) is a virtual L2 or L3 domain that belongs to a tenant. VNs are isolated from one another and may use overlapping addresses.

The network entity that sits on the edge of the Network Virtualization Overlays (NVO3) network is called Network Virtualization Edge (NVE), which implements network virtualization functions allowing for L2 and/or L3 tenant separation and hiding tenant addressing information (MAC and IP addresses). An NVE could be a virtual switch within a hypervisor, a physical switch or router, and a network service appliance or even be embedded within an End Station.

A VN may be identified by a virtual network identifier (VNID) (Figure 6.26), which is contained in a packet header. The encapsulated frame is delivered to the appropriate VN endpoint by the egress NVE using the VNID in the overlay encapsulation header.

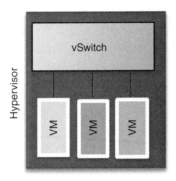

Figure 6.27. A hypervisor architecture.

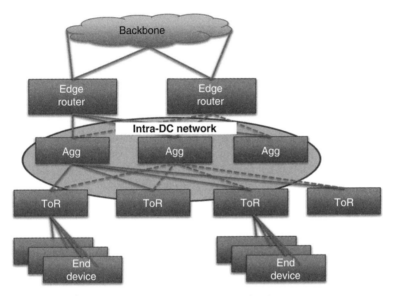

Figure 6.28. A generic architecture for data centers.

The network that provides the connectivity between NVEs is called underlay or underlay network that can use a completely different protocol (and address family) from that of the overlay. The underlying network can be completely unaware of the overlay packets. A tenant may have multiple instances where each instance is an instance of a virtual overlay network.

Figure 6.26 shows a DC reference model for network virtualization using layer 3 overlays where edge devices provide a logical interconnect between TESs that belong to specific tenant network.

An End Device can be a server blade with VMs and virtual switch, where the VM is the TES and the NVE functions may be performed by the virtual switch or the hypervisor or combination (Figure 6.27). The End Device can be a physical server with no VMs and virtual switch as well. As in the DC architecture depicted in Figure 6.28, the server is the TES and the NVE functions may be performed by the Top of Rack (ToR).

A TES attaches to an NVE, either directly or via a switched network.

The network virtualization functions of NVE allow for L2 and/or L3 tenant separation, tenant-related control plane activity, and service contexts from the Routed Backbone nodes. Backbone nodes utilize L3 techniques to interconnect NVE nodes in support of the overlay network. These devices perform forwarding based on outer L3 tunnel header and generally do not maintain per-tenant-service state albeit some applications such as multicast may require control plane or forwarding plane information that pertain to a tenant, group of tenants, tenant service, or a set of services that belong to one or more tunnels. When such tenant or tenant-service-related information is maintained in the core, overlay virtualization provides knobs to control that information.

In a DC, an End Device may be a server or server blade for computing, storage component, or a network appliance such as firewall and load balancer. Alternatively, the End Device may include software-based networking functions used to interconnect multiple hosts. An example of soft networking is the virtual switch in the server blades, used to interconnect multiple VMs. End Device may be single or multihomed to the ToR or End-of-Row (EoR) switches.

Conventional topology uses structured cable connections that are difficult to modify. Both EoR and ToR architectures are designed to facilitate changes required by virtualization.

A ToR is a hardware-based Ethernet switch aggregating all Ethernet links from the End Devices in a rack representing the entry point in the physical DC network for the hosts. ToRs may also provide routing functionality, virtual IP network connectivity, or layer 2 tunneling over IP for instance. ToRs are usually multihomed to switches in the intra-DC network.

The ToR topology is a bigger departure from the conventional architecture. It dedicates an Ethernet switch to every rack of servers. The ToR switch interconnects assets in each rack and provides a trunk connection to an aggregation point in the DC.

The EoR network topology dedicates an Ethernet switch to each row of equipment racks. The virtualized assets in each rack, in each row, are linked to a switch in the EoR rack. That switch also provides a trunk connection to a DC concentrator.

Like EoR, the ToR topology divides the switch fabric and the physical connections into two tiers. The difference is the granularity of the lower tier. Where EoR creates modularity in a row of racks, ToR creates modularity in each individual rack. Note that the ToR design does limit a server rack to only one switch. The aforementioned graphic shows two switches in a rack: one primary and one for redundancy. If the Ethernet switches are implemented as blades, there could be even more switches in a rack.

The EoR topology divides the switch fabric and physical connections from one tier into two, making the network more adaptable. EoR limits the length of the cables in the lower tier to the length of a row of racks. Shorter cables are generally easier to install and easier to change.

EoR topology confines the impact of asset reconfiguration to a row of racks instead of across an entire DC. EoR may reuse some elements of the existing physical network, although major changes and upgrades are likely.

In summary, ToR architecture has less structured cabling. It is modular and easier to change and expand. On the other hand, EoR architecture is less disruptive to infrastructure with fewer switches and trunk connections. It is easier to manage and support.

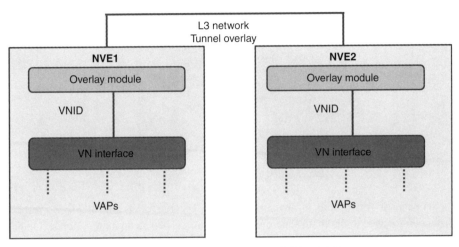

Figure 6.29. Generic reference model for NV Edge.

High-capacity intra-DC network consists of core switches aggregating multiple ToRs. Core switches are usually Ethernet switches but can also support routing capabilities. Connectivity to Internet and VPN customers is provided by DC GW, which is a router connected to the Internet and/or an IPVPN/L2VPN PE.

6.3.1.1 NVE
L3 network(s) or L3 VPN networks (either IPv6 or IPv4 or a combination thereof) provide transport for an emulated layer 2 created by NVE devices. Unicast and multicast tunneling methods are used to provide connectivity between the NVE devices. The NVE devices then present an emulated layer 2 network to the TESs at a VNI through Virtual Access Points (VAPs). The NVE devices map layer 2 unicast to layer 3 unicast point-to-point tunnels and may either map layer 2 multicast to layer 3 multicast tunnels or may replicate packets onto multiple layer 3 unicast tunnels.

The NVE is composed of a tenant-service instance that TESs interface with and an overlay module that provides tunneling overlay functions (e.g., encapsulation/decapsulation of tenant traffic from/to the tenant forwarding instance, tenant identification and mapping, etc.), as described in Figure 6.29.

NVE functionality could reside solely on the End Devices, on the ToRs, or on both the End Devices (i.e., spoke) and the ToRs (i.e., hub). TESs will interface with the tenant-service instances maintained on the NVE spokes, and tenant-service instances maintained on the NVE spokes will interface with the tenant-service instances maintained on the NVE hubs.

TESs are connected to the VNI through VAPs. The VAPs can be in reality physical ports on a ToR or virtual ports identified through logical interface identifiers that are VLANs and internal VSwitch Interface ID leading to a VM.

The VNI represents a set of configuration attributes defining access and tunnel policies and L2 and/or L3 forwarding functions.

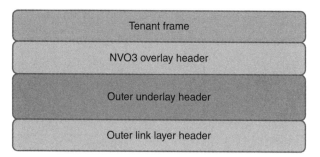

Figure 6.30. NVO3 encapsulated frame.

Per-tenant forwarding information base (FIB) tables and control plane protocol instances are used to maintain separate private contexts between tenants. Hence, tenants are free to use their own addressing schemes without concerns about address overlapping with other tenants.

6.3.1.2 Overlay Modules and VN Context

An overlay network is a layer of VN topology on top of the physical network. Overlay networks are widely used in telecommunications networks to support a broad set of services.

An NVO3 encapsulated frame is depicted in Figure 6.30. It consists of Tenant frame that is either Ethernet or IP based upon the VNI type, NVO3 overlay header containing VNI context information and other optional fields that can be used for processing this packet, outer underlay header that can be either IP or Multiprotocol Label Switching (MPLS), and outer link layer header that is specific to the physical transmission link used.

An NVO3 overlay header needs to be included after the underlay tunnel header to forward tenant traffic. The header contains a field that allows the encapsulated frame to be delivered to the appropriate VN endpoint by the egress NVE. The egress NVE uses this field to determine the appropriate VN context in which to process the packet. This field could be a VNID, which is unique to the administrative domain, or a locally significant identifier, which can be distributed by a control plane or configured via a management plane. If VNID is a global identifier, this field would be large enough to scale to hundreds of thousands of VNs.

In the data plane, each NVE, upon sending a tenant packet, must be able to encode the VN context for the destination NVE in addition to the L3 tunnel source address identifying the source NVE and the tunnel destination L3 address identifying the destination NVE. This allows the destination NVE to identify the tenant-service instance and therefore appropriately process and forward the tenant packet.

The overlay module provides tunneling overlay functions such as tunnel initiation and termination, encapsulation, and decapsulation of frames from VAPs and L3 Backbone. The module may forward IP traffic using tunnels. In a multitenant configuration, the tunnel aggregates frames from/to different VNIs. Tenant identification and traffic demultiplexing are based on the VN context such as VNID.

A unique VNID per-DC administrative domain is used to identify the related Tenant instances similar to IEEE VLAN or ISID tags to provide virtual L2 domains.

One VN context per VNI that is a per-tenant local value is automatically generated by the egress NVE and usually distributed by a control plane protocol to all the related NVEs similar to VRF MPLS labels in IP VPN. One VN context per VAP that is per-VAP local value is assigned and usually distributed by a control plane protocol similar to the use of per CE-PE MPLS labels in IP VPN.

Once the VN context is added to the frame, an L3 tunnel encapsulation is used to transport the frame to the destination NVE. Different IP tunneling options such as GRE, L2TP, and IPsec and tunneling options such as BGP VPN, PW and VPLS are available for both Ethernet and IP formats.

Control plane components that can be an on-net control protocol or a management control entity may be used to provide autoprovisioning/service discovery, address advertisement and tunnel mapping, and tunnel management.

As traffic reaches an ingress NVE, a lookup is performed to determine which tunnel the packet needs to be sent to. It is then encapsulated with a tunnel header containing the destination address of the egress overlay node. Intermediate nodes (between the ingress and egress NVEs) switch or route traffic based upon the outer destination address.

Mapping a final destination address to the proper tunnel is performed by NVEs. Populating these lookup tables can be control plane driven, management plane driven, or data plane driven. When a control plane protocol is used to distribute address advertisement and tunneling information, the autoprovisioning/service discovery could be combined with the address advertisement and tunnel mapping. Furthermore, a control plane protocol that carries both MAC and IP addresses eliminates the need for ARP and hence addresses one of the issues with explosive ARP handling.

A control plane protocol may be required to exchange tunnel state information for unicast or multicast tunnel such as active/standby status information between NVEs, up/down status information, pruning/grafting information for multicast tunnels, and so on.

The benefits of an overlay network may be summarized as:

- Unicast tunneling state management is handled at the edge of the network. Intermediate transport nodes are unaware of such state. This is not the case when multicast is enabled in the core network.
- Tunnels are used to aggregate traffic and hence offer the advantage of minimizing the amount of forwarding state required within the underlay network.
- Decoupling of the overlay addresses (MAC and IP) used by VMs from the underlay network. This offers a clear separation between addresses used within the overlay and the underlay networks, and it enables the use of overlapping address spaces by TESs.
- Support of a large number of VNIDs.

The issues with overlay networks may be summarized as:

- Overlay networks have no controls of underlay networks and lack critical network information. Overlays typically probe the network to measure link properties, such as available bandwidth or packet loss rate. It is difficult to accurately evaluate network properties. It might be preferable for the underlay network to expose usage and performance information.

- Miscommunication between overlay and underlay networks can lead to an inefficient usage of network resources.
- Fairness of resource sharing and collaboration among end nodes in overlay networks are two critical issues.
- When multiple overlays coexist on top of a common underlay network, the lack of coordination between overlays can lead to performance issues.
- Overlaid traffic may not traverse firewalls and NAT devices.
- Multicast service scalability. Multicast support may be required in the overlay network to address for each tenant flood containment or efficient multicast handling.
- Hash-based load balancing may not be optimal as the hash algorithm may not work well due to the limited number of combinations of tunnel source and destination addresses.

6.3.1.3 Data Path

When a frame is received by an ingress NVE from a tenant system over a local VAP, it needs to be parsed in order to identify which virtual network instance it belongs to. The parsing function can examine various fields in the data frame (e.g., VLAN ID) and/or the associated interface/port the frame came from [14].

Once a corresponding VNI is identified, a lookup is performed to determine where the frame needs to be sent. This lookup can be based on any combinations of various fields in the data frame such as destination MAC addresses, destination IP addresses, PCP coding, and DiffServ Code Point (DSCP) coding. The result of this lookup yields the corresponding information needed to build the overlay header, such as the destination L3 address of the egress NVE and a list of tunnels.

The egress NVE checks the context identifier and removes the encapsulation header and then forwards the original frame toward the appropriate recipient, usually a local VAP.

The NVE forwarding plane needs to identify VAP by using the local interface on which the frames are received, where the local interface may be an internal, virtual port in a virtual switch or a physical port on the ToR or sing the local interface and some fields in the frame header such as one or multiple VLANs or the source MAC.

VAPs are associated with a specific VNI at service instantiation time. A VNI identifies a per-tenant private context, that is, per-tenant policies and an FIB table to allow overlapping address space between tenants. There are different VNI types differentiated by the VN service they provide to tenant systems. Network virtualization can be provided by L2 and/or L3 VNIs.

Using a set of NVO3 tunnels, an L2 VNI can be formed where Ethernet multipoint service can be emulated such that TESs are interconnected by an emulated bridge. The bridge may use VLAN tags as a VAP.

Forwarding table entries that are populated by a control or management plane, or via data plane, provide mapping information between tenant system MAC addresses and VAPs on directly connected VNIs and L3 tunnel destination addresses over the overlay.

In the absence of a management or control plane, data plane learning can be used to populate forwarding tables. As frames arrive from VAPs or from overlay tunnels, the tenant system source MAC address is learned against the VAP or the NVO3 tunneling encapsulation source address on which the frame arrived. This implies that unknown unicast traffic be flooded (i.e., broadcast).

To deliver unknown unicast, broadcast, or multicast traffic, the NVE either supports ingress replication or multicast. For multicast, the NVE *must* have one or more multicast trees that can be used by local VNIs for flooding to NVEs belonging to the same VN. For each VNI, there is one flooding tree, and a multicast tree may be dedicated per VNI or shared across VNIs. In such cases, multiple VNIs may share the same default flooding tree.

An L3 VN providing virtualized IP routing and forwarding supports per-tenant forwarding instance with IP addressing isolation and L3 tunneling for interconnecting instances of the same VNI on NVEs. The inner TTL field is decremented by at least one as if the NVO3 egress NVE was one (or more) hop(s) away. The TTL field in the outer IP header is set to a value appropriate for delivery of the encapsulated frame to the tunnel exit point. Therefore, the default behavior could be that the overlay network looks like one hop to the sending NVE.

L2 and L3 VNIs can be deployed together. For example, an L2 VNI may be configured across a number of NVEs to offer L2 multipoint service connectivity, while an L3 VNI can be colocated to offer local routing capabilities and gateway functionality. Furthermore, in an NVE supporting integrated routing and bridging per tenant, an L2 VNI can be used to access to an L3 VNI on the NVE.

Class of Service (CoS) across or between overlay networks can be enforced by using CoS markings between TESs, overlay networks, and underlay networks where each networking layer enforces its own CoS policies independently. For example, PCP or DSCP can be used to indicate application-level CoS. On the other hand, NVE may classify packets based on TES CoS markings or other mechanisms such as DPI to identify CoS for services.

Depending on equipment used in physical network, the underlay network may use a different CoS set than the NVE CoS. The underlay CoS may also change as the NVO3 tunnels pass between different domains.

NVE service CoS may be provided through a QoS field, inside the NVO3 overlay header, as in the PCP field of service tag of 802.1p, DE bits in the VLAN and PBB ISID tags, and MPLS TC bits in the VPN labels.

6.3.1.4 LAG and ECMP

In order to increase bandwidth by load-balancing traffic over multiple paths and increase link availability, multipath over Link Aggregation Group (LAG) at layer 2 and Equal Cost Multipath (ECMP) paths at layer 3 can be used.

LAG and ECMP use a hash of various fields in the encapsulation (outermost) header(s) (e.g., source and destination MAC addresses for non-IP traffic, source and destination IP addresses, L4 protocol, L4 source and destination port numbers, etc.).

Furthermore, hardware deployed for the underlay network(s) will be most often unaware of the carried, innermost L2 frames or L3 packets transmitted by the TS. Thus, in order to perform fine-grained load balancing over LAG and ECMP paths in the underlying network, the encapsulation must result in sufficient entropy to exercise all paths through several LAG/ECMP hops. The entropy information may be inferred from the NVO3 overlay header or underlay header. If the overlay protocol does not support the necessary entropy information or the switches/routers in the underlay do not support parsing of the additional entropy information in the overlay header, underlay switches and routers should be programmable to select the appropriate fields in the underlay header for hash calculation based on the type of overlay header.

All packets that belong to a specific flow must follow the same path in order to prevent packet reordering.

All paths available to the overlay network need to be used efficiently. For example, different flows can be distributed as evenly as possible across multiple underlay network paths.

6.3.1.5 DiffServ and ECN Marking

When traffic is encapsulated in a tunnel header, the DSCP and Explicit Congestion Notification (ECN) markings can be set in the outer header and propagated to the inner header on decapsulation.

Per uniform model in [15], the inner DSCP marking can be copied to the outer header on tunnel ingress, and then outer header value will be copied back to the inner header at tunnel egress. ECN marking as defined in [16] may be performed here as well.

6.3.2 OAM

As we pointed out in the previous chapters, OAM consists of Fault Management, Configuration Management, Accounting Management, Performance Management, and Security Management. Fault Management consists of fault detection, fault verification, fault isolation, fault notification, alarm suppression, and fault recovery.

Fault detection deals with mechanism(s) that can detect both hard failures such as link and device failures and soft failures such as software failure, memory corruption, misconfiguration, and so on. After verifying that a fault has occurred along the data path, isolating the fault to the level of a given device or link is necessary. Therefore, a fault isolation mechanism is needed. A fault notification mechanism can be used in conjunction with a fault detection mechanism to notify the devices upstream and downstream to the fault detection point and NMS.

Fault notifications between layers are suggested [17]. The author does not believe in notifications from upper layer to lower layers, but from lower layers to upper layers. When there is a client/server relationship between two layered networks, the client layer should be notified of server layer failures. For example, NVO3 layer is a client of the outer IP server layer, while the inner IP layer is a client of the NVO3 server layer. The NVO3 layer should be notified of IP server layer failures. Similarly, the inner IP layer should be notified of NVO3 server layer.

Finally, fault recovery deals with recovering from the detected failure by switching to an alternate available data path (depending on the nature of the fault) using alternate devices or links. In fact, the controller can provision another VN, thus automatically resolving the reported problem.

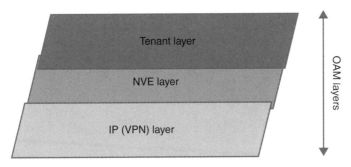

Figure 6.31. OAM layers in an NVO3 network.

In order to recover from failures quickly, redundant paths, redundant links, and redundant network elements (NEs) can be provisioned as appropriately by the controller. During failures, the traffic will be switched from failed primary to the secondary within 50 ms.

The controller may also directly monitor the status of VN components such as NVEs [18] in order to respond to their failures. In addition to fault notifications, the controller may deliver notifications to a higher-level orchestration component such as the one responsible for VM provisioning and management.

Performance Management deals with mechanism(s) that allows determining and measuring the performance of the network/services under consideration. Performance Management can be used to verify the compliance to both the service-level and network-level metric objectives/specifications. Performance Management typically consists of measuring performance metrics, for example, Frame Loss, Frame Delay, Frame Delay Variation (aka Jitter), Frame throughput, Frame discard, and so on, across managed entities when the managed entities are in available state. Performance Management is suspended across unavailable managed entities.

The L3 overlay network can be operated by a single service provider or span across multiple administrative domains as defined for Carrier Ethernet Networks [19]. While each of the layers is responsible for its own OAM, each layer may consist of several different administrative domains as shown in Figure 6.31.

The tenant layer could be IP/Ethernet and the IP (VPN) layer could be IP (VPN) over Ethernet, while network virtual edge will have to support both IP (VPN) layer and tenant layer.

At the L3 IP overlay network layer, IP (VPN) and/or Ethernet OAM mechanisms are used to probe link by link, node to node, and so on. OAM addressing here means physical node loopback or interface addresses.

At the NVE layer, NVO3 OAM messages are used to probe the NVE-to-NVE tunnels and NVE entity status. OAM addressing here means the physical node loopback together with the VNI.

At the Tenant layer, the IP and/or Ethernet OAM mechanisms are again used, but here, they are operating over the logical L2/L3 provided by the NVE through the VAP. OAM addressing at this layer deals with the logical interfaces on Vswitches and VMs.

The OAM layer below NVO3 will be supported by existing IP/L2 or IP/MPLS OAM methods and tools. An OAM domain can be defined as a closed set of NVEs and the tunnels that interconnect them. An NVE device dynamically discovers other NVE devices that share the same VNI within a given NVO3 domain.

Proactive connectivity monitoring is needed between two or more NVE devices that support the same VNIs within a given NVO3 domain. Automatic recovery from transmission facility failure by switchover to a redundant replacement facility may be triggered by OAM notifications.

Monitoring/tracing of all possible paths in the underlay network between a specified set of two or more NVE devices might differentiate equal cost paths that traverse LAG and/or ECMP.

Connectivity fault verification and isolation between two or more NVE devices that support the same VNI within a given NVO3 domain are needed.

VNI availability, VNI frame loss, VNI frame discard, VNI frame throughput, two-way frame delay, one-way frame delay, and delay variation between two NVE devices that support the same VNI within a given NVO3 domain are to be measured. Availability concept per VNI has not been defined yet. It is possible to adopt the availability concept for multipoint EVC as defined in [20] by replacing each UNI with a TES.

OAM frames will follow the same path as the NVO3 data frames. The QoS for OAM frames could be the same as the QoS for data frames.

NVO3 OAM frames need to be prevented from leaking outside their NVO3 domain. Similarly, NVO3 OAM frames from outside an NVO3 domain need to be prevented from entering the said NVO3 domain. The rules described in [19] can be applied here.

NVO3 OAM might leverage the OAM capabilities of the transport layer such as L3 or L2 and may adapt or interwork with its IP underlay OAM functions.

6.3.3 Security

Attacks for an NVO3 can be classified into two categories: inside attacks and outside attacks [21]. An attack is considered as an inside attack if the adversary performing the attack has got certain privileges in changing the configuration or software of an NVO3 device or a network device of the underlying network where the overlay is located upon and initiates the attack within the overlay security perimeter. In contrast, an attack is referred to as an outside attack if the adversary performing the attack (outside attacker or outsider) has no such privilege and can only initiate the attacks from compromised TSes. In a complex attack, inside and outside attacking operations may be performed in a well-organized way to expand the damages caused by the attack [12].

An outsider attacker can eavesdrop on the packets, replay the intercepted packets, and generate illegal packets and inject them into the network. With a successful outside attack, the attacker may be able to analyze the traffic pattern of a tenant or an end device, disrupt the network connectivity or degrade the network service quality, or access the contents of the data/control packets if they are not encrypted.

On the other hand, an inside attacker can perform any types of outside attacks from the inside or outside of the overlay perimeter using its privileges. For example, the inside attacker interferes with the normal operations of the overlay as a legal entity by sending packets containing invalid information or with improper frequencies, performs spoofing attacks and impersonates another legal device to communicate with victims using the cryptographic information it obtained, and accesses the contents of the data/control packets if they are encrypted with the keys held by the attacker.

For a secure NVO3, isolation of the VNs, spoofing detection, integrity protection and message origin authentication for the control packets, and protection of centralized servers to make sure that they will not become the bottleneck of the control plane especially under distributed denial-of-service (DDOS) attacks are required.

Basic security approaches are described below.

6.3.3.1 Securing the Communications between NVEs and TESs

If the network connecting the NVE and the TESs is potentially accessible to attackers, the security properties of data traffic such as integrity, confidentiality, and message origin authenticity can be provided by mechanisms such as IPsec, Secure Sockets Layer (SSL), and TCP Authentication Option (TCP-AO).

In order to secure the data/control traffic, automatic key management protocols such as Transport Layer Security (TLS) can be deployed where TESs belonging to different VNs use different keys to secure the control packet exchanges with their NVE.

In the colocated case, all the information exchanges between the NVE and the TESs are within the same device, and the TS traffic in different VNs needs to be isolated. In addition, the computing and memory resources used by the NVE, the hypervisor, and the TESs need to be isolated.

6.3.3.2 Securing the Communications within Overlays

In order to secure control plane, the integrity and origin authentication of the messages need to be guaranteed by encrypting signaling packets when the signaling messages are confidential. When the network devices exchange control plane packets, integrated security mechanisms or security protocols can be provided. Furthermore, cryptographic keys need to be deployed manually or dynamically generated by using certain automatic key management protocols such as TLS. The signaling messages belonging to different VNs can be secured by different keys.

When there are centralized servers providing mapping information within the overlay, it will be important to prevent a compromised NVE from impersonating the centralized servers to communicate with other NVEs by associating different NVEs with different keys when they exchange information with the centralized servers.

When there are a large amount of NVEs working within an NVO3 overlay, manual key management may become infeasible. Using automated key management solutions such as Extensible Authentication Protocol (EAP) [22] for NVO3 overlays becomes necessary. When an automated key management for NVO3 overlays is deployed, mutual authentication needs to be performed before two network devices such as NVEs in the overlay start exchanging the control packets.

The issues of DDOS attacks also need to be considered in designing the overlay control plane. For instance, in the VXLAN solution [23], an attacker attached to an NVE can try to manipulate the NVE to keep multicasting control messages by sending a large amount of ARP packets to query the inexistent VMs. In order to mitigate this type of attack, the NVEs should send signaling message in the overlay with a limited frequency.

6.3.3.3 Data Plane Security

When a data packet reaches the boundary of an overlay, it will be encapsulated and forwarded to the destination NVE through a proper tunnel [13]. An inside attacker compromising an underlying network device may intercept an encapsulated data packet transported in a tunnel, modify the contents in the encapsulating tunnel packet, and transfer it into another tunnel without being detected. When the modified packet reaches an NVE, the NVE may decapsulate the data packet and forward it into a VN according to the information within the encapsulating header generated by the attacker. Similarly, a compromised NVE may try to redirect the data packets within a VN into another VN by adding improper encapsulating tunnel headers to the data packets. Under such circumstances, in order to enforce the VN isolation property, signatures or digests need to be generated for both data packets and the encapsulating tunnel headers in order to provide data origin authentication and integrity protection. In addition, NVEs should use different keys to secure the packets transported in different tunnels.

6.3.3.4 Security Issues with VM Migration

The VM migration may also cause security risks. Because the VMs within a VN may move from one server to another that connects to a different NVE, the packets exchanging between two VMs may be transferred in a new path. If the security policies deployed on the firewalls of the two paths conflict or the firewalls on the new path lack essential state to process the packets, the communication between the VMs may be broken. To address this problem, one option is to enable the state migration and policy confliction detection between firewalls. The other one is to force all the traffic within a VN be processed by a single firewall. However, this solution may cause traffic optimization issues.

6.3.4 Applications

6.3.4.1 Basic VNs in a DC

A VN may exist within a DC [6]. The network enables a communication among TESs that are in a Closed User Group (CUG). A TES may be a physical server or VM on a server. A VN has a unique local or global VNID for switches/routers to properly differentiate it from other VNs. The CUGs are formed so that proper policies can be applied when the TESs in one CUG communicate with the TESs in other CUGs.

Figure 6.32 depicts two network virtual edges that may exist on a server or ToR. Each NVE may be the member of one or more VNs over L2 or L3 physical networks. The VN *T1* terminates on both NVE1 and NVE2, while the VN *Tn* terminates on NVE1 and the VN *Tm* at NVE2 only. If an NVE is a member of a VN, one or more virtual network instances (VNI) represented by routing and forwarding tables exist on the NVE. Each NVE has one overlay module to perform frame encapsulation/decapsulation and tunneling initiation/termination. In this scenario, a tunnel between NVE1 and NVE2 is necessary for the VN T1.

A TES attaches to a VN via a VAP on an NVE. One TES may participate in one or more VNs via VAPs; one NVE may be configured with multiple VAPs for a VN. Furthermore, if individual VNs use different address spaces, the TES participating in all of them will be configured with multiple addresses as well. In addition, multiple TESs may use one VAP to attach to a VN. For example, VMs are on a server and NVE is on ToR, and some VMs may attach to NVE via one VLAN.

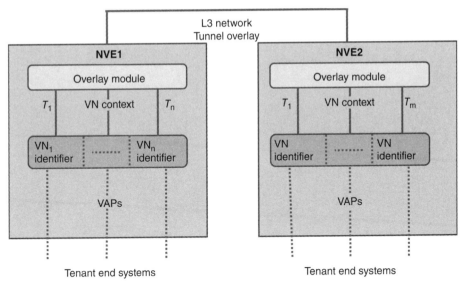

Figure 6.32. NVO3 for Tenant End System (TES) interconnection.

A VNI on an NVE is a routing and forwarding table that maintains the mapping of a TES and its attached NVE. The table entry may be updated by the control plane or data plane or management plane. It is possible that an NVE has more than one VNI associated with a VN.

When TES and NVE are colocated on a same physical device, the NVE is aware of the TES state at any time via internal application programming interface (API). When TES and NVE are remotely connected, a protocol is necessary for NVE to know TES state.

Individual VNs may use its own address space that is isolated from DC infrastructure, eliminating the route changes in the DC underlying network when VMs move.

If a VN spans across multiple DC sites, the tunnel between a pair of NVEs may in turn be tunneled over other intermediate tunnels over the Internet or other WANs, or the intra-DC and inter-DC tunnels are stitched together to form an end-to-end tunnel between two NVEs.

A user can access a VN consisting of NVE1 and NVE2 via the Internet by using IPsec tunnel as in Figure 6.33 where NVE2 is connected to TES running a VN gateway and NAT applications. The encrypted tunnel is established between the VN GW and the user machine or CPE at enterprise location. The VN GW provides authentication scheme and encryption.

6.3.4.2 DC VN and WAN VPN Interconnection

A DC provider and carrier may build a VN and VPN independently and interconnect the two. Figure 6.34 depicts an L3 overlay where a DC provider constructs an L3 VN between the NVE1 on a server and the NVE2 on the DC GW, while the carrier constructs an L3 VPN between PE1 and PE2 in its IP/MPLS network. An Ethernet interface physically connects the DC GW and PE2 devices. An Ethernet interface may be used between PE1 and CE to connect the L3 VPN and enterprise physical networks.

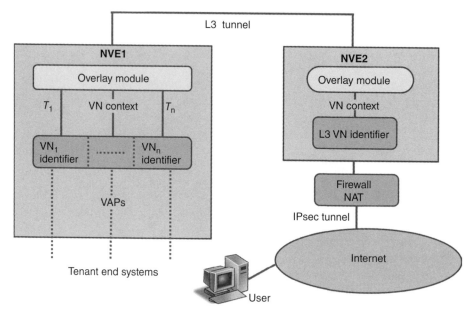

Figure 6.33. DC virtual network access via the Internet.

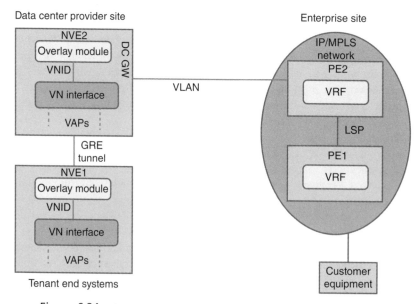

Figure 6.34. L3 VNI and L3 VPN interconnection across multinetworks.

This configuration allows the enterprise networks communicating to the L3 VN as if its own networks but not communicating with DC provider underlying physical networks as well as not other overlay networks in the DC. The enterprise may use its own address space on the L3 VN.

The DC provider can manage the VM and storage assignment to the L3 VN for the enterprise customer who can determine and run applications on the VMs. From the L3 VN perspective, an endpoint in the enterprise location appears as the endpoint associating to the NVE2. The NVE2 on the DC GW has to perform both the GRE tunnel termination [24] and the local VLAN termination and forward the packets in between. The DC provider and carrier negotiate the local VLAN ID used on the Ethernet interface.

This configuration makes the L3 VPN over the WANs that only has the reachability to the TES in the L3 VN. It does not have the reachability of DC physical networks and other VNs in the DC. However, the L3 VPN has the reachability of enterprise networks.

The eBGP protocol can be used between DC GW and PE2 for the route population in between. The eBGP, OSPF, or other can be used between PE1 and CE for the route population.

The L3 VN and compute resource are managed by the DC provider. The DC operator can place them at any location without notifying the enterprise and carrier because the DC physical network is completely isolated from the carrier and enterprise network. Furthermore, the DC operator may move the compute resources assigned to the enterprise from one server to another in the DC without the enterprise customer awareness, that is, no impact on the enterprise *live* applications running these resources.

6.3.4.3 Tenant VN with Bridging/Routing

For a tenant spanning multiple DCs, a DC provider may want to use L2 VN within a DC for its simplicity and L3 VN outside DCs for aggregation and scalability. A virtual gateway interface function (VNIF) can interconnect L2 VN and L3 VN in each DC (Figure 6.35).

The GW1 is the member of the L2 VN and L3 VN. The L2 VNI is the MAC/NVE mapping table and the L3 VNI is IP prefix/NVE mapping table. A packet coming to the GW1 from L2 VN will be decapsulated and converted into an IP packet and then encapsulated and sent to the site Z. The gateway uses ARP protocol to obtain MAC/IP mapping.

6.3.4.4 VDC

A DC provider may offer a VDC on top of a common network infrastructure for many customers and run service applications per-customer basis. The service applications may include firewall, gateway, DNS, load balancer, NAT, and so on.

A DC provider can construct several L2 VNs to group the end tenant systems together per application basis and create an L3 VNa for the internal routing. A server or VM runs firewall/gateway applications and connects to the L3 VN and Internet. A VPN tunnel can be built between the gateway and enterprise router. Users at Enterprise site can access the applications via the VPN tunnel and Internet via a gateway at the Enterprise site to access the applications via the gateway in the provider DC.

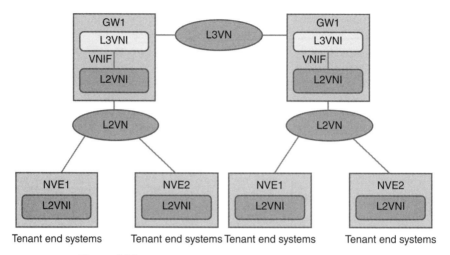

Figure 6.35. Tenant virtual network with bridging/routing.

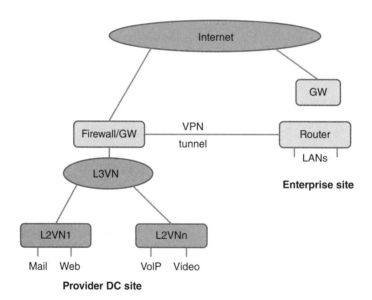

Figure 6.36. Virtual Data Center by using NVO3.

The enterprise operators can also use the VPN tunnel or IPsec over the Internet to access the VDC for the management purpose. The firewall/gateway provides application-level and packet-level gateway function and/or NAT function.

DC operators can configure the proper security policy and gateway function. DC operators may further set different QoS levels for the different applications for a customer.

This application requires the NVO3 solution to provide the DC operator an easy way to create NVEs and VNIs for any design and to quickly assign TESs to a VNI and easily configure policies on an NVE (Figure 6.36).

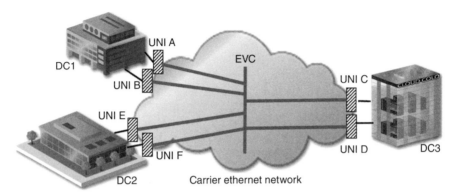

Figure 6.37. VN formed of EVCs where DCs and network are provided by one service provider.

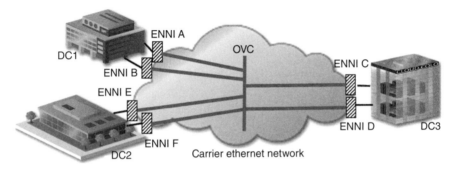

Figure 6.38. VN formed of EVCs where DCs and network are provided by different service providers.

6.3.4.5 EVC-Based VNs

VNs can be established by EVCs over Carrier Ethernet Networks as well as over non-Carrier Ethernet Networks as long as TES interfaces to networks via either UNI (Figure 6.37) or ENNI (Figure 6.38), and depending on if DCs and network are provided by one service provider or multiple.

6.4 NETWORK FUNCTIONS VIRTUALIZATION

In addition to the effort in IETF described earlier, service providers formed an industry specifications group within ETSI called the *Network Functions Virtualization* (NFV) Group [25]. The group is working on defining the requirements and architecture for the virtualization of network functions to:

- Simplify ongoing operations.
- Achieve high-performance, portable solutions.
- Support smooth integration with legacy platforms and existing EMS, NMS, Operations Support System (OSS), BSS, and orchestration systems.

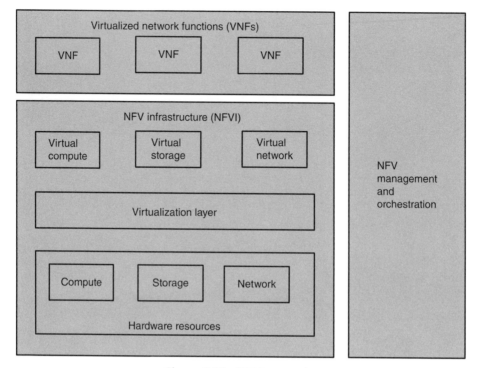

Figure 6.39. NFV framework.

- Enable an efficient migration to new virtualized platforms.
- Maximize network stability and service levels and ensure the appropriate level of resilience.

Service providers' networks are populated with a large and increasing variety of proprietary hardware appliances. Hardware-based appliances rapidly reach end of life, requiring much of the procure, design, integrate, deploy cycle to be repeated with little or no revenue benefit. Furthermore, hardware life cycles are becoming shorter as technology and service innovation accelerates, inhibiting the rollout of new revenue earning network services and constraining innovation [26].

NFV aims to address these problems by consolidating many network equipment types onto industry standard high-volume servers, switches, and storage, which could be located in DCs, network nodes, and customer premises.

A high-level NFV framework is depicted in Figure 6.39 [27]. Three working domains are defined:

- **Virtualized Network Function (NFV),** as the software implementation of a network function that is capable of running over the NFV Infrastructure (NFVI).
- **NFVI,** including the diversity of physical resources and how these can be virtualized. NFVI supports the execution of the Virtual Network Functions (VNFs).

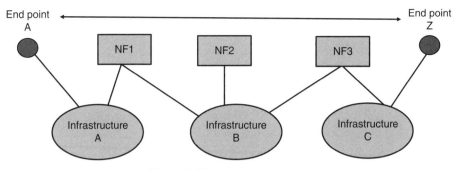

Figure 6.40. End-to-end service.

- **NFV Management and Orchestration,** which covers the orchestration and life cycle management of physical and/or software resources that support the infrastructure virtualization and the life cycle management of VNFs.

The relationships between VNFs are:

- **VNF Forwarding Graph (VNF-FG)** covers the case where network connectivity such as a chain of VNFs in a server tier with firewalls, NAT, and load balancer is important.
- **VNF Set** covers the case where the connectivity is not specified as in the case of virtualized residential gateway.

A network service is viewed as a forwarding graph of Network Functions (NFs) interconnected by supporting network infrastructure of a single or multiple operators. The underlying network function behavior contributes to the behavior of higher-level service. In other words, the network service behavior is a joint behavior of functional blocks such as NFs, NF Sets, and NF Forwarding Graphs.

The endpoints and network functions of network service (Figure 6.40) are represented as nodes and correspond to devices, applications, and/or physical server applications. An NFV Forwarding Graph can have network function nodes connected by logical links that can be unidirectional, bidirectional, multicast, and/or broadcast.

For example, in a Metro Ethernet Network depicted in Figure 6.41, CPE and edge router (PE) can be virtualized. In other words, some of CPE and PE functions can be implemented in software and reside in servers, as opposed to implementing them in proprietary hardware.

6.5 SOFTWARE-DEFINED NETWORKS

Software-defined networking (SDN) is defined by Open Networking Foundation (ONF) [28] as an emerging architecture that is dynamic, manageable, cost-effective, and adaptable. This architecture decouples the network control and forwarding functions enabling

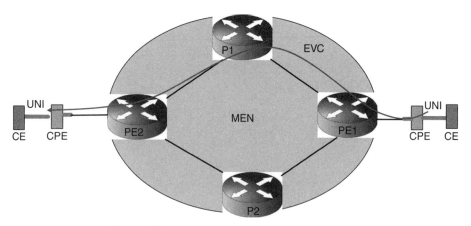

Figure 6.41. An example Metro Ethernet Network.

the network control to become directly programmable and the underlying infrastructure to be abstracted for applications and network services.

The switching plane can be heterogeneous, composed of NEs from multiple vendors, and it can provide distinct services with different characteristics, configurations, and controls at the packet and/or optical layers. Abstracting the control plane from the NEs allows network-platform-specific characteristics and differences that do not affect services to be hidden. In addition, software-defined networking (SDN) is based on the principle that applications can request needed resources from the network via interfaces to the control plane. Through these interfaces, applications can dynamically request network resources or network information that may span disparate technologies. For instance, the application layer can dynamically request and obtain network resources at the packet flow, circuit, or even optical level based on application layer requirements. Current SDN implementations focus on Ethernet switching primarily for DC resource optimization.

Today's network infrastructure is complex and proprietary despite of substantial effort for standardizations in numerous standards organizations. The networks are too difficult to program. Rate of innovation is lower compared to other areas such as applications. Complex functions such as routing and traffic engineering are built into special hardware and software. Although standard protocols such as SNMP and CLI are used for management interfaces, the management is pretty much different for each device. Therefore, OPEX and CAPEX are high for today's networks. It is hard to adapt to new application requirements and security threats and scale.

Business pressures are forcing broadband providers to reconsider how high-speed Internet services are managed and priced. Faster connection speeds and rapidly growing consumer demand for bandwidth-intensive applications and services are driving an endless cycle of investments to expand network capacity. At the same time, per-subscriber usage is increasing across all demographics, while per-subscriber revenue growth is lagging behind increasing capital and operating expenses, resulting in lower profit margins and less budget for additional equipment and bandwidth. The challenge for broadband providers is to establish new service management and pricing models that satisfy the needs of all subscribers while ensuring long-term business sustainability.

As communications networks have grown in size and complexity, streamlining the architecture and implementation to reduce costs, simplify management, improve service provisioning time, and improve resource utilization has become increasingly important. Ideally, an application or service could be completely decoupled from the underlying network infrastructure, but this is not always realistic. Parameters that affect application performance, such as bit rate, packet loss, and latency, are closely tied to the underlying network. To meet application performance objectives, it becomes necessary for the application or its proxy to ensure that the underlying network is aware of the application requirements and provides the necessary services. A network control plane component is often used to find and configure the required resources in the network, as well as map the application traffic to these resources. There are many approaches for applications to request and receive a service from a network. In all cases, the application interacts with some sort of a network control plane.

The transport network with no visibility into IP traffic patterns and application requirements often does not need to support dynamic services. They remain largely static and are provisioned manually where bringing up a new circuit to support a service can take weeks or days.

There is a great deal of optimism that SDN will make networks more flexible, dynamic, and cost-efficient while greatly simplifying operational complexity. Vendors have begun unveiling its open network environment, extending network capabilities, and extracting greater intelligence from network traffic through programmatic interfaces.

With SDN, IT can orchestrate network services and automate control of the network according to high-level policies, rather than low-level network device configurations. By eliminating manual device-by-device configuration, IT resources can be optimized to lower costs and increase competitiveness.

IT can create VNs to handle the real-time demands of video or voice applications, ensuring that users have the quality experience they demand. IT can gain unified control over wireless and wired LANs, making it easier for mobile users to stay seamlessly connected to their applications and services, no matter where they are. SDN principles allow service providers greater control to run cost-optimized and service-optimized converged packet-circuit networks, where they have maximum flexibility in choosing the correct mix of technologies depending on service needs. With common control over packets and circuits, carriers can innovate outside the box by designing networking applications specifically taking advantage of the strengths of both kinds of switching technologies.

The rise of cloud services and virtualization and large-scale computing and storage in huge DCs require on-demand availability of additional network capacity (i.e., scalability) and rapid service creation and delivery. SDN is expected to automate service provisioning at least and help service providers to deliver services much quicker.

The infrastructure consists of both physical and VN devices such as switches and routers. These devices implement the OpenFlow protocol as a standards-based method of implementing traffic forwarding rules. The control layer consists of a centralized control plane for the entire network to provide a single centralized view of the entire network. The control layer utilizes OpenFlow to communicate with the infrastructure layer.

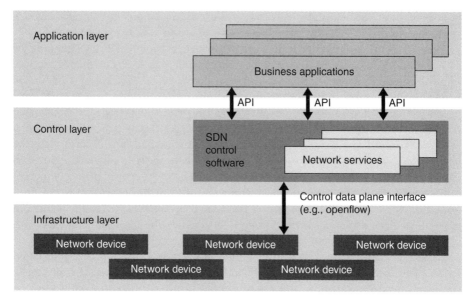

Figure 6.42. SDN layers.

The application layer consists of network services, orchestration tools, and business applications that interact with the control layer. These applications leverage open interfaces to communicate with the control layer and the network state (Figure 6.42).

IP and transport networks today do not interact. As a result of this lack of interaction, the Internet core cannot benefit from more scalable circuit switches, nor take advantage of dynamic circuit switching. Dynamic circuits can recover faster from failures and provide bandwidth-on-demand with guaranteed low latency and jitter. These are hard to provide in today's networks.

The OpenFlow™ protocol is a foundational element for building SDN solutions. Virtualized networks, enabled by the OpenFlow protocol, should make it easier for organizations to configure and manage their applications and resources across campus networks. For example, a retailer can provision payment devices using virtualized networks. Compliance with Payment Card Industry (PCI) security regulations may become easier.

SDN and OpenFlow meet the goal of helping users develop applications that manage and control the network independent of the underlying topology or switch vendor while providing access to resources on a per-flow basis.

The purpose of SDN is to reduce the complexity of network operation, increase robustness, maximize interaction between layers, and maximize network resources and efficiency and minimize time to revenue for new applications and services. The SDN objectives may be listed as:

- Lower layer network abstraction
- Partitioning of resources
- Network automation

- Application-to-network relationship
- Provides access to the forwarding plane of network devices
- Reduce DC complexity
- Provide more robust resiliency and network flexibility
- Optimize the use of the network resources for applications and higher-layer platforms
- Accelerate application innovation
- Flexible control of flows passing through the network
- Forwarding of packets according to software-defined rules
- Load balancing of packets according to software-defined heuristics combined with hardware flow hashing
- Flexible modification of the frame (NAT, tunneling, tag rewrites) to aid in interfacing between hosts, local networks, and external network interfaces
- Intelligent flow classification (security, virtual domaining, etc.) that is software defined and is done in parallel with forwarding and frame modifications
- Vendor-independent interface to the switching elements
- Ability to work beside existing protocols (hybrid switch)
- Ability to overlay SDN

According to Plexxi and Lightspeed Ventures [29], companies are already making buying decisions based on transitioning their networks to SDN. 30–40% of total networking spend could be influenced by SDN over the next 6 years, resulting in a total SDN market of $35B by 2018 [30]. Despite the shift to software-defined networking, hardware will continue to play a significant role in the network infrastructure. By 2018, 46% of overall DC network spending will be on SDN-enabled optical, switching, and routing hardware.

6.5.1 Standards

SDN standards are being set by multiple organizations. The ONF is dedicated to promoting and standardizing software-defined networking, including the OpenFlow protocol. It is founded in March 2011 by Deutsche Telekom, Facebook, Google, Microsoft, Verizon, and Yahoo. Currently, there are more than 50 members. The new ONF-Optical Transport WG (OTWG) will address SDN and OpenFlow™ control capabilities for optical transport networks (OTNs).

Open Networking Research Center (ONRC) and ON.Lab seek to *open up the Internet infrastructure for innovations* and enable the larger network industry to build networks that offer increasingly sophisticated functionality yet are cheaper and simpler to manage than current networks. ON.Lab is a sister organization where ONRC develop, distribute, and support open-source SDN tools and platforms for the larger community.

OpenStack Foundation promotes the development, distribution, and adoption of the OpenStack cloud OS.

ETSI Industry Specification Group (ISG) for NFV was initiated by seven of the world's leading telecoms network operators for virtualization of network functions. AT&T, BT, Deutsche Telekom, Orange, Telecom Italia, Telefonica, and Verizon have been joined by 52 other network operators and telecoms equipment vendors. ITU also works on SDN. In November 2012, the ITU drafted a new resolution on software-defined networking (SDN) that instructs ITU-T Study Group 13 (Future Networks including mobile and NGN) to expand and accelerate its work in the SDN domain.

Internet Research Task Force Software-Defined Networking Research Group (IRTF SDNRG) investigates SDN from various perspectives and provides a forum for researchers. SDNRG provides objective definitions, metrics, and background research with the goal of providing this information as input to protocol, network, and service design to SDOs and other standards-producing organizations such as the IETF, ETSI, ATIS, ITU-T, IEEE, ONF, MEF, and DMTF.

IETF is also another standards organization working on SDN.

6.5.2 SDN Building Blocks

SDN building blocks [31, 32] and the author's view of layers involved in SDN are depicted in Figure 6.43.

The application layer sits on top of service layers. The service layer interacts with an orchestrator that can interact with multiple controllers via a northbound API. The northbound interface between applications/services allows the applications to authenticate and learn of which objects they have authorization to manipulate or to interact with objects belonging to controlling software. The SDN Orchestrator requests object

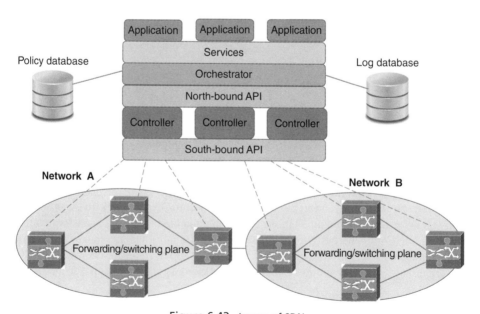

Figure 6.43. Layers of SDN.

models from each of the controlling software it is responsible for managing and manipulates them. The Orchestrator can be centralized or distributed as well.

Reference [33] proposes an SDN Orchestrator interface between Policy DB and Orchestrator for network policy and authentication and authorization. If there are multiple orchestrators, an SDN Orchestrator to Orchestrator Interface is needed to allow interaction between orchestrators.

Reference [33] also proposes an SDN Orchestrator interface between Orchestrator and logging system for SDN Orchestrators to collect and communicate the relevant activity logs to support log-consuming applications to operate in multi-SDN Orchestrator environment.

SDN controller instructs the switches as to what actions they should take via the southbound API. OpenFlow is rapidly becoming the dominant way for an SDN controller to communicate with switches. The equipment shown in Figure 6.43 can either be pure OpenFlow switches, whereby the only function that they provide is to forward packets, or they can be OpenFlow-hybrid switches that support both OpenFlow-enabled flow forwarding and traditional Ethernet switch bridging and routing.

The SDN Control Interface is used between the Orchestrator and an application entity, Policy DB, Location Services, Plug-In, or another Orchestrator in order to send commands, receive replies, and emit notifications.

The author believes that it would take a long time to see a network with all SDN supporting equipment. As a result, SDN-enabled devices will have to coexist with legacy systems.

In the following sections, we will describe OpenFlow and examples for SDN networks.

6.5.3 OpenFlow

OpenFlow is a fundamental networking technology that enables innovative network solutions based on centralized, controller-based management of multiple physical network devices using *forwarding instruction sets* rather than traditional system-level networking protocols. OpenFlow has been in development since 2007, led by Stanford University and the University of California at Berkeley.

It became a standard that is now defined by the ONF since 2011. OpenFlow hides the complexity of the individual components of network devices. OpenFlow centralizes the control of those devices in a virtualized manner. This should simplify network management for network managers. The growing OpenFlow ecosystem now includes routers, switches, virtual switches, and access points from a range of vendors, like HP.

OpenFlow is one of the candidate protocols to convey specific configuration information toward devices. As such, OpenFlow is one possible component of the global SDN toolkit. The OpenFlow provides software control of the flow tables that instruct switches how to handle traffic within an SDN-based network. OpenFlow provides access to the data plane of the network and allows software to determine the path that data packets or flows will take.

Since the transition to SDN or OpenFlow will take some time, OpenFlow must coexist with traditional networks. For the transition to be smooth, OpenFlow must be supported by hybrid switches that also support traditional L2/L3 switching and IP routing.

The OpenFlow protocol uses a standardized instruction set that enables an OpenFlow controller to send a common set of instructions to any OpenFlow-enabled switch.

OpenFlow is an open-standards way of virtualizing the network. Network managers can specify different policy rules for different groups of devices and users, which create multiple virtualized networks regardless of the physical network connections. This allows network managers to easily customize and manage these virtualized networks to ensure proper policies such as forwarding path, QoS, and security.

OpenFlow is designed to be programmable. The OpenFlow instruction set allows network managers to try new ideas or create new protocols to solve problems specific to their organizations' network needs. This allows experimentation with new services and protocols on a real-world network that cannot be simulated in a test lab.

In a classical router or switch, the fast packet forwarding (data path) and the high-level routing decisions (control path) occur on the same device. An OpenFlow Switch separates these two functions. The data path portion still resides on the switch, while high-level routing decisions are moved to a separate controller, typically an external server. The OpenFlow Switch and Controller communicate via the OpenFlow protocol, which defines messages, such as packet-received, send-packet-out, modify-forwarding-table, and get-stats.

The data path of an OpenFlow Switch presents a clean flow table abstraction; each flow table entry contains a set of packet fields to match and an action (such as send-out-port, modify-field, or drop). When an OpenFlow Switch receives a packet it has never seen before, for which it has no matching flow entries, it sends this packet to the controller. The controller then makes a decision on how to handle this packet. It can drop the packet, or it can add a flow entry directing the switch on how to forward similar packets in the future.

OpenFlow allows new and innovative controller-based solutions across a wide range of networking environments for applications such as VM mobility, high-security networks, and next-generation IP-based mobile networks.

OpenFlow Switch Specification Version 1.3.0 [29] covers the components and the basic functions of the switch and the OpenFlow protocol to manage an OpenFlow switch from a remote controller. OpenFlow components are depicted in Figure 6.44.

An OpenFlow Switch consists of one or more flow tables and a group table, which perform packet lookups and forwarding, and an OpenFlow channel to an external controller (Figure 6.44). The switch communicates with the controller and the controller manages the switch via the OpenFlow protocol.

Using the OpenFlow protocol, the controller can add, update, and delete flow entries in flow tables, both reactively (in response to packets) and proactively. Each flow table in the switch contains a set of flow entries where each flow entry consists of match fields, counters, and a set of instructions to apply to matching packets.

Matching starts at the first flow table and may continue to additional flow table. Flow entries match packets in priority order, with the first matching entry in each table being used. If a matching entry is found, the instructions associated with the specific

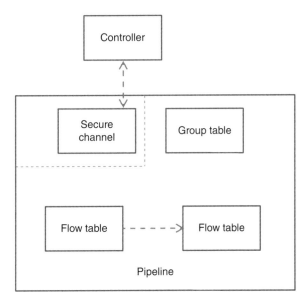

Figure 6.44. Main components of an OpenFlow switch.

flow entry are executed. If no match is found in a flow table, the outcome is based on configuration of the table-miss flow entry. For example, the packet may be forwarded to the controller over the OpenFlow channel, may be dropped, or may continue to the next flow table.

Instructions associated with each flow entry either contain actions or modify pipeline processing. Actions included in instructions describe packet forwarding, packet modification, and group table processing. Pipeline processing instructions allow packets to be sent to subsequent tables for further processing and allow information, in the form of metadata, to be communicated between tables. Table pipeline processing stops when the instruction set associated with a matching flow entry does not specify a next table; at this point, the packet is usually modified and forwarded.

Flow entries may forward to a physical port or a logical port defined by the switch or a reserved port. Reserved ports may specify generic forwarding actions such as sending to the controller, flooding, or forwarding using non-OpenFlow switch processing. The switch-defined logical ports may specify LAGs, tunnels, or loopback interfaces.

Actions associated with flow entries may also direct packets to a group, which represents sets of actions for flooding, as well as more complex forwarding semantics (e.g., multipath, fast reroute, and link aggregation). Groups also enable multiple flow entries to forward to a single identifier (e.g., IP forwarding to a common next hop), allowing common output actions across flow entries to be changed efficiently.

OpenFlow packets are received on an ingress port and processed by the OpenFlow pipeline that may forward them to an output port. The ingress port can be used when matching packets. The OpenFlow pipeline can decide to send the packet on an output port using the output action.

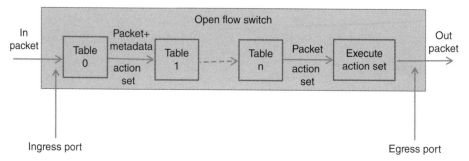

Figure 6.45. Packet flow in an OpenFlow switch.

An OpenFlow physical port is a switch-defined port corresponding to a hardware interface of the switch such as Ethernet switch port. The OpenFlow switch may be virtualized over the switch hardware where an OpenFlow physical port may represent a virtual slice of the corresponding hardware interface of the switch.

An OpenFlow logical port is a switch-defined port that does not correspond directly to a hardware interface such as LAGs, tunnels, and loopback interfaces. Logical ports may include packet encapsulation and may map to various physical ports. The processing done by a logical port is transparent to OpenFlow processing. However, a logical port interacts with OpenFlow processing associated OpenFlow physical port.

A packet associated with a logical port may have an extra metadata field called Tunnel-ID associated with it, compared to a physical port. When a packet is received on a logical port, both its logical port and its underlying physical port are reported to the controller.

OpenFlow-compliant switches come in two types: OpenFlow-only and OpenFlow-hybrid. OpenFlow-only switches support only OpenFlow operation, where all packets are processed by the OpenFlow pipeline and cannot be processed otherwise. OpenFlow-hybrid switches support both OpenFlow operation and L2 Ethernet switching operation, VLAN isolation, L3 routing (IPv4 routing, IPv6 routing, etc.), ACL, and QoS processing. They provide a classification routing traffic to either the OpenFlow pipeline or the normal pipeline. For example, a switch may use the VLAN tag or input port of the packet or IP address to decide whether to process the packet using one of the pipelines or direct all packets to the OpenFlow pipeline.

The flow tables of an OpenFlow switch are sequentially numbered, starting at zero. Pipeline processing always starts at the first flow table: the packet is first matched against the flow entries of flow table zero. Other flow tables may be used depending on the outcome of the match in the first table.

When a packet arrives at an OpenFlow switch, the header fields are compared to flow table entries (Figure 6.45). If a match is found, the packet is either forwarded to specified port(s) or dropped, depending on actions stored in the flow table. A flow entry can only direct a packet to a flow table number that is greater than its own flow table number.

When an OpenFlow switch receives a packet that does not match the flow table entries, it encapsulates the packet and sends it to the controller. The controller then

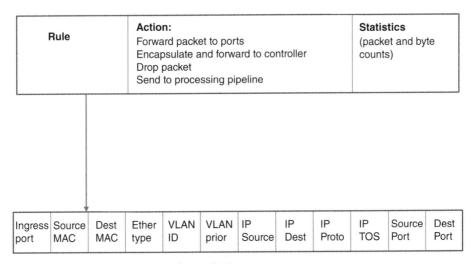

Rule	Action: Forward packet to ports Encapsulate and forward to controller Drop packet Send to processing pipeline	Statistics (packet and byte counts)

Ingress port	Source MAC	Dest MAC	Ether type	VLAN ID	VLAN prior	IP Source	IP Dest	IP Proto	IP TOS	Source Port	Dest Port

Figure 6.46. Flow header.

decides how the packet should be handled and notifies the switch to either drop the packet or make a new entry in the flow table to support the new flow. Figure 6.46 shows the 12-tuple of header fields that is used to match flows in the flow table.

Flow Matching Fields are defined in [34] as shown in Table 6.3.

An OpenFlow controller should be able to match on any field (Figure 6.46), support for OpenFlow actions, and discover the network using the Link Layer Discovery Protocol (LLDP). If a packet does not match a flow entry in a flow table, this is a table miss. The behavior on a table miss depends on the table configuration. A table-miss flow entry in the flow table may specify how to process unmatched packets. The options include dropping them, passing them to another table, or sending them to the controller over the control channel via packet-in messages.

A flow table consists of flow entries. Each flow table entry contains Match Fields, Priority, Counters Instructions, Timeouts, and Cookie. Match Fields consist of the ingress port and packet headers and optionally metadata specified by a previous table. Priority is matching precedence of the flow entry. Counters are for counting matching packets. The action set or pipeline processing are modified by Instructions. Timeouts set the maximum amount of time or idle time before flow is expired by the switch.

The OpenFlow channel is the interface that connects each OpenFlow switch to a controller. Through this interface, the controller configures and manages the switch, receives events from the switch, and sends packets out the switch.

A typical OpenFlow controller manages multiple OpenFlow channels, each one to a different OpenFlow switch. An OpenFlow switch may have one OpenFlow channel to a single controller, or multiple channels for reliability, each to a different controller.

The OpenFlow channel may be composed of multiple network connections to exploit parallelism. The OpenFlow switch always initiates a connection to an OpenFlow controller. The channel messages are formatted according to the OpenFlow protocol and usually encrypted using TLS.

TABLE 6.3.   OpenFlow Matching Parameters, Actions, and Statistics

Switch input port
Switch physical input port
Metadata passed between tables
Ethernet destination address
Ethernet source address
Ethernet frame type
VLAN ID
VLAN priority
IP DSCP (6 bits in ToS field)
IP ECN (2 bits in ToS field)
IP protocol
IPv4 source address
IPv4 destination address
TCP source port
TCP destination port
UDP source port
UDP destination port
SCTP source port
SCTP destination port
ICMP type
ICMP code
ARP opcode
ARP source IPv4 address
ARP target IPv4 address
ARP source hardware address
ARP target hardware address
IPv6 source address
IPv6 destination address
IPv6 Flow Label
ICMPv6 type
ICMPv6 code
Target address for ND
Source link layer for ND
Target link layer for ND
MPLS label
MPLS TC
MPLS BoS bit
PBB ISID
Logical port metadata
IPv6 extension header pseudofield

Cookie is an opaque data value chosen by the controller and may be used by the controller to filter flow statistics, flow modification, and flow deletion. A flow table entry is identified by its match fields and priority: the match fields and priority taken together identify a unique low entry in the flow table. The flow entry that wildcards all fields and has priority equal zero is called the table-miss flow entry.

The OpenFlow protocol supports three message types: controller-to-switch, asynchronous, and symmetric, each with multiple subtypes. Controller-to-switch messages are initiated by the controller and used to directly manage or inspect the state of the switch. Asynchronous messages are initiated by the switch and used to update the controller of network events and changes to the switch state. Symmetric messages are initiated by either the switch or the controller and sent without solicitation.

An OpenFlow controller typically manages an OpenFlow switch remotely over one or more networks using TCP/IP. It may be a separate dedicated network or an in-band controller connection.

For a connection setup, the switch needs to establish communication with a controller at a user-configurable or fixed IP address of a user-specified port. If the switch knows the IP address of the controller, the switch initiates a standard TLS or TCP connection to the controller. Traffic to and from the OpenFlow channel is not run through the OpenFlow pipeline. Therefore, the switch must identify incoming traffic as local before checking it against the flow tables.

In the case that a switch loses contact with all controllers, as a result of echo request timeouts, TLS session timeouts, or other disconnections, the switch should immediately enter either *fail secure mode* or *fail stand-alone mode*, depending upon the switch implementation and configuration. In *fail secure mode*, packets and messages destined to the controllers are dropped. Flow entries continue to expire according to their timeouts in *fail secure mode*. In *fail stand-alone mode*, the switch acts as a legacy Ethernet switch or router that is available on Hybrid switches. Upon connecting to a controller again, the controller then has the option of deleting all flow entries, if desired.

The first time a switch starts up, it will operate in either *fail secure mode* or *fail stand-alone Mode*, until it successfully connects to a controller.

The switch may establish communication with a single controller or may establish communication with multiple controllers for redundancy. The handover between controllers is entirely managed by the controllers themselves, which enables fast recovery from failure and also controller load balancing.

An OpenFlow switch may provide per-port or per-flow queuing. Each queue is described by a set of parameters such as minimum and maximum data rates guaranteed.

6.5.4 Representational State Transfer API

An API is essentially a definition of how software components can interact with each other. If your router or switch has *an API*, it means that vendor has put into place some kind of mechanism where you can send it commands through more automated means than CLI (like a script).

In order to speak devices that are not running OpenFlow, the applications need to build languages. In OpenDaylight, these languages are modularized and placed in a repertoire of mechanisms that can be used to configure network devices. These languages are commonly referred to as *southbound* APIs for the SDN controller, because they are below the controller, which is an abstracted entity that sits above the physical or virtual infrastructure.

To instruct the controller to configure these devices, a *northbound* interface or API, which provides a list of vendor-agnostic base network functions, is established. The controller interprets these commands into a language that each infrastructure node can understand. This API is commonly referred to as a *northbound* API.

This architecture is fairly common among SDN initiatives, and just as common is the reference to a Representational State Transfer (REST) API.

A REST API, or an API that is RESTful, which means it adheres to the constraints of REST, is not a protocol, language, or established standard. It is essentially six constraints that an API must follow to be RESTful [35]. The point of these constraints is to maximize the scalability, independence, and interoperability of software interactions. These six constraints are:

1. **Client/Server:** This relationship must exist to maximize the portability of server-side functions to other platforms. With SDN, this usually means that completely different applications, even in different languages, can use the same functions in a REST API. The *applications* would be the client, and the controller would be the *server*.

2. **Stateless:** All state is kept client side, and the server does not retain any record of client state. This results in a much more efficient SDN controller.

3. **Caching:** Just like *cookies* in a web browser, it's a good idea for the client to maintain a local copy of information that is commonly used. This improves performance and scalability, because it decreases the number of times a business application would have to query the network's REST API. Some functions should not be cacheable, however, and it is up to the REST API to define what should be cached.

4. **Layered System:** Many times, a system of applications is composed of many parts. A REST API must be built in a way that a client interacts with its neighbor that could be a server, load balancer, and so on and doesn't need to see *beyond* that neighbor. By providing a REST API northbound, we don't have to teach our applications how to speak southbound languages like SNMP, SSH, NETCONF, and so on.

5. **Uniform Interface:** Independent from the information retrieved, the method by which it is presented is always consistent. For instance, a REST API function may return a value from a DB. It does not return a DB language, but likely some kind of open markup like JSON or XML (more on this later). That markup is also used when retrieving something *entirely* different, say, the contents of a routing table. The information is different, but it is presented in the same way.

6. **Code on Demand:** This is actually an optional constraint of REST, since it might not work with some firewall configurations, but the idea is to transmit working code inside an API call. If none of an API's functions did what you wanted, but you knew how to make it so, you could transmit the necessary code to be run server side.

The purpose here is to maximize the usefulness of an API to provide services to a large number and variety of clients, which in the case of SDN is likely to be our business logic applications or cloud orchestration such as OpenStack.

6.5.5 SDN Application in Optical Networks

An SDN architecture for optical networks can be designed as depicted in Figure 6.47 where the control plane is implemented in an OSS acting as a centralized controller. The controller communicates with the optical systems hardware using a standardized interface. From this interface, the details of hardware implementation can be abstracted. Applications can interact with the network through the controller using an API to coordinate network resources and optimize overall network performance. An SDN controller generally provides two APIs, one for applications to request resources from the network and the other to connect directly to the NEs or to an EMS that controls the network.

OpenFlow provides a standardized interface to the hardware for Ethernet, IP, and MPLS networks, but does not cover circuit- or wavelength-based equipment. It needs to be expanded for optical networks.

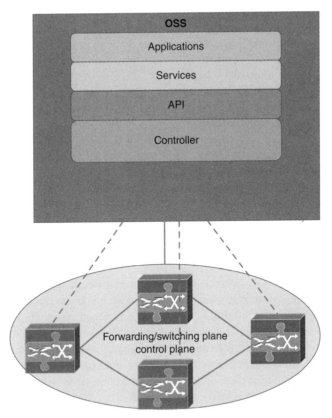

Figure 6.47. SDN architecture for optical networks.

Over the optical network, various services can be offered such as private line. Applications running over these services can dynamically request network resources at wavelength, circuit, or flow granularity via network coordination or orchestration function.

The centralized OSS can support an application interface to support application-driven network resource allocation, while the controller function can be still embedded in the NEs as in optical networks with a generalized MPLS (GMPLS) dynamic control plane with integrated control and switching planes. In these networks, a dynamic control plane is distributed to NEs to expedite the circuit design process and improve provisioning times.

To set up a connection, the network establishes communication paths among the application's endpoints. The communication paths are selected to satisfy jitter, packet loss, and delay and resiliency requirements of the application. This connection may be a label-switched path (LSP) on an MPLS network, an EVC mapped into an OTN or Synchronous Optical Networking (SONET) circuit, or an optical wavelength service.

Without dynamic allocation of resources, resources must be allocated considering the worst-case usage of an application, leading to underutilized resources. On the other hand, dynamic allocation of resources could assign unused resources of an application or a service to other applications or services.

Switching can be supported in both the electrical and optical domains. Electrical switching can be at the frame granularity for packet services or at the time slot level for TDM circuits with guaranteed bandwidths. The characteristics of the underlying optical paths are often factored into the selection of the TDM or packet circuit.

Optical architectures can be constructed to switch fibers or ports that may contain multiple channels over a wide spectrum or to provide wavelength-selective switching with fixed or flexible slices of spectrum. Wavelength switching is typically implemented as a reconfigurable optical add/drop multiplexer (ROADM) that can add, drop, or express wavelengths through the node.

A wavelength service can be added at an ingress node, switched through the network, and dropped at an egress node. This SDN architecture allows automated setup of a wavelength between two endpoints, but the control structure is generally slow and requires minutes to complete the setup of the wavelength path [28]. The circuit reach is limited by the optical performance of the transceiver and network impairments but can be extended by signal regeneration. Power levels can be increased to minimize amplifier transients.

Wavelength networks can be made more dynamic by adding fast wavelength tunability to either the transmitter or receiver to implement subwavelength switching. All-optical subwavelength switching allows sharing of wavelengths in the time domain so that the same transmitter or receiver (depending on the implementation) can communicate with multiple nodes. Transmitter or receiver tunability allows the network to be reconfigured based on traffic demands if a wavelength is assigned based on the traffic destination or source. The advantage of these architectures is that transceiver resources are not fixed between two endpoints but can be dynamically reused based on traffic demands.

The network needs to be reconfigured or reprogrammed based on application layer traffic demands and performance requirements. Under the control of an SDN controller, in order to meet application needs, one can maximize network resource utilization and achieve the desired performance, by having all the tunable parameters available at the application layer.

SDN control mechanisms and controllable network attributes depend on the underlying network technology. Packet services can be connectionless or connection oriented. A connectionless packet service such as IP routing can use a connection-oriented service. Connection-oriented technologies such as GMPLS can provide transport services to applications. An LSP between two endpoints with SLA constraints such as protection, loss, delay, jitter, and bandwidth to satisfy the needs of applications running over it can be established by the NEs in the path in a distributed control plane environment or by a centralized SD N controller.

In a distributed control plane, one endpoint of the connection (or circuit) is told what connection it needs to set up and the attributes of the connection in terms of bandwidth, protection, loss, delay, and jitter objectives. If there are multiple independent endpoints in the network setting up connections at the same time, it is possible that the network may not be utilized in the best possible way. Some links may be more loaded than others. Link utilization may also be fragmented such that no single link may be able to satisfy a new demand even though the network is underutilized. There are also cases where the service cannot be satisfied by an NE because it or the network lacks the necessary capabilities to learn or disseminate relevant topology information.

An SDN controller must be aware of the network topology, link attributes such as bandwidth and delay, and dependencies between links such as shared risk link groups. Topology information can be stored in the form of a DB or learned from the network using routing protocols and updated based on network events such as link failures or link additions. Delay information can be measured or computed and can be part of the DB or topology information.

An SDN controller, using topology information, can compute a path on a network that satisfies connection, circuit, or wavelength constraints by optimizing globally the placement of connections on the network as the computation is done centrally for all as opposed to being distributed where every connection is set up independently. An SDN controller can move connections to new paths while satisfying their constraints in order to satisfy the constraints of new connection(s) that would have been blocked otherwise. It can also compute diverse paths irrespective of whether they share the same endpoints.

When global network resource optimization is required or there are interdependent connection constraints across nodes, the SDN controller approach has benefits over a control plane that is solely distributed to NEs due to its global knowledge of application requirements and network resources.

Furthermore, an SDN controller can compute a path across Internet Gateway Protocol (IGP) areas satisfying SLA constraints, configure the performance monitoring capabilities of these paths, and collect them. If the performance objectives are violated due to degraded links, congestion, or increased delay, the path can be recomputed to avoid problem segments. It can coordinate the application network needs across disparate technologies, each with its own controllable attributes, to realize a network service.

While centralized controllers provide for global network optimization and coordination among connections, they may be slower to react to network changes (e.g., link failure) than a distributed control plane. This can stem from the time to propagate the change from the network to the SDN controller and from the response time at the SDN controller, which is potentially processing events or requests pertaining to multiple NEs, unlike a control plane processor dedicated per NE.

In order for SDN to scale and response to network events, the deployment of multiple geographically distributed CPUs forming an SDN control cluster is likely. Coordination within the cluster becomes necessary to maintain decision centralization and control across operational or authoritative domain boundaries controlled via individual SDN controllers to establish services that cross these domains.

While centralized SDN control can address the coordination of application requirements with the network, large networks will still need to rely on a distributed embedded control plane to provide fast reaction to network events. This embedded control plane is more scalable and leverages mature and reliable technologies that were developed to replace early centralized control approaches. Connectionless IP networks are a good example of truly distributed control.

To implement large scalable networks that are coupled to the application layer, it becomes essential that SDN-based centralized control and distributed embedded control coexist on the same network. SDN controllers that belong to different vendors need to interoperate.

6.5.6 SDN Application in Virtualized Centers

The emergence of cloud-based computing changed the networking service models. Essentially, cloud computing is to utilize centralized processing, storage, and computing to create a new service layer on top of the network. The services can be modeled as IaaS, PaaS, SaaS, and so on. Interactive services that combine voice, video, and data could replace separated voice and video services. Enterprise services through private cloud may replace the traditional ISP-initiated VPNs. This change has been validated by service offerings from Google, Amazon, and Apple in recent years and likely will continue to proliferate in years to come.

The cloud-based services require the networking resources to be available in anywhere and anytime. Service providers and enterprises are increasingly offering services and applications from a group of DCs. For cloud service providers, networking is a utility business. The network types (circuit or packet) and networking technologies (Ethernet, IP, or MPLS) do not make any differences, as long as the network can reliably transport user data at competitive price.

SDN enables cloud service providers to have a uniform application interface to the underlying networks, so that they can optimize the use of the networking resources. SDN does not define and dictate the DC interior architecture and management. Instead, SDN is to interface with the DCs to make use of the network resources.

Cloud services can be roughly divided in storage, computing, and networking. SDN is only responsible for the networking portion. It enables the cloud service providers to provision network resources from the application layer. The operation itself must subject to proper business arrangement between DC and network service providers. The SDN resource programming can only take place on abstract level.

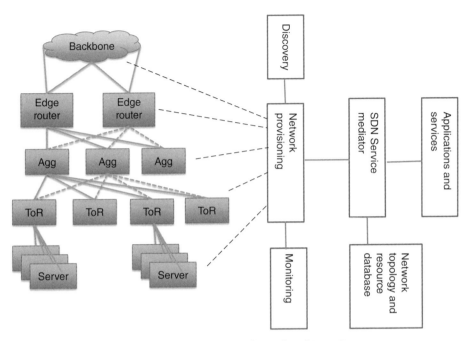

Figure 6.48. SDN architecture for a Virtual Data Center.

For a VDC [33], an overlay network can be constructed for a group of users where each user can be represented by an individual VM. All users in the same group will share a common VN as we discussed in section 6.3. SDN shields underlying network from users and aggregates the user traffic over a selected set of network connections (Figure 6.48).

The Service Mediator interfaces with network edge nodes and provisions the network tunnels. When the application is to set up a logical connection between two VMs, the Service Mediator will identify the user network identification (e.g., VLAN or VXLAN) and select the network tunnel to use. The application is to program the ToR switches and initiate the logical connections. All VM traffic will be aggregated through selected network connections for latency and bandwidth guarantees.

Another example is to use L3 virtualization using SDN for DCs [33] (Figure 6.49).

VM users may trigger the discovery process and connect to BGP gateways when connecting to other users. The SDN Service Mediator correlates the routing information among BGP GWs. This runs at IP layer and reuses much of the existing and proven routing techniques. The underlying solution is identical to that of MPLS VPN.

6.5.7 SDN Application in Carrier Ethernet Services

MEF has established physical layer-agnostic Carrier Ethernet Services that are being widely used in the industry as described in Chapter 4. MEF needs to define the interface between application/service layer and controller, controller functions for underlying physical network that may or may not be Carrier Ethernet, and the interface between

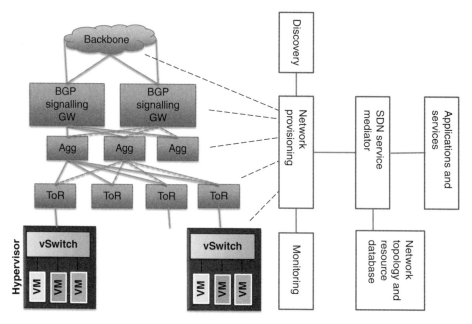

Figure 6.49. SDN with L3 virtualization.

Controller and physical network by extending OpenFlow architecture. In fact, for Carrier Ethernet, it is very likely that we need to separate application and service layers since MEF already defined services and define the interface between application and service layers.

MEF developed specifications to address network interfaces (UNI and ENNI), connections (EVCs), fault and performance management, and QoS that can be considered as part of the Infrastructure layer. In Information model, some of service OAM functionalities and bandwidth profiles can be considered as Control Plane functionalities. On the other hand, MEF-defined services, as we described in Section 4.2, can be considered as part of the service layer. The mapping of MEF developed specifications and corresponding SDN layers is depicted in Figure 6.50 [36]. More work needs to be done for the adaptation of Carrier Ethernet to SDN.

The key objective of applying SDN concept to Carrier Ethernet Service would be rapid service provisioning and circuit turnup as indicated in [37]. In order to accelerate the service provisioning, service providers need to have an accurate view of network topology and resources and be able to modify the resources on demand.

According to [37], Transition Networks suggest to define service templates for predefined CE2.0 services (i.e., EPL, EVPL, EP-LAN, and EVP-LAN), create a catalog of service offerings with various features, and push service attributes and configurations to devices forming the physical network.

Once the service is provisioned, the service provider runs a service activation testing (SAT) before handing it over a customer. The SAT may be performed and measurements are collected by a centralized system.

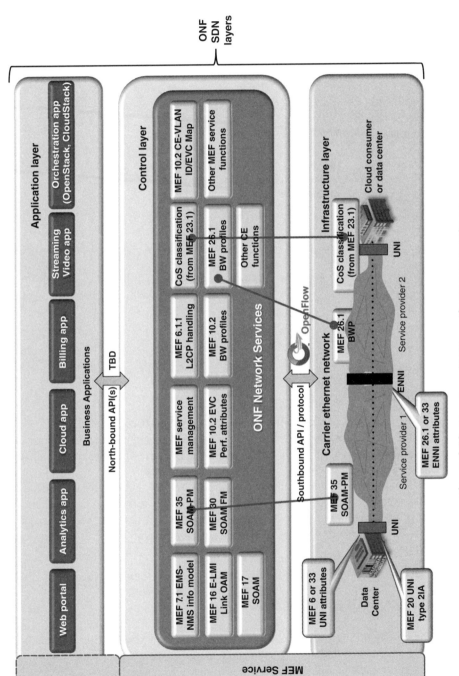

Figure 6.50. Mapping of MEF specifications to SDN layers.

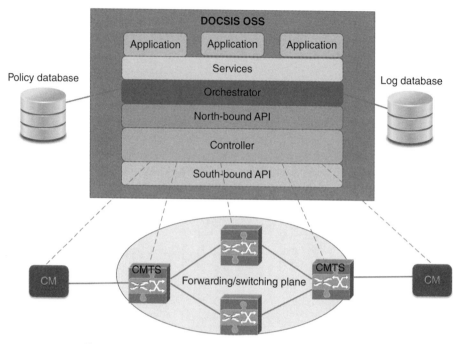

Figure 6.51. A likely SDN architecture for DOCSIS networks.

6.5.8 SDN Application in DOCSIS Networks

MSOs have been provisioning millions of cable modems (CMs) automatically via DOCSIS back-office systems and DOCSIS autoprovisioning methods, as we mentioned in the previous chapters. Another key component of the DOCSIS system is cable modem termination system (CMTS); its provisioning is semiautomatic.

SDN should expand this autoprovisioning beyond CM and CMTS and include autoprovisioning of end-to-end connections within MSO networks. In order to accomplish this, CM, CMTS, and other NEs within the DOCSIS network need to speak OpenFlow. The controller functions and APIs can be built into DOCSIS back-office systems (Figure 6.51).

6.6 SELF-MANAGED NETWORKS

Virtualization, Cloud, and SDN concepts that we have described earlier are mainly focused on operational efficiency and maximum utilization of network resources. Further operational efficiencies can be achieved with self-managed networks. Network resources and services should be automatically provisioned, and faulty components should be automatically identified and fixed by the network itself.

The future networks are very likely to be self-managed. Autoconfiguration of NEs such as CM and CMTS is being practiced by MSOs using DOCSIS back-office systems. Similar procedures are also used by DPoE networks (see Section 4.3) [38] for autoconfiguration of NEs and services. The addition of self-managed capabilities would result in ultimate operational efficiency in the networks.

Self-managed Internet and mobile systems have been extensively studied by ETSI and 3GPP standards organizations [34, 39–45]. Despite this, the fixed network industry is focused on autoconfiguration [46] and monitoring of network resources and services, isolating problems when there are failures, and fixing them by sending technicians to the sites most of the time or downloading certain configuration files remotely for configuration-related problems. The concept of network identifying problems by itself and fixing them and sending technicians to the failure site only when there is single point of hardware failures (i.e., there is no hardware redundancy) [47] is not practiced. Tools for self-managed networks are not developed either. In this section, we will not discuss the autoconfiguration, but the fault management of centrally self-managed networks and distributed self-managed networks [47, 48].

Each self-managed NE (sNE) in a network monitors its hardware and software resources periodically, runs diagnostic tests during failures in a hierarchical fashion and identifies problems if they are local to the sNE and fixable by the sNE, and reports failures and fixes to a centralized self-managing network management system (sNMS)[3] to be accessed by network operators, field technicians, customers, and other sNEs on the path of a given connection. If the problem is not locally fixable by the sNE, the sNMS will run its own Rule-Based Logic to determine if the problem is fixable remotely by the sNMS. If it is not, a notification is sent to a network operator or field technician to fix the problem.

If the network is divided into autonomous regions where each region has its own regional network management system (i.e., distributed network management), then problems that are not fixable by sNEs will be sent to the associated self-managing regional network management system (sRNMS) first. If the sRNMS cannot fix it, then the problem is escalated to the sNMS. This hierarchy of fixing entities in the self-managed networks is necessary to minimize the time needed for fault isolation and fixing and to optimize fault management architecture in NE and NMS.

The failure type, if the problem is fixable locally by sNE, remotely by sRNMS or sNMS, or remotely by a technician, and the estimated fix time are communicated with a newly defined message format. The hierarchy of fixing failures is network architecture dependent, as discussed in Sections 6.6.6.1–6.6.6.6.

This concept is applied to VNs and SDN. sNE and sNMS architectures for centrally and distributedly managed VNs and SDN are described.

It is expected that the concept of self-managed networks to change how networks operate today reduces the cost of operation (OPEX) and network provisioning and maintenance intervals dramatically. Self-managed networks should accelerate realizations of cloud services and software-defined networks (SDNs) as well.

---

[3]Since the focus in this section is element and network-level fault management, we use NMS to represent an operation support system. This NMS could be an OSS, as described in Chapter 5.

6.6.1 NE and NMS Architectures for Centrally Self-Managed Networks

Centrally self-managed network architectures consisting of sNEs and sNMS are depicted in Figures 6.52 and 6.53. An sNE (Figure 6.54) consists of intelligent agents. Each self-managing agent (i.e., intelligent agent) monitors the entity that it belongs to, runs diagnostic tests to identify problems during failures, initiates a failure message, fixes problems, and initiates fix reporting to the central self-management system, sNMS. The message indicating that the fixing entity is sNE is communicated to other NEs, sNMS, field technicians, and customers (if desired). If the problem is determined to be not fixable locally after two–three tries or without a try, depending on the problem, a message is sent to sNMS by the sNE indicating that the fixing entity is unidentified.

An sNE consists of an intelligent NE (iNE) and one or more intelligent agents of each type (Figure 6.52). The agents are one or more intelligent Hardware Maintenance Agent(s) (iHMA(s)), intelligent Operating System Maintenance Agent(s) (iOMA(s)), intelligent Application Maintenance Agent(s) (iAMA(s)), and intelligent Capacity Management Agent(s) (iCMA(s)), depending on the implementation.

iHMA periodically monitors hardware entities such as CPU, memory, physical ports, communication channels, buffers, backplane, power supplies, and so on and initiates predefined maintenance actions during hardware failures. iOMA periodically monitors OS and initiates predefined maintenance actions during OS failures. iAMA periodically monitors application software and protocol software and initiates predefined maintenance actions during application and protocol software failures. iCMA

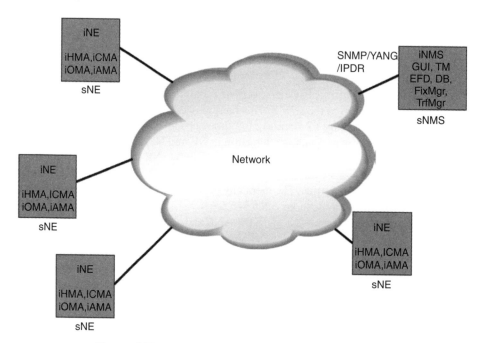

Figure 6.52. Centrally self-managed network architecture.

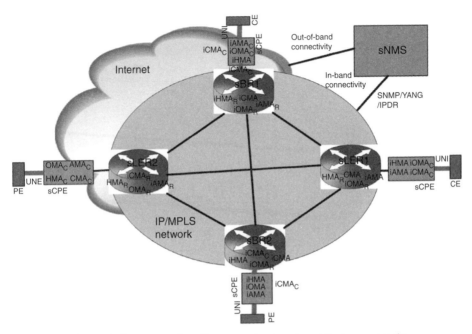

Figure 6.53. A centrally self-managed network architecture example.

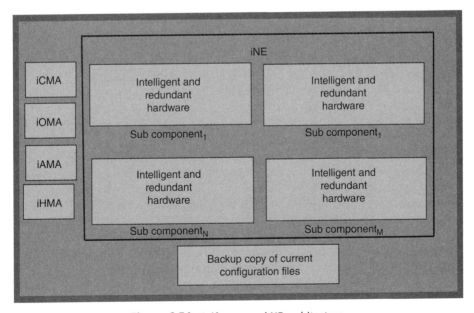

Figure 6.54. Self-managed NE architecture.

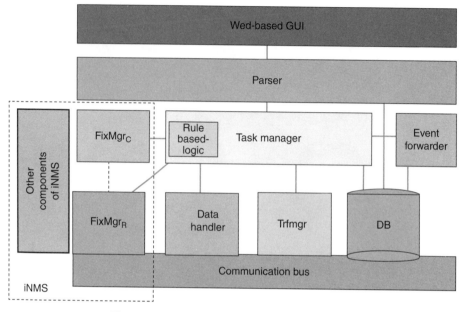

Figure 6.55. Architecture of self-managing NMS.

periodically monitors system capacity, load, and performance and collects measurements. During failures, iCMA initiates predefined maintenance actions.

In addition to the aforementioned intelligent agents above, the sNE needs to be designed as an iNE with redundant hardware and software components as depicted in Figure 6.54, where each hardware and software component is capable of running its own diagnostics and identifying faulty subcomponents.

On the other hand, sNMS consists of intelligent NMS (iNMS) that mainly deals with remote fixes, Task Manager (TM) to manage tasks to be executed and copies of software modules for each type of iNE; Traffic Manager (TrfMgr) to deal with network-level traffic management issues such as routing policies, load balancing, connection admission control (CAC), and congestion control; Event Forwarding Discriminator (EFD) to forward failure and fix notifications to network operators and customers; DB to store data; and graphical user interface (GUI) (Figures 6.55 and 6.56).

Web-based GUI provides human and machine interfaces, opens and closes sessions with its clients, performs initial authentication, validates all submitted data from the clients, processes chart data generated by parser, and generates flash-based charts and graphs for its clients. Parser processes GUI templates. DB stores GUI events and data collected. TM prioritizes and schedules execution of the tasks including repair and configuration activities that can be performed remotely using a Rule-Based Logic module. Data Handler collects end-to-end connection-related measurements and NE capacity measurements and stores them in DB to support TrfMgr.

There are no changes introduced to interfaces between NMS and network for self-management. The well-known protocols such as SNMP, IP Detail Record (IPDR)

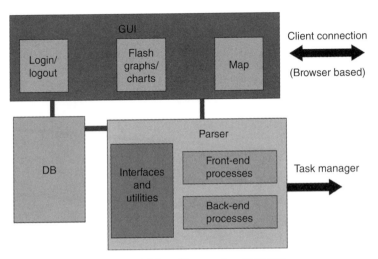

Figure 6.56. Self-managing NMS GUI.

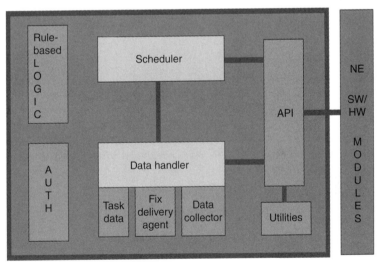

Figure 6.57. Intelligent agent architecture.

for usage information, Network Configuration Protocol (NETCONF) for manipulating configuration data and examining state information, and YANG modeling can be employed.

The intelligent agent architecture is depicted in Figure 6.57. Its Rule-Based Logic module determines problems and initiates fixes if the problems are local to sNE, initiates tests for the fixes, determines if the fix procedure or a step or some of the steps are to be repeated, and initiates a message to all related parties about the fixes. If the problem

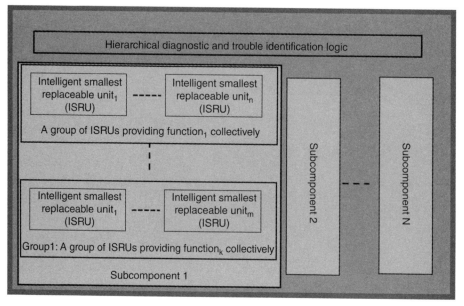

Figure 6.58. Intelligent NE architecture.

is not local to sNE, the sNE informs all related parties including sNMS for its conclusion, which is that the fixing entity is unidentified. If the result of diagnostics cannot identify the failed component that is inconclusive, that is conveyed as well.

The Scheduler module determines the priority and order of the tasks for each functional entity within sNE. An API provides interface to various types of software and hardware entities within sNE. The Data Handler module collects necessary data for the sNE, such as relevant measurements, performs the fix, and keeps the data associated with the task. The AUTH module authenticates local user access and remote user access from sNMS interface to sNE agents. The Utilities module supports various file operations.

6.6.2 iNE and iNMS Architectures

iNE (Figure 6.58) consists of intelligent subcomponents such as chips, OS, and protocol software that are capable of periodic self-checking, declaring a failure when it is unable to perform its functions, running diagnostics and identifying whether the faulty entity is within the subcomponent or not, and escalating the diagnostics to the next level in the hierarchy when the diagnostics are inconclusive.

When there is a failure, if failed entity is unidentified as a result of the diagnostic tests run by the intelligent subcomponents, iNE is able to run diagnostics for a predefined set of subcomponents that are collectively performing a specific function. A predefined set of subcomponents can be a collection of components that are contributing

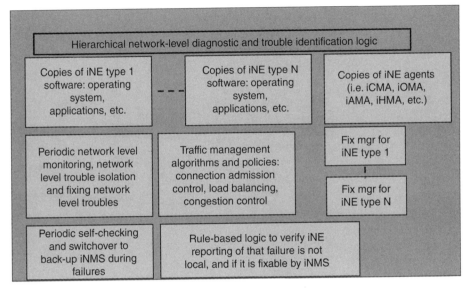

Figure 6.59. Intelligent NMS architecture.

to the realization of a main function such as packet forwarding, deep packet inspection, event forwarding, and so on.

If the diagnostic tests run for a predefined set of subcomponents cannot identify the failed entity, iNE is able to run diagnostics at NE level to determine the failure. After the failure is identified to the smallest replaceable hardware (e.g., chips, wires connecting chips, backplane, etc.) and/or software entity (e.g., kernel, log, protocol software, EFD, etc.), the responsible agents determine if the failure is fixable and initiate a message to related parties with estimated fix time to repair. If iNE diagnostics are inconclusive, that is communicated as well (see Section 6.6.6).

On the other hand, iNMS (Figure 6.59) periodically monitors the network that sNMS is managing, identifies network-level failures, estimates and communicates the fix time to related parties, and fixes them. When sNE reports that the failure is not local (i.e., either tests are inconclusive or sNE is not capable of fixing it), Rule-Based Logic of sNMS verifies if sNE failure is not local.

sNMS is redundant where the active sNMS is protected by a standby sNMS. Active and standby sNMS units perform periodic self-checking. When the active sNMS fails, the standby sNMS takes over the responsibilities. It is certainly possible to configure both units of redundant sNMS as active.

The TM of sNMS manages tasks to be executed by the iNMS. The Rule-Based Logic determines if the problem is remotely fixable by the iNMS.

iNMS includes a Fix Manager (FixMgr) for each sNE type to fix sNE problems remotely; stores software modules specific to sNEs and network-level traffic management algorithms such as routing policies, load balancing, CAC, and congestion control; and executes the algorithm when needed. Furthermore, iNMS holds a copy of each sNE agent and remotely loads into sNE when needed.

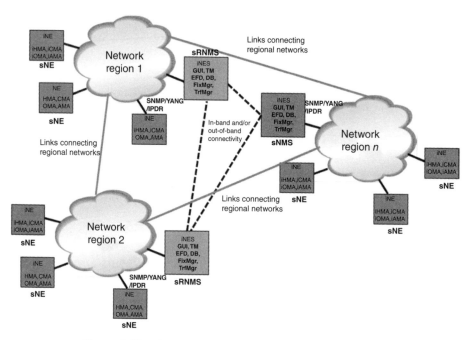

Figure 6.60. Distributed self-managed network architecture.[4]

6.6.3 NE and NMS Architectures for Distributed Self-Managed Networks

In a distributed architecture (Figure 6.60), the network is divided into subnetworks (i.e., regional networks), where each subnetwork has its own sRNMS. The sRNMS architecture is identical to sNMS. It provides all the centralized management functions for its own subnetwork and informs sNMS and other sRNMSs about possible failures and fixes. Informing the sRNMs of other regions is necessary to ensure the coordination of fixes and failures impacting other regions. The sRNMS can choose to inform sNEs in its domain.

End-to-end network-level monitoring and end-to-end network-level capabilities beyond regional boundaries are left to sNMS. Some of the end-to-end network-level capabilities are CAC, load balancing, and congestion control at network level.

6.6.4 NE and NMS Architectures for Self-Managed VNs

A VN can be created over physical networks based on different technologies such as IP, MPLS, SONET, Ethernet, wireless, and so on, as depicted in Figure 6.61 [49]. In order for a VN to monitor itself and fix faulty components during failures, we need to add self-managed VN infrastructure in addition to self-managed network infrastructure

---

[4] In this architecture, sNMS plays a role of sRNMS for its own subnetwork and a role of central sNMS for the entire network.

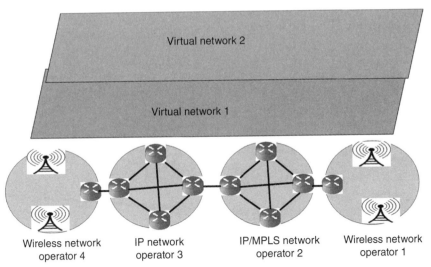

Figure 6.61. Virtual networks over an underlay network of multiple technologies.

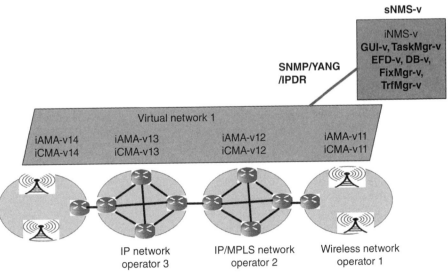

Figure 6.62. Centrally self-managed virtual network architecture.

associated with each underlay network. The existence of the self-managed VN infra-structure is independent of self-managed underlay physical network described in the previous sections.

Each self-managed VN has one or more Intelligent Application Maintenance Agent(s) (iAMA-v(s)) and one or more Intelligent Capacity Management Agent(s) (iCMA-v(s)) as depicted in Figures 6.62 and 6.63. The iAMA-v monitors application software and protocol software and initiates predefined maintenance actions during

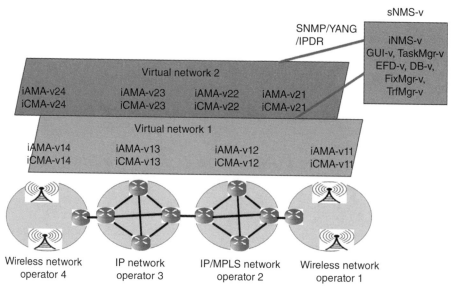

Figure 6.63. Centrally self-managed multiple virtual networks.

failures. The iCMA-v monitors VN capacity, system load, and performance; collects system measurements; and initiates predefined actions during failures.

Each agent monitors the entity that it belongs to, reports failures to the central management system of the virtual network (sNMS-v) in a centrally managed network, fixes problems, and reports fixes to sNMS-v. If a problem is determined to be not fixable after two or three fix tries or without a try, the agent sends a message with unidentified fixing entity to SNMS-v. The fix is provided by SNMS-v or a field technician if the responsible agent cannot.

A distributedly managed VN is depicted in Figure 6.64. In the distributed architecture, each VN has its own regional management system (sRNMS-v). The sRNMS-v provides all centralized management functions for its VN and informs the sNMS-v about failures and fixes.

End-to-end network-level activities beyond VN boundaries will be left to sNMS-v. These activities can be end-to-end CAC, load balancing, and congestion control.

6.6.5 NE and NMS Architectures for Self-Managed SDN

In software-defined networks (SDNs) (Figure 6.65), a network is abstracted into two separate layers, namely, Control Layer (Control Plane) and Infrastructure Layer (Data Plane). The Control Plane manages switch and routing tables, while the Data Plane performs the layer 2 and 3 filtering, forwarding, and routing. The layer 2 and 3 filtering can be part of Control Plane as well. APIs are built on top of control plane to support various applications.

The application software by periodic keep-alive messages, run diagnostic tests during failures, identify failed module(s) during application software failures, and

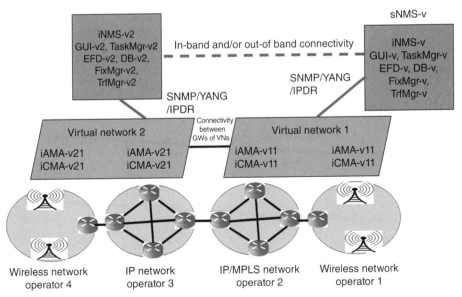

Figure 6.64. Distributed self-managed virtual network architecture.

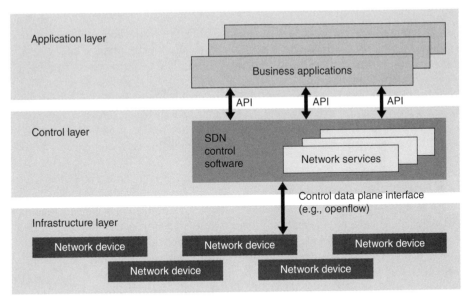

Figure 6.65. SDN architecture.

fix them. Similarly, each Control Plane protocol has an Intelligent Control Plane Management Agent (iCPA$_{CL}$) to monitor control plane software by periodic keep-alive messages, run diagnostic tests during failures, identify failed module(s), and fix them.

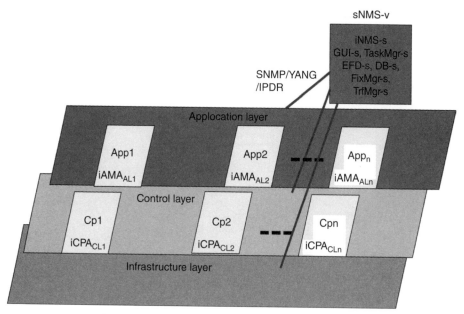

Figure 6.66. Centrally self-managed SDN architecture.

A centrally self-managed SDN is depicted in Figure 6.66 where failures and their resolutions are communicated to a centralized management system sNMS-s by the agents. Failures that are not fixed by iAMA$_{AL}$s and iCPA$_{CL}$s are either fixed by the sNMS-s or by field technicians.

An sNE architecture supporting SDN is depicted in Figure 6.67. The main difference between the sNE in Figure 6.54 and the sNE in Figure 6.67 is that the SDN supporting NE has application and control plane agents, iAMA$_{AL}$s and iCPA$_{CL}$s, to deal with SDN layer issues, in addition to the self-managed architecture components defined for an sNE.

In the distributed self-managed SDN as depicted in Figure 6.68, each layer has its own sNMS, sNMS-s$_{AL}$, and sNMS-s$_{CL}$.

The sNMS-s$_{AL}$ and sNMS-s$_{CL}$ provide all the centralized fault management functions for application and control layers and inform sNMS-s about possible failures and fixes. End-to-end network-level issues such as those associated with CAC, load balancing, and end-to-end congestion control that may require the knowledge of the availability of resources at all layer and issues involved in multiple layers are left to sNMS-s to resolve. In order to resolve these issues, the sNMS-s might need to work with central sNMS.

6.6.6 In-Band Communications of Failure Types, Estimated Fix Time, and Fix

In today's networks, failures related to ports and connections are mostly reported to NMS via SNMP traps or in-band communications to NEs via Alarm Indication Signal (AIS), Remote Defect Indicator (RDI), Connectivity Check Message (CCM)-related

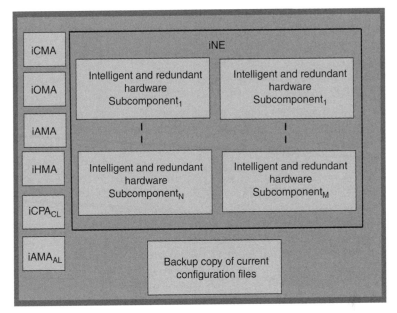

Figure 6.67. Self-managed SDN NE architecture.

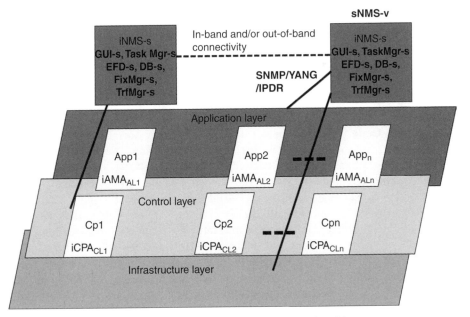

Figure 6.68. Distributed self-managed network architecture.

IFG	P	SFD	DA	SA	L/T	fNE ID	fComp ID	Op code	Failure code	Fix code	Fix time	PAD(25 bytes- all zeros)	CRC

IFG: Interframe Gap, 12 bytes
P/SFD (Preamble/Start of Frame Delimiter)-8 Bytes (P-7 bytes, SFD-1 byte)
L/T (Length/Type): Length of frame or data type , 2 bytes (0x8808)
CRC: 4 bytes
DA: 01:80:C2:00:00:02 (6 bytes)-Slow protocol multicast address
fNE ID: 6 bytes, Failed sNE Identifier
fComp ID: 4 bytes, Failed Component Identifier
Op Code: 2 bytes-0x0202 for Disabled and 0x0303 for Enabled status
Failure Code: 4 bytes
Fix Code: 1 byte identifying fixing entity, sNE (x00), sNMS (x01), sRMS (x02), sNMS-v (x03), sRNMS-v (x04), sNMS-s (x05), sRNMS-s (x06), field technician (x07), unidentified entity or inconclusive diag(x08)
Fix Time: 4 bytes indicating fix time in seconds by NE, NMS, or field technician

Figure 6.69. Frame format for self-managed Ethernet networks.

events such as Loss of Continuity (LoC) [50], and so on. These alarms and traps identify failed NE, port, or connection, but don't identify the component contributing to the failure. For self-management, it is necessary to identify faulty components, the entity that can fix the failure, and the estimated time for fix and communicate that to all parties involved (i.e., NEs and NMS and field technicians, customers), so that working NEs can store (if desired) data routed to the failed NE(s) for the duration of fix and reroute traffic around the failed NE(s) or port(s). This section introduces a concept of informing NEs, NMS, field technicians, and users (if desired) about the type of failures, the hierarchy for fixing the failures (i.e., whether it is locally fixable by sNE or remotely fixable by sRNMS or sNMS or locally fixable by a field technician), and a frame format for in-band communications.

Figure 6.69 depicts a proposed Ethernet frame for Ethernet networks to carry all the information described earlier for Carrier Ethernet Networks. Similar messages can be created for other types of networks such as IP, MPLS, and IMS.

For Ethernet networks, slow protocol multicast address is used to inform sNEs, sNMS, and field technician devices connected to the network. **fNE ID** indicates MAC address of the failed NE. **fComp ID** indicates the failed component identifier within the sNE. **Op Code** indicates whether the sNE or port is operationally disabled or enabled. This operational status is disabled during failures and becomes enabled after the failure is fixed. **Failure Code** indicates the failure type. If the failure type is unidentified through diagnostics, Failure Code will be unidentified or inconclusive or not local to sNE. **Fix Code** identifies repairing entity whether it is sNE, sNMS, or a filed technician. It is possible to allocate six bytes to Fix Code field to indicate MAC address of the fixing entity. It is possible to identify the failure type and not being able to fix it. In this case, fixing entity is unidentified. It is also possible that both failure code and fix code are unidentified. **Fix time** indicates the estimated time in seconds for repair that is filled by the repairing entity. In order for sNE and sNMS to provide the estimated fix time, the fix time must be for each type of failure needs to be stored in sNE and sNMS or sRNMS.

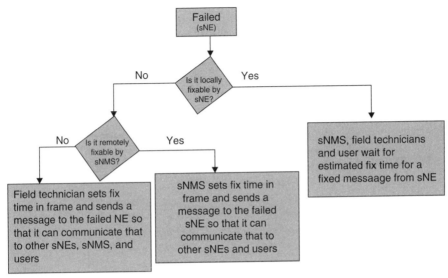

Figure 6.70. Centrally self-managed network architecture.

Given sNMS interface use network management protocols such as SNMP, the information in the message (Figure 6.68) needs to be conveyed to sNMS via an SNMP trap. Similarly, the SNMP trap from sNMS needs to be converted into an in-band message to convey the information to sNEs.

6.6.6.1 Failure Fixing Hierarchy in Centralized Networks

In a centrally self-managed network, when there is a failure, sNE determines if the failure is local to sNE or not. If the failure is local, then sNE informs other sNEs, sNMS, field technicians, and customers about the failure type and fix time (Figure 6.70). If sNE decides that the failure is not local to sNE, then sNE escalates the problem to sNMS. The sNMS verifies that it is not local to sNE and determines if it can fix the problem. If sNMS can fix the problem, sNMS communicates the failure type and fix time to NEs, field technicians, and customers. If sNMS determines the failure is not fixable, sNMS escalates the problem to field technicians. The field technician communicates the fix time to NEs, sNMS, and customers. After the fix is completed, the fixing entity initiates a self-managed notification with Enabled status (i.e., Opcode is set to Enabled) to other sNEs, sNMS, and customers. Both sNMS and field technicians use one of the sNEs to send notifications to the remaining interested parties.

The sNMS and field technician communicate failures and fixes via an sNMS. If there is a node failure (i.e., sNE completely fails due to a power failure for example), neither sNMS nor filed technicians will be able to communicate with the sNE. Therefore, sNMS and field technicians would use another sNE to communicate the failure.

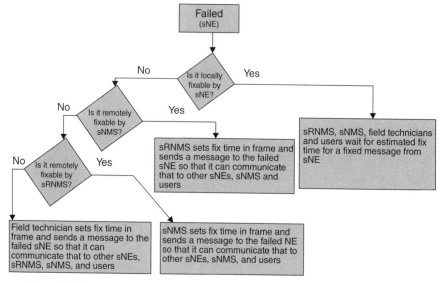

Figure 6.71. Process flow for fixing failures in distributed self-managed network.

6.6.6.2 Failure Fixing Hierarchy in Distributed Networks

In a distributed self-managed network, when there is a failure, sNE determines if the failure is local to sNE or not. If the failure is local, then sNE informs other NEs in that region, sRNMS, field technicians, and customers about the failure type and fix time (Figure 6.71). If sNE decides that the failure is not local to sNE, then sNE escalates the problem to its sRNMS. The sRNMS conveys the information to sNMS. The sRNMS verifies that it is not local to sNE and determines if it can fix the problem. If sRNMS can fix the problem, sRNMS communicates the failure type and fix time to sNEs in that region, other sRNMSs, field technicians, and customers. If sRNMS determines the failure is not fixable, sRNMS escalates the problem to sNMS; otherwise, sRNMS sends a self-managed notification to sNEs in the region, field technicians, and customers indicating the failure type and fix time. After the fix is completed, sRNMS sends a self-managed frame with Enabled status to sNEs, sNMS, and field technicians.

If sNMS can fix the problem, sNMS communicates the failure type and fix time to sNEs, sRNMS, field technicians, and customers. If sNMS determines the failure is not fixable, the sNMS escalates the problem to field technicians. The field technician communicates the fix time to sNEs, sNMS, sRNMS, and customers. After the fix is completed, the field technician sends a self-managed notification with Enabled status (i.e., Opcode is set to Enabled) to sNEs, sNMS, and customers.

6.6.6.3 Failure Fixing Hierarchy in Centralized VNs

In a centrally managed VN, when there is a failure in VN, iAMA or iCMA monitoring failed VN determines if the failure is fixable or not and communicates that to sNMS-v.

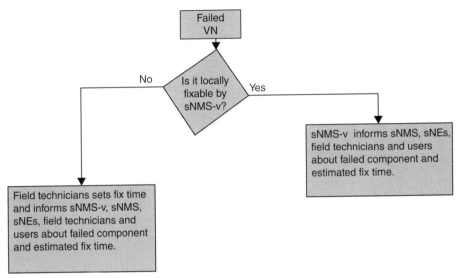

Figure 6.72. Process of fixing failures in centrally self-managed virtual network.

If the failure is fixable, then sNMS-v informs sNMS, sNEs (Figure 6.72), field technicians, and customers about the failure type and fix time.

If sNMS-v determines that the failure is not fixable, then it escalates the problem to sNMS. If sNMS can fix the problem, sNMS communicates the failure type and fix time to sNMS-v, field technicians, and customers; otherwise, it escalates the problem to field technicians. The field technician communicates the fix time to sNMS-v, sNMS, and customers. After the fix is completed, the field technician initiates a self-managed notification with Enabled status (i.e., Opcode is set to Enabled) to sNMS-v, sNMS, and customers. The sNMS may inform sNEs.

6.6.6.4 Failure Fixing Hierarchy in Distributed VNs

In a distributed self-managed VN, when there is a failure in VN, iAMA or iCMA monitoring failed entity determines if the failure is fixable or not and communicates that to sRNMS-v (Figure 6.73). If the failure is fixable, then sRNMS-v informs other sRNMS-v, sNMS, field technicians, and customers about the failure type and fix time.

If sRNMS-v determines that the failure is not fixable, then it escalates the problem to sNMS. If sNMS-v can fix the problem, then the sNMS communicates the failure type and fix time to sRNMS-v, sNMS, field technicians, and customers; otherwise, the sNMS escalates the problem to field technicians. The field technician communicates the fix time to sRNMS-v, sNMS-v, sNMS, and customers. After the fix is completed, the field technician initiates via the responsible agent a self-managed notification with Enabled status (i.e., Opcode is set to Enabled) to sRNMS-v, sNMS-v, sNMS, and customers.

Here, it is assumed that the central sNMS also helps to resolve VN issues if it is designed to do so.

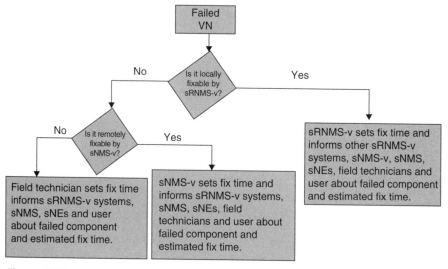

Figure 6.73. Process of fixing failures in self- and distributedly managed virtual network.

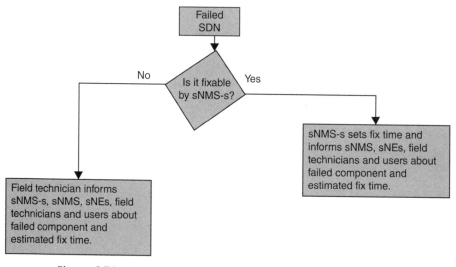

Figure 6.74. Process for fixing failures in centrally self-managed SDN.

6.6.6.5 Failure Fixing Hierarchy in Centralized SDNs

In a centrally self-managed SDN, when there is a failure in SDN, $iAMA_{AL}$ or $iCPA_{CL}$ monitoring failed entity determines if the failure is fixable or not and communicates that to the SDN controller, sNMS-s. If the failure is fixable, then sNMS-s informs sNMS, field technicians, and customers about the failure type and fix time (Figure 6.74). In turn, the sNMS may inform sNEs.

If sNMS-s determines that the failure is not fixable, then it escalates the problem to sNMS. If sNMS can fix the problem, sNMS communicates the failure type and fix time

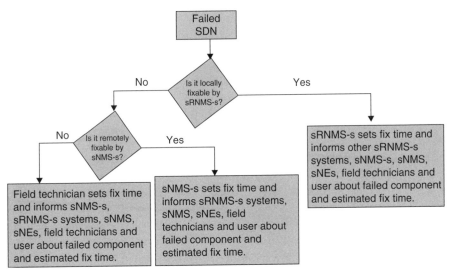

Figure 6.75. Process of fixing failures in distributed self-managed SDN.

to sNMS-s, NEs, field technicians, and customers; otherwise, sNMS escalates the problem to field technicians. The field technician communicates the fix time to sNMS-v, sNMS, and customers. After the fix is completed, the field technician sends a self-managed notification with Enabled status (i.e., Opcode is set to Enabled) to sNMS-s, sNMS, and customers. The sNMS may inform sNEs.

6.6.6.6 Failure Fixing Hierarchy in Distributed SDNs

In a distributed self-managed SDN, when there is a failure in SDN, $iAMA_{AL}$ or $iCPA_{CL}$ monitoring failed entity determines if the failure is fixable or not and communicates that to sRNMS-s. If the failure is fixable, then sRNMS-s informs other sRNMS-s, sNMS, field technicians, and customers about the failure type and fix time (Figure 6.75). The sNMS may inform sNEs after receiving the message from sRNMS-s.

If sRNMS-s determines that the failure is not fixable, then it escalates the problem to sNMS-s. If sNMS can fix the problem, the sNMS communicates the failure type and fix time to sRNMS-s, sNEs, field technicians, and customers; otherwise, the sNMS-s escalates the problem to field technician. The field technician communicates the fix time to sRNMS-s, sNMS-s, sNMS, and customers. After the fix is completed, the field technician sends a self-managed frame with Enabled status (i.e., Opcode is set to Enabled) to NEs, sRNMS-s, sNMS-s, sNMS, and customers.

6.7 CONCLUSION

In this chapter, cloud, virtualization, SDN, and self-managed concepts, architectures, and their building blocks are described as possible future directions for MSO networks and services.

Application examples are given. MSOs are already implemented cloud-based services using VMs. Initiatives for virtualization of some of network functions, separation of data and control plane functions, and further automation of OAMPT functions using SDN concept are underway.

Cloud-based applications operate as regular applications, mainly using web services platforms. Cloud computing services are merging as infrastructure services. Multiprovider and multidomain resources are involved. The Intercloud Architecture addresses problems with multidomain heterogeneous cloud-based applications integration and inter-provider and interplatform interoperability.

Software-defined networking (SDN) architecture decouples the network control and forwarding functions, enabling the network control to become directly programmable and allowing application/service layer to be aware of network infrastructure, and vice versa. As a result, service providers expect to provision services quickly, while users expect to have more control of the services that they are receiving. Switching and routing equipment are expected to be simplified, while control functions are centralized.

The transition from non-SDN networks to SDN networks for service providers might take a long time. This process can be somewhat shortened if switching and routing equipment vendors follow open standards.

Self-managed network concept, NE and NMS architectures, and fault management communication mechanisms for centrally and distributedly self-managed networks are introduced. The concept is applied to VNs and SDN.

Cloud, virtualization, SDN, and self-managed concepts introduce substantial changes to equipment, network, and operations systems. It may take a decade for these four concepts to be fully implemented and used in today's networks.

REFERENCES

1. Fraunhofer Institute for Open Communication Systems, Cloud Concepts for the Public Sector in Germany—Use Cases, December 2011.
2. NIST SP 500-292 NIST Cloud Computing Reference Architecture, September 2011.
3. ESG White Paper, Cloud Computing: Networking and WAN Optimization.
4. Gartner Says Worldwide Cloud Services Market to Surpass $109 Billion in 2012, September 2012, http://www.gartner.com/newsroom/id/2163616 (accessed August 19, 2014).
5. Gartner Says Worldwide Public Cloud Services Market to Total $131 Billion, Feb 28, 2013, http://www.gartner.com/newsroom/id/2352816
6. Alan Murphy F5 White Paper, Virtualization Defined—Eight Different Ways, 2010.
7. L. Yong, et al., "Use Cases for DC Network Virtualization Overlays," draft-mity-nvo3-use-case-04, October 2012.
8. Y. Demchenko, et al., Intercloud Architecture for Interoperability and Integration, SNE technical report SNE-UVA-2012-05 Release 1, Draft Version 0.5, September 2012.
9. MEF White Paper, Carrier Ethernet for Delivery of Private Cloud Services, February 2012.
10. Y. Demchenko, et al., Intercloud Architecture Framework for Heterogeneous Multi-Provider Cloud based Infrastructure Services Provisioning. International Journal of Next-Generation Computing, July 2013.

11. Fluke Networks White Paper, Connectivity in the Virtualized Datacenter: How to Ensure Next-Generation Services, May 2010.

12. VmWare White Paper, Virtualization Overview, 2006.

13. Lasserre, et al., "Internet-Draft Framework for DC Network Virtualization," draft-lasserre-nvo3-framework-03.txt, July 2012.

14. N. Bitar, et al., "NVO3 Data Plane Requirements," draft-ietf-nvo3-dataplane-requirements-01. txt," July 2013.

15. D. Black, RFC2983 "Differentiated Services and Tunnels," October 2000.

16. B. Briscoe, RFC6040 "Tunnelling of Explicit Congestion Notification," November 2010.

17. P. Ashwood-Smith, et al., "NVO3 Operational Requirements," draft-ashwood-nvo3-operational-requirement-03, July 2013.

18. Marc Lasserre, et al., "Framework for DC Network Virtualization," draft-ietf-nvo3-framework-05. txt, January 2014.

19. MEF 30.1, Service OAM Fault Management Implementation Agreement: Phase 2, April 2013.

20. MEF 10.3, Ethernet Services Attributes Phase 3, October 2013.

21. S. Hartman, et al., "Security Requirements of NVO3," draft-hartman-nvo3-security-requirements-02, January 2014.

22. J. Vollbrecht, RFC4137 "State Machines for Extensible Authentication Protocol (EAP) Peer and Authenticator," August 2005.

23. M. Mahalingam, et al., "A Framework for Overlaying Virtualized Layer 2 Networks over Layer 3 Networks," draft-mahalingam-dutt-dcops-vxlan-02.txt, August 2012.

24. M. Riegel, RFC4197 "Use of Provider Edge to Provider Edge (PE-PE) Generic Routing Encapsulation (GRE) or IPin BGP/MPLS IP Virtual Private Networks," January 2007.

25. NFV Members: http://portal.etsi.org/NFV/NFV_List_members.asp (accessed August 19, 2014).

26. Margaret Chiosi, et al., Network Functions Virtualisation—Introductory, White Paper Issue 1 Network Functions Virtualisation, http://portal.etsi.org/NFV/NFV_White_Paper.pdf, October 2012 (accessed August 19, 2014).

27. ETSI GS NFV 002 v1.1.1 Network Function Virtualisation (NFV); Architectural Framework, 2013-10.

28. S. Gringeri, et al., "Extending Software Defined Network Principles to Include Optical Transport," IEEE Comm. Magazine, March 2013.

29. ONF, OpenFlow Switch Specification Version 1.3.0, June 25, 2012.

30. Matthew Palmer, http://www.sdncentral.com/market/sdn-market-sizing/2013/04, April 24, 2013 (accessed August 19, 2014).

31. T. Nadeau, et al., "Framework for Software Defined Networks," draft-nadeau-sdn-frame-work-00, April 2012.

32. Ashton, Metzler Associates, "Ten Things to Look for in an SDN controller," May 2013.

33. P. Pan, et al., "Software-Defined Network (SDN) Problem Statement and Use Cases for Data Center Applications," draft-pan-sdn-dc-problem-statement-and-use-cases-02.txt, March 2012.

34. ETSI GS AFI 002 V1.1.1 Autonomic network engineering for the self-managing Future Internet; Generic Autonomic Network Architecture, 2013-04.

35. R.T. Fielding, "Architectural Styles and the Design of Network-based Software Architectures," Doctoral dissertation, University of California, Irvine, 2000.

36. MEF: "Carrier Ethernet and SDN Part 2: Practical Considerations", July 2014.

37. S. Hubbard, "Automating Carrier Ethernet 2.0 Service Delivery and Preparing for SDN-Enabled Services," September 2013.

38. E. Malette and M. Hajduczenia, "Automating provisioning of Demarcation Devices in DOCSIS® Provisioning of EPON (DPoE™)," IEEE Comm. Magazine, September, 2012.

39. ETSI TS 132 507 V9.1.0 (3GPP TS 32.507 version 9.1.0 Release 9) Technical Specification UMTS; LTE; Telecommunication management; Self-configuration of network elements; IRP: SOAP SS, 2010–07.

40. ETSI TS 132 526 V10.4.0 (3GPP TS 32.526 version 10.4.0 Release 10) Digital cellular telecommunications system (Phase 2+);UMTS; LTE; Telecommunication management; Self-Organizing Networks; Policy Network Resource Model Integration Reference Point; SS definitions, 2013–10.

41. ETSI TS 132 501 V11.0.0 (3GPP TS 32.501 version 11.0.0 Release 11) UMTS; LTE; Telecommunication management; Self-configuration of network elements; Concepts and requirements, 2012–09.

42. ETSI TS 132 506 V10.0.0 (3GPP TS 32.506 version 10.0.0 Release 10) Technical Specification Digital cellular telecommunications system (Phase 2+);UMTS;LTE; Telecommunication management; Self-Configuration of Network Elements Integration Reference Point; SS definitions, 2011-05.

43. ETSI TS 132 506 V11.0.0 (3GPP TS 32.506 version 11.0.0 Release 11) Digital cellular telecommunications system (Phase 2+);UMTS;LTE; Telecommunication management; Self-Configuration of Network Elements Integration Reference Point; SS definitions, 2012-10.

44. ETSI TR 136 902 V9.3.1 (3GPP TR 36.902 version 9.3.1 Release 9) Technical Report LTE; E-UTRAN; Self-configuring and self-optimizing network use cases and solutions, 2011-05.

45. ETSI GS AFI 001 V1.1.1 Group Specification Autonomic network engineering for the self-managing Future Internet (AFI); Scenarios, Use Cases and Requirements for Autonomic/Self-Managing Future Internet, 2011-06.

46. Keller, Alexander et al., (Eds.), "Self-Managed Networks, Systems, and Services Second IEEE International Workshops," SelfMan 2006, Dublin, Ireland, June 16, 2006, Proceedings.

47. M. Toy, Self-Managed Networks, Comcast internal document, November, 2012.

48. M. Toy, Self-Managed Carrier Ethernet Networks, April 2014, MEF Meeting in Budapest, self-managed-networks-comcast-mtoy.pdf, https://wiki.metroethernetforum.com/pages/viewpageattachments.action?pageId=39193359.

49. N. Wang, et al., "A Two-Dimensional Architecture for End-to-End Resource Management in Virtual Network Environments," IEEE Network, September 2012, 26 (5), 8–14.

50. ITU-T Y.1731, "OAM functions and mechanisms for Ethernet based networks," 2008.

51. CEF, CEF 1.0: CEF Reference Architecture, Draft, October 2014.

INDEX

Note: Page numbers in *italics* refer to Figures; those in **bold** to Tables

AaaS *see* Applications as a Service (AaaS)
accounting management, 5, 85–6, 286
Active Queue Management (AQM), 5, 13, 14, 22, 88
adaptive bit-rate (ABR) video, 34, 35, 37, 39, 40, *42*, 43, 48
Ad Decisioning System (ADS), 40–41, 43
Address Resolution Protocol (ARP), 21, 210, 283, 289, 293
Alternate Content Decisioning System *see* Placement Opportunity Information System (POIS)
application-level gateway (ALG), 178–9
Application Programming Interfaces (APIs), 41, 51, 79, 98, 185, 189–90, 192, 199, 200, 206, 209–10, 213–14, 216–18, 223, 291, 302–3, 309–11, 318, 328
Applications as a Service (AaaS), 198, 223
AQM *see* active queue management (AQM)
automatic channel selection (ACS), 73
AWS Elastic Computing or ECS, 223
AWS marketplace services, 223

backup primary channels, 5, 88
Baseline Privacy Interface Plus (BPI+), 100, 101, 103–5
Baseline Privacy Key Management (BPKM), 101
best-effort (BE) service, 14, 21
billing systems and formats
  for business services
    business services order management, 236, *236*
    DOCSIS-based provisioning, 235

e-Care (self-service) platforms, 242
electronic data interchange (EDI) standard, 236
residential-type products and services, 235
SOAP or RESTful services, 236
for residential services
  business and service management layer, *97*
  charges, 99
  downstream interfaces, 98
  function, 99
  product catalog, 99
  telephone number (TN) inventory system, 98
  upstream interfaces, 98
Breakout Gateway Control Function (BGCF), 29
business network architectures and services
  Business Real-Time Communications Services, 7
  business voice, 171–80
  Carrier Ethernet, 6–7
  Cloud Ethernet Forum (CEF), 7
  DOCSIS Provisioning of EPON (DPoE), 161–7
  EPON Protocol over Coax (EPoC), 168–71
  Metro Ethernet services, 7, 8
  Operations, Administration, Maintenance, Provisioning, and Troubleshooting (OAMPT), 7–8
  Private Branch Exchanges (PBXs), 6–7
  "Business Real-Time Communications Services," 171–2

Cable Networks, Services, and Management, First Edition. Edited by Mehmet Toy.
© 2015 The Institute of Electrical and Electronics Engineers, Inc. Published 2015 by John Wiley & Sons, Inc.

Business Support Systems (BSS)
administrative subsystems, 189–90
benefits of outsourcing billing, 190
functional systems, 189
service creation, 190
business voice
communications service portfolio, 171–2
ESG support, Broadband Forum TR-069, 180
Hosted IP-Centrex service, 172, *172*
IMS enhancements, 173–4
PacketCable Enterprise SIP Gateway, 178–80
SIP Forum SIPconnect1.1, 176–8
SIP-PBX registration/routing, 174–6
SIP Trunking service, 173, *173*

CableLabs DOCSIS, 187
cable modem (CM)
at customer location, 3
DOCSIS, *164,* 318
management, *86,* 86–7, **87**
upstream and downstream profiles, 19–20
cable modem termination system (CMTS)
customer premise equipment (CPE)
provisioning, 219
DOCSIS 3.0 management, 56, 104
DOCSIS 3.1 management, 5–6, 12–13
DOCSIS Provisioning of EPON
(DPoE) System, 163
extended message integrity check (MIC), 104
management, 87, **88,** *88,* 89, 95
management information base
(MIBs), **104,** 104–5, **238, 240**
modular *see* modular cable modem
termination system (M-CMTS)
multiple system operators (MSOs)
networks, 318–19
performance metrics collection, 95–6
residential network architectures
and services, 2–3
"CableMover" program, 199, 200
Cable Telecommunications and Marketing
Association (CTAM), 199
cable TV (CATV) system, 1, 11, 38,
89–90, 94, 184, 203
Campaign Management System (CMS), 41
Carrier Ethernet
cloud services
Cloud Brokers, 266, 267
Cloud Carriers, 266, *266*

E-LAN services, 269–70
Ethernet Private Line (EPL), 267–8
E-Tree services, 269
delivery, 7
EVCs, 153, **154**
revenue, 135
SDN application, 315–16, *317*
CCAP *see* Converged Cable Access
Platform (CCAP)
CDNs *see* content distribution
networks (CDNs)
class of service (CoS), 137–8, 142–3, 147,
149, **149, 150, 151, 152, 153,** *158,*
231, 233, 285
Closed User Group (CUG), 290
Cloud Computing Reference
Architecture (CCRA), *254*
cloud computing services
actors, 252, **255**
advantages, 250, 251
architecture, 252, 255–9
Carrier Ethernet, 266–70
characteristics, 247–8
Cloud Computing Reference
Architecture (CCRA), *254*
cloud-enabled DCs, 251
Cloud Provider, 256
compute elements, 250
connectivity fabric, 250
deployment models, 249–50
elastic cloud infrastructure, 251–2
Infrastructure as a service (IaaS), 252
interactions, actors, 256, **256**
intercloud architecture, 270–274
layers, *263*
network-centric cloud architecture, 252
network optimization, 248
orchestration, 250–251
physical resource layer, 263
resource abstraction and control layer, 262–3
security, 265
service delivery models, 248–9
Software as a Service (SaaS), 257–8
standards, 252, **253**
system components, 262
technical and business requirements, 251
usage scenario (carriers, brokers
and auditors), 256, **256,** 257
Cloud Ethernet Forum (CEF), 7

cloud services
   cloud based applications, 9
   computing technologies, 9
   management
      accounting and billing, 264
      business support, 263
      components, 263, *264*
      contract management, 264
      customer management, 263–4
      description, 263
      Infrastructure as a Service (IaaS), 265
      inventory management, 264
      metering, 264
      monitoring and reporting, 264
      pricing and rating, 264
      rapid provisioning, 264
      reporting and auditing, 264
      resource changing, 264
   multiple system operators (MSOs), 8
   provisioning, 223
   virtualization, 9
CMS *see* Campaign Management System
         (CMScoarse wavelength division
         multiplexing (CWDM), 191–3, 218, 242
code verification certificate (CVC), 103, 239
command-line interface (CLI)
   manual service provisioning, 206, *207*
   M2M communication, 207
   NE (human-to-machine)
         communication (GUI), 207–8
   scripting, 209, 309
   and SNMP, 216, 298
   Unix CLI, 207, 208
committed information rate (CIR), 21,
         145, 146, 149–50
Common Alerting Protocol (CAP), 43
Common Encryption (CENC) Standard, 39
Common Object Request Broker Architecture
         (CORBA), 185, 216
common off-the-shelf (COTS) hardware, 49
community cloud
   description, 249–50
   management, 260
   on-site, *261*
configuration management, 5, 13, 56, 68, 85, 188
consecutive high loss intervals (CHLIs), 233
constant bit rate (CBR), 16
consumer-owned and managed
         (COAM), 35, 39–40, 44

content distribution networks (CDNs)
   architecture and services
      common off-the-shelf (COTS)
            hardware, 49
      content caching and delivery, 47, 49–50
      content publishing, 49
      delivery components, 47
      Edge Web Cache (EC), 45–7, *46*
      Network Digital Video Recorder
            (NDVR), 48
      quadrature amplitude modulation
            (QAM)-based STBs, 49
      service components, 46
      storage components, 47
      transcoding *vs.* storage, 51
      usage-based billing, 47, *48*
      video-on-demand (VOD) services, 48
      web-site graphics, 46
   management and reporting
      "Big Data" projects, 52–3
      content management system , 51
      hierarchical cache algorithms, 51, *52*
      video content, 45, 48–9
Content Information Service (CIS), 41, 43
content management system , 47, 51
Converged Cable Access Platform (CCAP)
   architectures and services
      "all IP" service offering, 54
      broadcast and narrowcast, 55
      data and video applications, 53
      displacement or augmentation, 65–9
      distributed architectures, 61–4
      EPON linecards, 54
      "future proofing" capabilities, 54
      migration and deployment
            considerations, 64–5
      mixed access technologies, 58–9
      modular head-end architecture, 56–8
      multiple program transport stream
            (MPTS) video, 55
      multiple system operators (MSOs)
            core network, 3, 12–13
      operational models, 55
      radio frequency over glass (RFoG), 59–61
      service group spanning, 54
      system capacity and density, 53
      virtual node splitting, 54
   business services, 2
   DOCSIS 3.0/3.1, 12

Converged Cable Access Platform
(CCAP) (*cont'd*)
FCAPS (Fault, Configuration, Accounting,
Performance, and Security) model, 4
management approach, 5
migration
development and planning, 69
MHA support, 64–5
technology and HFC evolution, *66*
phases
convergence, 67
enhancements, 67–8
initial deployment, 66–7
maturation, 68–9
security, 241
"CPE WAN Management Protocol
(CWMP)", 178–9, 180
Customer Account System (CAS), 91, 98, 99
customer-edge (CE) equipment, 153, 211, 221–3
Customer Premises Equipment (CPE)
autoprovisioning, 4, 85
DPoE ONU
configuration file, 219
de-coded Type–Length–Value
(TLV), 219, **220**
full coded Type–Length–Value
(TLV), 220, **221**
hexadecimal representation, 219, *219*
TLV and sub-TLV, 220, **222**
inventory system, 90, 97
RFoG system, 6
single points of failure (SPOFs), 219
UNI-C functions, 140
Customer Service Representative
(CSR), 189, 201, 204
CVC *see* code verification certificate (CVC)

Data over Cable Service Interface
Specification (DOCSIS)
3.0/3.1 *see* DOCSIS 3.0/3.1
based services, 200
billing systems and formats, 235
cable modem (CM), *164*, 318
network security administration, 85
Provisioning of EPON *see* DOCSIS
Provisioning of EPON (DPoE)
service and device management, 85
software-defined networking
(SDN), 318, *318*

delay measurement message
(DMM), *229*, 232–4, *233*, 243
delay measurement response (DMR), *229*,
232–4, *233*, 243
DiffServ Code Point (DSCP), 20, 143, 150,
150, 151, 284, 285, 286, **308**
Digital Rights Management (DRM)
Adobe Access, 39
business rule management, 40
business support systems (BSS), 40, 189
FairPlay DRM, 39
PlayReady DRM, 39
Widevine, 39
Digital Video Broadcasting—Cable 2
(DVB-C2), 18
Digital Video Broadcasting—Satellite 2
(DVB-S2), 18
"Distributed CCAP" concept
challenges, 62, 63–4
distributed architecture, 61, *62*
multiple system operators (MSOs), 62–3
distribution system (DS), 72
DOCSIS *see* Data over Cable Service
Interface Specification (DOCSIS)
DOCSIS 3.0/3.1
capacity evolution
channel bandwidths, 17–18
frequency bands, 16–17
modulation techniques and FEC, 18
subcarrier bit loading, 18–19
upstream and downstream
profiles, 19–20
evolution
active queue management (AQM)
algorithm, 13
advanced bandwidth allocation
method, 14
channels upstream and downstream, 14
data-over-cable reference
architecture, 14, *15*
first-generation, 16
forward error correction (FEC) and
modulation techniques, 14, 18
physical layer (PHY) capabilities, 16
QoS support, 13–14
radio frequency over glass (RFoG), 16
high-speed data (HSD) or broadband
services, 12, *13*
management model, 13

scheduling services
  best-effort (BE) service, 21
  buffer bloat, 22
  committed information rate (CIR), 21
  non-real-time polling service (nrtPS), 21
  Proportional Integral Controller-
    Enhanced (PIE) algorithm, 22
  real-time polling service (rtPS), 21
  unsolicited grant service (UGS), 21
  unsolicited grant service with activity
    detection (UGS-AD), 21
security
  Baseline Privacy Key Management
    (BPKM), 101
  CableLabs, 100
  CM cloning, 104–5
  configuration file TLV encodings, 102
  device authentication and traffic
    encryption, 100–103
  DOCSIS certificate PKI hierarchy, 100, *101*
  early authentication and
    encryption (EAE), 101
  encrypted traffic, 101
  features, 100
  key encryption key (KEK), 101
  provisioning, 104
  security association identifier (SAID), 101
  service and network management, 88–9
  software download, 103
  service flows (SF), 20–21
  SNMP MIBS
    certificate management, 101, **102**
    managing key exchange, 101, **102**
DOCSIS Light Sleep Mode (DLS), 5, 88
DOCSIS MPEG-TS (D-MPT), 57
DOCSIS Operations and Support System
    Interface (OSSI), 190
DOCSIS Provisioning of EPON (DPoE)
  architecture and services
    CM configuration file, 165
    downstream encryption, 164
    four-step device provisioning process, *165*
    internet data service, 167–8
    logical link identifier (LLID), 165
    PON topology, 161, *162*
    service provisioning concepts, 166–7
    stages, 164–5
    standards, 167
    Trivial File Transfer Protocol (TFTP), 165

ELINE and ELAN services, 202
Ethernet transport, 202
ONU (D-ONU), 201, 219–22
security
  certificate public key infrastructure
    (PKI), 237, *237*
  certification authority (CA) certificate, 237
  device authentication and traffic
    encryption, 237–8
  ONU cloning, 240
  provisioning, 239–40, **240**
  secure software download (SSD)
    mechanism, 239
  SNMP MIBS for Certificate
    Management, 238, **238**
Document Schema Description
    Languages (DSDL), 213
Document-Type Definition (DTD), 213
Domain Name System (DNS), 50, 106,
    109–10, 113, 185, 200, 207, 217,
    276, 293
DownstreamExternal PHY Interface
    (DEPI), 57, 65–7, 69
DPoE *see* DOCSIS Provisioning
    of EPON (DPoE)
dual band–dual concurrent (DBDC), 73
DVB-C2 *see* Digital Video
    Broadcasting—Cable 2 (DVB-C2)
DVB-S2 *see* Digital Video
    Broadcasting—Satellite 2 (DVB-S2)
DWDM, 191–2, 218
Dynamic Host Configuration Protocol
    (DHCP), 91, 101, 104, 106, 110–111,
    114, 123, 165, 167, 185, 205, 207,
    210–211, 217, 219, 240

early authentication and encryption
    (EAE), 101, 238
Edge Resource Management
    (ERM) protocol, 56
Edge Web Cache (EC), 45, *46*
EDNS-client-subnet, 50
E-LAN services
  Cloud Consumers, 270
  EP-LAN services, 270
  E-Tree service, 155
  multipoint L2 VPNs, 155
  from SP, *160*
  UNIs, 155

element management systems (EMSs)
  definition, 187
  equipment manufacturers, 8, 183, 186–7
  network element (NE), 186
  organizations and protocols, 187
  service order management systems, 198
  vendor-specific, 213, 218
  virtual machines (VMs), 187, 188
  web interface or client application, 188
embedded Digital Multimedia
        Adaptor (eDVA), 201
embedded Multimedia Terminal Adapter
        (eMTA), 23–4, 91, 201
ENNI *see* External Network-to-Network
        Interface (ENNI)
EP-LAN *see* Ethernet private LAN (EP-LAN)
EPOC *see* EPON Protocol over Coax (EPoC)
EPON Protocol over Coax (EPoC)
  Bridge FCU (B-FCU), 169
  CableLabs EPoC system, 169
  coaxial network, 168
  coax network units (CNUs), 168–9
  migration strategies, 170, *171*
  orthogonal frequency division
        multiplexing (OFDM), 169
  P2MP topology, 168
  Repeater FCU (R-FCU), 169
  security, 240–241
Equal Cost Multipath (ECMP), 285–6, 288
Ethernet
  access
    DOCSIS Provisioning of EPON (DPoE),
        2, 7–8, 61, 134, 161–8, 187, 237–40
    EPON Protocol over Coax (EPoC), 2,
        7–8, 134, 168–71, 240–241
    Ethernet passive optical network
        (EPON), 54, 58–61, 66–8, 134, 162–4,
        169–70, 191, 193, 201, 218, 242
  Carrier *see* Carrier Ethernet
  frame
    broadcast service frame, 139
    canonical format indicator (CFI), 137–8
    class of service (CoS)/priority
        information, 138
    C-Tag and S-Tag, *139*
    cyclic redundancy check (CRC), 137
    ETH layer PDUs, 138
    IEEE 802.3–2002/2005 compliant
        frame formats, 138, *138, 139*

Layer 2 Control Protocol (L2CP)
        frame, 139
    multicast service frame, 139
    service frame disposition, 139
    unicast service frame, 139
    VLAN tag, 138, *138*
  Metro *see* Metro Ethernet
  services, 201–2
  transport of
    IP/MPLS, 154, 202
    L2VPN, 201, 227
    VPLS, 154, 202, 283
    VPWS, 202
Ethernet private LAN (EP-LAN),
        155–7, 168, 270, *271,* 316
Ethernet Private Line (EPL) services
    Access EPL service, 157
    automatic service provisioning, 209, *209*
    Cloud Consumers, 267, *268*
    Ethernet Virtual Private Line
        (EVPL) service, 267
    service frame transparency, 142
Ethernet Virtual Connection (EVC)
    bundling map, 142
    description, 140
    multipoint-to-multipoint
        (P2MP) EVC, 141, *154*
    point-to-point EVC, 141, 142, *142, 154*
    subscriber flows, 136
    UNI/EVC attributes, 142–4
Ethernet virtual private (EVP-LAN),
        155, *157,* 168, 316
E-Tree services, 142, **154,** 155–6,
        269, *269, 270*
EVC *see* Ethernet Virtual
        Connection (EVC)
EVP-LAN *see* Ethernet virtual private
        (EVP-LAN)
Explicit Congestion Notification (ECN)
        marking, 286, **308**
Extensible Authentication Protocol
        (EAP), 71, 237, 238, 289
eXtensible Markup Language (XML)
    APIs, *213,* 213–14
    HTML, 213
    REST and RESTful APIs, 214
    SOAP, 192
eXtensible Messaging and Presence
        Protocol (XMPP), 214

External Network-to-Network Interface (ENNI), *145,* 145–6, 147, 150, 157–8, 159, 161, 202, 295, 316

Fault, Configuration, Accounting, Performance, and Security (FCAPS) model, 4, 85, 131
fault management (FM)
    for business services
        alarm indication signal (AIS), 226
        for business services, 242–3
        continuity check messages (CCMs), 225, *225*
        dying gasp or link fault, 225
        IEEE 802.1ag SOAM framework, 228–9
        Link OAM use case, 224, *224*
        Linktrace Message (LTM), 226, *227*
        Lock Message (LCK), *227,* 227–8
        loopback messages (LBMs), 226, *226*
        Test Message (TST), 227, *227*
    for residential services
        alarm surveillance, 93
        correction, 93–4
        description, 85
        localization, 93
        testing, 94
FCAPS *see* Fault, Configuration, Accounting, Performance, and Security (FCAPS) model
fiber-to-the-home (FTTH) networks, 168, 170, *171,* 181
Fibre Channel over Ethernet (FCoE), 251
File Transfer Methods
    FTP, 21, 49, 98, 192, 208
    HTTP or HTTPS, 14, 37–8, 39, 50, 53, 110, 129, 179, 192, 206, 208, 211, 214, 216
    SFTP, 192, 208
    Trivial File Transfer Protocol (TFTP), 91, 103, 110–111, 131, 165, 167, 185, 192, 208, 219, 239
forward error correction (FEC) code, 14, 16, 18
frequency division duplexing (FDD), 16
FTTH networks *see* fiber-to-the-home (FTTH) networks

Graphical User Interface (GUI)
    DWDM, 218
    element management systems, 186–7
    order entry systems, 201
    provisioning interfaces, 207–10

provisioning protocols, 211–12
    web-based, 322

hackers, 100, 103–5, 132
HDS *see* HTTP Dynamic Streaming (HDS)
High-efficiency WLAN (HEW), 71
high loss intervals (HLIs), 233
High Speed Data (HSD) Internet or Broadband service, 1–2, 11–12, *13,* 48, 53, 76, 78, 94, 170, 178
HLS *see* HTTP Live Stream (HLS)
Home Subscriber Server (HSS) database function, 25–6, 29–34, 125, 127–8, 172–4, 176
Hosted IP-Centrex service, 172, *172*
"hosted PBX," 195
Hotspot 2.0, 71
HTTP Dynamic Streaming (HDS), 37, 44
HTTP Live Stream (HLS), 37, 40, 44
hybrid cloud
    description, 250
    IaaS and PaaS, 272
    management, 261–2, *262*
hybrid-fiber/coax (HFC) cable network, 2

IaaS *see* Infrastructure as a Service (IaaS)
IEEE 802.1ag SOAM framework
    DMM/DMR, 229, *229*
    hierarchical OAM domains, 228, *228*
    maintenance entity group end point (MEP), 228–9
    maintenance entity group intermediate point (MIP), 229
    MEG levels, 229, **229**
IETF Trivial File Transfer Protocol (TFTP), 192
IMS *see* IP Multimedia Subsystem (IMS)
IMS Authentication and Key Agreement (AKA), 123, 126, 241
in-band communications
    Ethernet frame, 332, *332*
    Failure Code, 332
    failure fixing hierarchy
        centralized networks, 333, *333*
        centralized SDNs, *336,* 336–7
        centralized VNs, 334–5, *335*
        distributed networks, 334, *334*
        distributed SDNs, 337, *337*
        distributed VNs, 335, *336*
    Fix Code, 332

incorporated IP Version 6 (IPv6),
13–14, 50, 143, 150, 167,
281, 306
Independent Rate of Return (IRR)
calculations, 197
Industry Specification Group (ISG), 302
Infrastructure as a Service (IaaS)
Cloud Consumer, 257, 259
Cloud IaaS, 249
Cloud Provider, 257, 259
cloud services, 223, 272–4, 278
example services, *258*
growth rate, 252
intercloud control and management, *274*
management, 265
multiple system operators (MSOs), 198
prices, 249
service management, 199
service order security, 204, 265
VMs, 265
initial filter criteria (IFC), 32–4
intercloud architecture, cloud
computing services
components, 272–3
IaaS and PaaS, communication, 274, *274*
IT infrastructure disaster recovery, 271
layers, 273
multilayer cloud service model
(CSM), 273, *273*
network interconnecting
infrastructure, *271, 272*
Intercloud Control and Management
Plane (ICCMP), 272, 273
Intercloud Federation Framework
(ICFF), 272, 273
Intercloud Operation Framework
(ICOF), 272–3
Interconnection Border Control Function
(IBCF), 25, 29, 33, 174
internet access, 135, 153, 156, 189,
190, 193, 199–201, 203, 207,
266–7
Internet Engineering Task Force
(IETF), 23–4, 26–7, 32, 53,
93, 95, 121, 123, 126, 129, 192,
213, 215
Internet Gateway Protocol (IGP), 313
Internet Protocol High-Speed Data
(IPHSD) service, 167–8

Internet Protocol TeleVision (IPTV) system
advertising insertion
adaptive bit-rate (ABR) video,
40, *42*, 43–4
ad-viewing experience, 43
asset, 40
"Digital Program Insertion Cueing
Message for Cable," 41
linear or on-demand video, 41
Manifest Manipulator queries Ad
Decisioning System, 43
SCTE-130, 41
video content, 40
architecture and services
adaptive bit-rate (ABR) video, 35, 37
closed captioning and accessibility, 44
delivery architecture, components, 34
digital rights management (DRM), 38–40
"download-to-go" service, 39
generic HTTP web servers, 38
local linear video, 35, *36*
multiple system operators
(MSOs) goals, 34
national linear video, 35, *36*
Online Multimedia Authorization
Protocol (OMAP), 39
pay-per-view (PPV), 35, *36*
pay-TV service, 34
satellite-based reception sites, 35, *36*
"scrambling and conditional
access (CA)", 38
"SmartTVs," 35
sports blackouts, 44–5
video content protection, 39
video file segmentation
and delivery, 37, *38*
video-on-demand (VOD)
programming, 35, *36*
linear and on-demand video, 12
Internet Protocol (IP) traffic, 2, 12,
201, 282, 299
Internet Protocol version 4 (IPv4), 50,
143, 150, 167, 281
Internet Research Task Force Software-Defined
Networking Research Group (IRTF
SDNRG), 302
Interrogating Call Session Controller Function
(I-CSCF), 29, 31–2, 34, 125–7
intrusion detection systems (IDS), 186, 188

IP Detail Record/Streaming Protocol
(IPDR/SP), 6, 86–7, 95, *96,* 132
IP Multimedia Subsystem (IMS)
architecture and services
application layer, 29
half-call model, 27
overview, *28,* 28–9
public user identity, 26
Real-time Transport Protocol (RTP), 27
Session Initiation Protocol
(SIP), 12, 26–7
traditional telephony, 27
registration, 26
security, 241
session establishment, 26–7
subscription, 27
use cases
I-CSCF queries, 31, 32
operator networks, 33–4
P-CSCF, 30–33
REGISTER requests, 30–31
registration message sequence, 30
S-CSCF queries, 31–2
session establishment, 32, *33*
IPTV *see* Internet Protocol TeleVision
(IPTV) system
ISG *see* Industry Specification Group (ISG)

JAIN *see* JAVA Application Program
Interface (API)'s for Integrated
Networks (JAIN)
JAVA Application Program Interface (API)'s
for Integrated Networks (JAIN), 185
JavaScript Object Notation
(JSON), 214–15, *215*
JSON *see* JavaScript Object Notation (JSON)

LAG *see* Link Aggregation Group (LAG)
LAN, 1, 2, 12, 135, 155–7, 159, *160,*
161, 168, 178–81, 201–2, 251,
269–70, *271*
Layer 2 Virtual Private Network
(L2VPN), 57, 193, 201, 206–7, 242
Link Aggregation Group (LAG), 145,
285–6, 288, 305, 306
logical link identifier (LLID), 165,
169, 201, 238
long-term evolution (LTE), 18
low-density parity check (LDPC), 18

MAC *see* Move–Add–Change (MAC)
machine-to-machine (M2M), 206–7, 215
media gateway (MG), 24
media gateway (MGW), 24, 29
Media Gateway Controller (MGC), 24
Media GCF (MGCF), 29
Media Presentation Description (MPD), 37
media terminal adapter (MTA), 23,
24, 59, 106, 109–16
MEF *see* Metro Ethernet Forum (MEF)
Metro Ethernet
access EPL and EVPL, 156–8
basic network reference model, 136, *136*
Carrier Ethernet Network (CEN), 136
characteristic information
(ETH_CI) *see* Ethernet
class of service
color-aware mode/color-blind mode, 150
committed information rate
(CIR), 149–50
DSCP—CoS mapping, 150, **150**
excess information rate (EIR), 149–50
parameters (H, M, and L classes), 148–9
PCP and DSCP mapping
(H, M, and L classes), 150, **151**
PCP—CoS mapping, 149, **149**
performance tiers (PTs), 152
service frame delivery performance, 150
SLAs, **152–3**
Ethernet Virtual Connection (EVC),
136, 141–2
External Network-to-Network
Interface (ENNI), 145, *145*
Network Interface Device (NID), 159, *159*
operator virtual connections
(OVCs), 145–6, 159–61
protocol stack, 137, *137*
services
Carrier Ethernet services, 153, **154**
E-Line service, 154
Ethernet private LAN (EP-LAN), 155, *156*
Ethernet private line (EPL), 154
Ethernet virtual private line
(EVPL), 154, *155*
E-Tree service, 155–6, *157*
EVP-LAN, 155, *156*
I-NNI, 153
transport layer (TRAN layer), 137
UNI/EVC attributes, 142–4

Metro Ethernet (*cont'd*)
  User Network Interface (UNI),
    136, 140–141
  virtual UNI/remote UNI
    (VUNI/RUNI), 147–8
Metro Ethernet Forum (MEF)
  Carrier Ethernet definition, 135
  Cloud Broker definition, 266
  EP-LAN services, 270
  EP-Tree services, 269, *269*
  Ethernet Cloud Carrier, 266
  Ethernet Private Line (EPL) services, 267–8
  EVP-Tree services, 269
  mapping, 316, *317*
  Network as a Service (NaaS) concept, 248
MHA *see* modular head-end
    architecture (MHA)
mobile virtual network operator
    (MVNOs), 203
mobile wireless services, 203
modular cable modem termination
    system (M-CMTS), 56, *57*, 64–7
modular head-end architecture (MHA)
  cable modem termination system
    (CMTS), *58*
  definition, 56
  DownstreamExternal PHY Interface (DEPI)
    tunnel modes, 57
  edge quadrature amplitude
    modulation (QAM), 56, *58*
  Edge Resource Management
    (ERM) protocol, 56
  modular cable modem termination
    system (M-CMTS), 56, 57
  Packet Stream Protocol (PSP), 57
monthly recurring cost (MRC), 196
Move–Add–Change (MAC), 7, 14, 20, 63,
    94–6, 101, 104–5, 109–10, 135,
    137–41, **139**, 146, 165, 169–70, 184,
    237, 242, 278, 283–5, 293
MPEG-DASH, 37, 39, 44, 50
Multimedia over Coax Alliance (MoCA), 18
Multiple Dwelling Units
    (MDU), 55, 61–2, 64, 66
multiple program transport stream
    (MPTS) video, 55
multiple system operators (MSOs), 1, 2,
    3, 8, 12–13, 34, 62–3, 85, 193,
    198, 318–19

Multi-Point Control Protocol
    (MPCP), 164, 168–9
multiuser MIMO (MU-MIMO), 70–71

NaaS *see* Network as a Service (NaaS)
national linear video, 3, 35
NETCONF *see* Network Configuration
    (NETCONF) Protocol
Network as a Service (NaaS), 248, 272
Network Configuration (NETCONF)
    Protocol, 206, 211, 214, 215,
    216, 218, 310, 323
Network Digital Video Recorder (NDVR), 48
network element (NE), 5–6, 8, 9, 22,
    45, 85–6, 183, 218
Network Functions Virtualization (NFV)
  CPE and PE functions, Metro Ethernet
    Network, 297, 298
  end-to-end service, 297, *297*
  framework, 296, *296*
  NFV Infrastructure (NFVI), 296
  Virtual Network Functions (VNFs), 296
  VNFs relationships, 297
  working domains, 296–7
Network Interface Devices
    (NIDs), 159, *159, 218
Network Management System (NMS), 8,
    163, 183, 186, 187–8, *188,* 319
network operations centers (NOCs), 4–6,
    85–6, 93–4, 189, 204
networks, 190–191
network systems provisioning, 217–18
Network Virtualization Edge (NVE)
  data plane, 282
  forwarding information base
    (FIB) tables, 282
  multicast, 285
  network virtualization functions, 280
  NVO3 OAM messages, 287
  proactive connectivity monitoring, 288
  service CoS, 285
  tenant-service, 281
  tenant virtual network with bridging/
    routing, *294*
  and TESs communication, 289
  underlay/underlay network, 279
  unicast and multicast tunneling
    methods, 281
  virtual switch, 278

Network Virtualization Overlays
(NVO3) network
attacks, 288
encapsulated frame, 282, *282*
Extensible Authentication
Protocol (EAP), 289
inside attacks and outside attacks, 288–9
OAM frames, 288
OAM layers, 287, *287*
overlay header, 282, 286
overlays, 289
server layer, 286
Tenant End System (TES)
interconnection, *291*
tunneling, 285
Virtual Data Center, 294
"Next Generation Corporate Network"
(NGCN), 173–4, 176
NFV *see* Network Functions
Virtualization (NFV)
non-real-time polling service (nrtPS), 21
non-recurring costs (NRC), 196
null data packet (NDP) approach, 70

OAMPT *See* Operations, Administration,
Maintenance, Provisioning, and
Troubleshooting (OAMPT)
OID, 207
ONF *see* Open Networking Foundation (ONF)
ONRC *see* Open Networking Research
Center (ONRC)
OpenFlow, 300, 303–9
OpenFlow protocol, 10, 299–301,
304, 307, 309
Open Mobile Alliance (OMA), 185, 216
Open Networking Foundation (ONF), 9,
297, 301, 302, 303, *317*
Open Networking Research Center
(ONRC), 301
operation, administration, and maintenance
(OAM), 7, 135, 140–141, 145, 164,
167, 202, *224,* 224–5, *228,* 228–9,
230, 234, 242–3, 286–8, 316
Operations Support Systems (OSS), 87, 89,
184–5, 186, *188,* 189–91, 197, 200,
202, 204–6, 211, 217–19, 242
operational systems and management
architectures *see also* Operations,
Administration, Maintenance,

Provisioning, and Troubleshooting
(OAMPT)
for business services
business services operations,
requirements, 184
Business Support Systems (BSS), 185
control and centralization, 195
distributed systems, 194–5
3GPP IMS-based services, 185
networks, 190–191
Operations Support Systems
(OSS), 184–5
operations, 194
protocols, 192
SDNs and OAMPT, 196
security, 193–4
for residential services
accounting management, 85–6
cable modem management, *86,* 86–7, **87**
CMTS/CCAP management, 87, **88,** *88*
configuration management, 85
DOCSIS service and device
management, 85
DOCSIS 3.1 service and network
management, 88–9
fault management, 85
performance management, 86
quality-of-service (QoS), 86
recommendation, 85
security management, 86
telecommunications operators, 85
subsystems
Business Support Systems
(BSS), 189–90
cable-specific systems, 190
Element Management Systems
(EMSs), 186–7
intrusion detection systems (IDS), 188
Network Management Systems
(NMSs), 187–8
operations support systems, 189
systems, 191
Operations, Administration, Maintenance,
Provisioning, and Troubleshooting
(OAMPT)
for business services
billing systems and formats, 235–6
equipment manufacturers, 8
fault management, 224–9

Operations, Administration, Maintenance,
    Provisioning, and Troubleshooting
    (OAMPT) (cont'd)
    operations systems see operational
        systems and management
        architectures
    performance management, 230–235
    provisioning see provisioning process
    security, 237–41
    service orders, 196–205
    service providers, 8
  for residential services
    accounting management, 5
    "Billing" system, 6
    billing systems and formats, 97–9
    capabilities, 84
    configuration management, 4–5
    Customer Premises Equipment (CPE), 4
    DOCSIS 3.0 and 3.1 networks, 5–6
    fault management, 4, 5, 93–4
    FCAPS model, 4
    maintenance, 4
    network administration, 4
    network operations center (NOC), 4–5
    operational systems and management
        architectures, 85–9
    performance management, 5–6, 94–6
    residential services, 85
    security management, 5, 100–105
    service order process, 4
    service orders, 89–91
operator virtual connections (OVCs), 145,
    145–7, 146, 157–9, 161, 202,
    205–6, 228
opportunistic key caching (OKC), 72
Optical Beat Interference (OBI), 16, 59
optical distribution network (ODN), 60
Optical Line Terminal (OLT), 7, 54, 60,
    161, 163, 169–70
Optical Network Units (ONUs), 3, 7,
    59–60, 161, 164–7, 169–70,
    201, 219–22, 240
optical transport networks (OTNs), 301, 312
Order Entry system, 6, 97, 197, 201, 204
Order management systems, 197–8
Orthogonal Frequency Division Multiple
    Access (OFDMA), 13–14, 18, 95–6
orthogonal frequency division multiplexing
    (OFDM), 13, 14, 16, 65, 95, 96, 169

OSSI see DOCSIS Operations and Support
    System Interface (OSSI)
"over-the-top" (OTT) services, 54–5,
    203–4, 206

PaaS see Platform as a Service (PaaS)
PacketCable
    architecture and services
        PacketCable 1.5, 22–5
        PacketCable 2.0, 24–6
        service providers (SP), 22
        Voice over Internet Protocol
            (VoIP) technology, 22
    Enterprise SIP Gateway
        hosted IP-Centrex deployment, 178, 178
        network address translator (NAT)
            traversal and access control rules, 178
        provisioning gateway hierarchy
            diagram, 179, 180
        SIP Endpoint Test Agent (SETA), 179
    project, 1, 2–3, 12, 22
    security
        assertion, and configuration and
            management, 130–131
        PacketCable 1.5, 106–17
        PacketCable 2.0, 117, 121–2
        signaling, 128–30
    user and UE authentication
        access domain reference points
            description, 123
        IMS registration message flow, 124
        IMS specifications, 1258
        integrity and optional confidentiality, 123
        SIP authentication, 126–8, 127
        steps, 125–6
        Universal Integrated Circuit
            Card (UICC), 122
PacketCable 1.5 see also PacketCable
    architecture and services
        call management server (CMS), 23
        endpoint device (E-MTA), 23
        IP-based replacement, 22
        IP Multimedia System (IMS),
            innovations, 24–5
        media gateway controller (MGC), 24
        network call signaling (NCS), 23
        provisioning procedures, 24
        Quality of Service (QoS) model, 24
        Real-Time Transport Protocol (RTP), 24

Session Initiation Protocol (SIP), 23
  "softswitch-based" architecture, 23
  traditional telephony features, 22
security
  file MIB objects, configuration, **111**
  interfaces, **106**
  Kerberos, 116
  key distribution center (KDC), 106
  MTA provisioning process, 106
  operator guidance, 105
  PacketCable PKI, 116–17
  post-MTA provisioning security flows, **112**
  provisioning flows, **107–9**
  provisioning steps and, 106, 110, 111
  securing call signaling flows, **112,** 112–13
  security MIBs, **118–20**
  ticket expiration time (TicketEXP), 113–14
  ticket grace period (TicketGP), 114–15
PacketCable 2.0 *see also* PacketCable
  architecture and services
    advanced communications services, 24
    advanced multimedia services, 22
    enhancements, 25–6
    IP Multimedia System (IMS)
      architecture, 24–5
    Session Initiation Protocol (SIP)
      procedures, 26
  security
    MIB objects, 117
    PKI certificates, **117**
    reference architecture, 121–2
Packet Stream Protocol (PSP), 57
Passive Optical Network (PON)
    system, 3, 7, 59, 60, 153–4,
    161, *162,* 164, 170
Passpoint, 71
Passpoint certified, 71
Payment Card Industry (PCI), 300
pay-per-view (PPV) system, 3, 35, 98
"perfect" clones, 105
performance management (PM)
  for business services
    availability performance, 230, 233
    frame delay (FD) measurement, 230,
      231–3, *232, 233*
    frame loss ratio (FLR), 230–231, *231, 232*
    inter-frame delay variation
      (IFDV), 230, 233
    Metro Ethernet services, 230

performance monitoring solution
    framework, *234,* 234–5
  resiliency performance, 230, 233
  for residential services
    cable plant segmentation, 94–5
    CM performance metric collection, 96
    CMTS and CCAP performance
      metrics collection, 95–6
    definition, 94
    description, 86
    IPDR/SP network model, *96*
    trending analysis, 94
PERL, 185, 207
Placement Opportunity Information
    System (POIS), 41, 43, 45
Platform as a Service (PaaS)
  Cloud Consumer, 257, *258*
  Cloud Provider, 249, 257
  hybrid cloud, 272
  intercloud architecture, 274, *274*
  multiple system operators (MSOs), 198
PNM *see* proactive network
    maintenance (PNM)
primary service flows, 21
Private Branch Exchanges
    (PBXs), 6, 134, 173–9, 195
private clouds
  description, 249
  management, 260, *261*
Proactive Network Maintenance
    (PNM), 5, 88, 93
Product Catalogue, 6, 97, 99
Proportional Integral Controller-Enhanced
    (PIE) algorithm, 22
Provide Backbone Transport (PBT), 7, 135
Provider Backbone Bridge
    (PBB), 7, 135, 285
provisioning process *see also* Operations,
    Administration, Maintenance,
    Provisioning, and Troubleshooting
    (OAMPT)
  application systems, 223
  BSS/OSS stack, 91
  cloud services, 223
  CM provisioning and software
    updates, 91–2
  customer equipment (CE), 221–3
  customer premises equipment
    (CPE), 219–21

interfaces
command-line interface (CLI), 206–8
Graphical User Interface (GUI), 208–9
scripting and automation, *209,* 209–10
NE provisioning, 218
network systems, 217–18
protocols
and combinations, 211
Common Object Request Broker
Architecture (CORBA), 216
Extensible Markup Language
APIs, 213–14
JavaScript Object Notation
(JSON), 214–15
MPT systems, 211–12, *212*
Network Configuration Protocol
(NETCONF), 215
Representational State Transfer (REST
and RESTful) Architecture APIs, 216
Simple Network Management Protocol
(SNMP), 212–13
Simple Object Access Protocol
(SOAP), 214
YANG, 215
SDN, 216–17
self-service, 223–4
selling telecom services, 210–211
server, 218
service orders, 205
Proxy Call Session Controller Function
(P-CSCF), 29–33, 123, 125–6,
130, 174, 241
public cloud
description, 249–50
management, 260, *260*
PYTHON, 185

quality assurance (QA) team, 69
Quality of Service (QoS), 3, 5–7, 13–14,
20–22, 24–5, 51, 63, 72, 78, 80, 86,
94–5, 96, 121, 135, 162, 166, 264,
285, 288, 294, 304, 306, 316

radio frequency over glass (RFoG) system
architecture, 59–60
and Converged Cable Access
Platform (CCAP), 59
DOCSIS 3.1 specification, 16
and EPON, 61, *61*

Radio resource management/self-organizing
network (RRM/SON), 72
3rd Generation Partnership Project
(3GPP), 1, 12, 24, 26, 30, 37,
69, 117, 126, 174–6, 185, 319
real-time Polling Service (rtPS), 21
Real-time Transport Protocol
(RTP), 24, 27, 29, 178, 179
reconfigurable optical add/drop
multiplexer (ROADM), 312
remote procedure call (RPC), 215
remote UNI (RUNI), 147–8
Representational State Transfer
(REST and RESTful)
Architecture APIs, 185, 192,
206, 211, 214–17, 310
residential network architectures
and services
Cable Modem (CM), 3
cable modem termination
system (CMTS), 2–3
content distribution networks
(CDNs), 2, 3, 45–53
Converged Cable Access Platform
(CCAP), 2, 3, 53–65
DOCSIS 3.0/3.1, 2, 3, 12–22
High Speed Data (HSD) Internet
or Broadband service, 2
Internet Protocol TeleVision (IPTV)
system, 34–45
IP Multimedia Subsystem (IMS), 2, 26–34
linear and on-demand video, 3
national VOD programming, 3
OAMPT (Operations, Administration,
Maintenance, Provisioning,
Troubleshooting), 4–6
packet cable, 22–6
PacketCable 1.5, 3
PacketCable 2.0, 3
packet-switched network, 2
Passive Optical Network (PON) system, 3
Voice over Internet Protocol (VoIP), 2, 3
Wi-Fi, 4, 69–78
REST APIs *see* Representational State
Transfer (REST and RESTful)
Architecture APIs
RESTful APIs *see* Representational State
Transfer (REST and RESTful)
Architecture APIs

RFoG *see* radio frequency over glass (RFoG) system
RF-over-Glass Optical Network Unit (R-ONU), 59

SaaS *see* Software as a Service (SaaS)
SAMIS *see* Subscriber Accounting Management Interface Specification (SAMIS)
SAV *see* source address verification (SAV)
SCTE-130, 41
SDN *see* software-defined networking (SDN)
Secure Sockets Layer (SSL), 215, 289
secure software download (SSD) mechanism, 91, 103, 239
security *see also* Operations, Administration, Maintenance, Provisioning, and Troubleshooting (OAMPT)
  for business customers, 193
  Converged Cable Access Platform (CCAP), 241
  in DOCSIS 3.0/3.1, 100–105
  DOCSIS Provisioning of EPON (DPoE), 237–40
  encryption, 100
  EPON Protocol over Coax (EPoC), 240–241
  Ethernet, 193
  hackers, 100
  IP Multimedia Subsystem (IMS), 241
  management, 86
  Metro Ethernet services, 193–4
  multiple system operators (MSOs), 193
  network, 242
  PacketCable security *see* PacketCable
  servers and devices, 100
  spoofing servers, 100
self-managed networks
  architectures
    Data Handler module, 324
    intelligent agents, 320, *320, 321*
    intelligent Capacity Management Agent(s) (iCMA(s)), 320, 322
    Rule-Based Logic module, 323, *323*
    Scheduler module, 324
  in-band communications, 330–337
  iNE and iNMS architectures, 324–5
  NE and NMS architectures
    centrally self-managed networks, 320–324
    distributed self-managed networks, 326, *326*

self-managed SDN, 328–30, *330*
self-managed VNs, 326–8, *328, 329*
self-managing network management system (sNMS), 319
self-managing regional network management system (sRNMS), 319
self-managing regional network management system (sRNMS), 319, 326, 328, 332, 334, 335, 337
self-service, 189–90, 200, 204, 223–4, 242, 247, 266
server provisioning, 217, 218
service delivery, 3, 7, 13, 22, 24, 34, 53, 171–2, 185, 198, 201–2, 230, 234, 248
service design, 196, 198, 205, 302
service flows (SF), 14, 20–22, 96, 201, 204
Service Level Agreement (SLA), 87, 135, 152, **152, 153,** 189, 197, 224, 230, 235, 251, 256, 264, 272, 313
Service OAM (SOAM) framework, 202, 226–7, 228–9, 231, 233, 242–3
service order management, 197–8, 199
Service Order Manager, 6, 97
service orders *see also* Operations, Administration, Maintenance, Provisioning, and Troubleshooting (OAMPT)
  access and beyond, 203
  defining services, 203–4
  management, 197–8
  product management, 196–7
  security, 204
  service management across operators, 199
  service type
    Ethernet network to network interface, 202
    Ethernet services, 201–2
    fixed wireless services, 202
    inter-carrier Ethernet, 202
    internet access, 200–201
    mobile wireless services, 203
    voice services, 201
  standards, 198–9
  wholesale markets, 204–5
Serving Call Session Controller Function (S-CSCF), 29–34, 125–30
Session Initiation Protocol (SIP), 12, 23, 26–7, 126–8, **127,** 173, *173,* 176–80
set-top-box (STB), 16–17, 24–5, 35, 38, 49–50, 59, 66, 68, 90

signaling gateway (SG), 24, 71
signaling gateway (SGW), 24, 29
signal-to-noise ratio (SNR), 18, 70, 95
Simple Network Management
        Protocol (SNMP)
    management information base
        (MIB), 212, 213
    management of NEs, 212
    servers, 185
    SNMPv1 and SNMPv2, 212
    traps, 192
Simple Object Access Protocol
        (SOAP), 187, 192, 206, 211,
        214, 216, 236
"single points of failure" (SPOFs), 219
SIP Endpoint Test Agent (SETA), 179
SIP Forum SIPconnect1.1, 176–8
SIP Trunking service, 173, *173*
small and medium size enterprises
        (SMEs), 2, 6, 134
Society of Cable Telecommunications
        Engineers (SCTE), 41, 60
Software as a Service (SaaS)
    cloud services, 248
    consumers, 257
    example services, *258*
    multiple system operators (MSOs), 198
software-defined networking (SDN)
    building blocks, *302,* 302–3
    carrier Ethernet services, 315–17, *317*
    Cloud services, 8–10
    definition, 297
    DOCSIS networks, 318, *318*
    layers, 300, *300*
    and OAMPT, 196
    objectives, 300–301
    OpenFlow, 299, 303–9
    optical networks, 311–14
    per-subscriber usage, 298
    provisioning, 216–17
    representational state transfer
        API, 309–11
    SDN Control Interface, 303
    standards, 301–2
    switching plane, 298
    virtualized centers, 314–15
source address verification
        (SAV), 104, **104**, 240, **240**
source-specific multicast (SSM), 14

southbound APIs, 309
sRNMS *see* self-managing regional
        network management system
        (sRNMS)
standalone Digital Multimedia
        Adapter (sDVA), 201
Subscriber Accounting Management
        Interface Specification
        (SAMIS), 5, 85–6
Synchronous Optical Networking
        (SONET) circuit, 137, 218, 312
synthetic loss message (SLM), *231,*
        233, 234, 243
synthetics loss reply (SLR), 231,
        231, 233–4

TCL, 185, 207
TCP Authentication Option
        (TCP-AO), 289
Telecommunications Managed Network
        (TMN), 131, 185, 242
Telecommunications Management Forum
        (TM Forum), 198–9
TFTP *see* Trivial File Transfer
        Protocol (TFTP)
time and frequency division multiplexing
        (TaFDMA) design, 16
time-varying control (TVC)
        values, 103, 239
Transport Layer Security (TLS), 128–9,
        238, 289, 307, 309
Trivial File Transfer Protocol (TFTP), 91,
        103, 110–111, 131, 165, 167, 185,
        192, 208, 219, 239
TVC values *see* time-varying control
        (TVC) values

UNI *see* User Network Interface (UNI)
unified communications services
        (UCS), 185
Universal Edge QAM (UEQAM), 56,
        57, 65–8
unsolicited grant service with activity
        detection (UGS-AD), 21
user-based security model (USM), 131
User Network Interface (UNI), 136,
        140–141, 142–4, 221
USM *see* user-based security
        model (USM)

VACM *see* view-based access
        control model (VACM)
value-added services (VAS), 185
VDC *see* Virtual Data Center (VDC)
Video Ad Serving Template (VAST), 41
video encoders, 37, 41
Video Multiple Ad Playlist (VMAP), 41
video-on-demand (VOD) services, 3,
        35, 48–9, 67
Video Player-Ad Interface Definition
        (VPAID), 41
view-based access control model
        (VACM), 131
virtual cable modem (vCM), 163–5,
        164, 190, 239
Virtual Data Center (VDC)
    DC provider, 293
    NVO3, 294, *294*
    users, Enterprise site, 293
virtual gateway interface function
        (VNIF), 293
virtual infrastructure, 9, 275, 277, 310
virtualization *see also* Network Functions
        Virtualization (NFV)
    application server, 276
    architecture
        Backbone nodes, 280
        data centers, 279, *279*
        data path, 284–5
        DiffServ and ECN marking, 286
        End Device, 279, 280
        EoR topology, 280
        high-capacity intra-DC network, 281
        LAG and ECMP, 285–6
        Network Virtualization Edge (NVE), 281–2
        Network Virtualization Overlays
            (NVO3) network, 278
        overlay modules and VN context, 282–4
        ToR, 280
        underlay/underlay network, 279
        Virtual Data Center (VDC), 277–8
        virtual network (VN), 278
        virtual network identifier
            (VNID), 278, *278*
    basic VNs, DC, 290–291, *291, 292*
    blade servers, 277
    DC VN and WAN VPN
        interconnection, 291, *292,* 293
    dual-core processor technology, 277

    EVC-based VNs, 295, *295*
    hardware, 275
    management, 275
    multiple OSs, 277
    network, 275
    OAM, 286–8
    OS virtualization, 276
    security
        communications, NVEs and TESs, 289
        communications within overlays, 289
        data plane security, 290
        inside attacks and outside attacks, 288–9
        VM migration, 290
    server with hypervisor layer, *275*
    service, 276–7
    storage, 275–6
    tenant VN with bridging/routing,
        293, *294*
    VDC, 293–4, *294*
virtual machine images (VMIs), 195
Virtual Machines (VM), 248–52, 257, 259,
        262, 265, 276–83, 287, 289–91, 293,
        304, 315, 338
virtual network instances (VNI), 257, 281–5,
        287–8, 290–291, *292,* 293–4
virtual NID (vNID), 159, 278, *279,* 282–3
Virtual Private Network (VPN), 57, 61, 135,
        155, 190, 193, 195, 201, 204, 206–7,
        223, 242, 281, 283, 285, 287, 291–4,
        314–15
virtual UNI (VUNI), 147–8
VNF Forwarding Graph (VNF-FG), 297
VOD *see* video-on-demand (VOD) services
Voice over Internet Protocol (VoIP), 1,
        2, 3, 12, 22, 148, 178, 180, 190,
        194, 213, 242
VoIP *see* Voice over Internet
        Protocol (VoIP)

wave-division multiplexing
        (WDM) system, 60–61
wholesale markets, 204–5
Wi-Fi
    architecture and services
        air interfaces, 70–71
        cable Wi-Fi, 76–7
        community Wi-Fi, 73–4, **75**
        handoff algorithms, 71–2
        home Wi-Fi networks, 72–3

Wi-Fi (*cont'd*)
    Hotspot 2.0, 71
    public Wi-Fi hotspot, 75–6
    radio resource management/self-organiz-
        ing networks (RRM/SON), 72
    multiple system operators (MSOs), 2
    service providers (SP), 4

Wi-Fi alliance (WFA), 71
Wireless Service Locator
        (WSL) tool, 77, *78*

XaaS, 198

"YANG models," 215–16, 323

IEEE Press Series on
Networks and Services Management

The goal of this series is to publish high quality technical reference books and textbooks on network and services management for communications and information technology professional societies, private sector and government organizations as well as research centers and universities around the world. This Series focuses on Fault, Configuration, Accounting, Performance, and Security (FCAPS) management in areas including, but not limited to, telecommunications network and services, technologies and implementations, IP networks and services, and wireless networks and services.

Series Editors:
Thomas Plevyak
Veli Sahin

1. *Telecommunications Network Management into the 21st Century*
   Edited by Thomas Plevyak and Salah Aidarous
2. *Telecommunications Network Management: Technologies and Implementations: Techniques, Standards, Technologies, and Applications*
   Edited by Salah Aidarous and Thomas Plevyak
3. *Fundamentals of Telecommunications Network Management*
   Lakshmi G. Raman
4. *Security for Telecommunications Network Management*
   Moshe Rozenblit
5. *Integrated Telecommunications Management Solutions*
   Graham Chen and Qinzheng Kong
6. *Managing IP Networks Challenges and Opportunities*
   Thomas Plevyak and Salah Aidarous
7. *Next-Generation Telecommunications Networks, Services, and Management*
   Edited by Thomas Plevyak and Veli Sahin
8. *Introduction to IT Address Management*
   Timothy Rooney
9. *IP Address Management: Principles and Practices*
   Timothy Rooney
10. *Telecommunications System Reliability Engineering, Theory, and Practice*
    Mark L. Ayers
11. *IPv6 Deployment and Management*
    Michael Dooley and Timothy Rooney
12. *Security Management of Next Generation Telecommunications Networks and Services*
    Stuart Jacobs
13. *Cable Networks, Services, and Management*
    Mehmet Toy